Biogeographic regionalization of the Neotropical region

Neotropical Biogeography

Regionalization and Evolution

CRC Biogeography Series

Series Editor
Malte C. Ebach
University of New South Wales
School of Biological
Earth and Environmental Sciences
New South Wales
Australia

Handbook of Australasian Biogeography, *Malte C. Ebach*

Biogeography and Evolution in New Zealand, *Michael Heads*

Neotropical Biogeography
Regionalization and Evolution

Juan J. Morrone
Professor of Biogeography, Systematics
and Comparative Biology
Museo de Zoología "Alfonso L. Herrera"
Facultad de Ciencias, UNAM

CRC Press
Taylor & Francis Group
Boca Raton London New York

CRC Press is an imprint of the
Taylor & Francis Group, an **informa** business

CRC Press
Taylor & Francis Group
6000 Broken Sound Parkway NW, Suite 300
Boca Raton, FL 33487-2742

First issued in paperback 2020

© 2017 by Taylor & Francis Group, LLC
CRC Press is an imprint of Taylor & Francis Group, an Informa business

No claim to original U.S. Government works

ISBN 13: 978-0-367-65805-2 (pbk)
ISBN 13: 978-1-138-03248-4 (hbk)

Library of Congress Cataloging-in-Publication Data

Names: Morrone, Juan J.
Title: Neotropical biogeography : regionalization and evolution / Juan J. Morrone.
Description: Boca Raton : CRC Press, 2017. | Includes bibliographical references.
Identifiers: LCCN 2016045862| ISBN 9781138032484 (hardback : alk. paper) |
ISBN 9781315390666 (ebook)
Subjects: LCSH: Biogeography--Central America. | Biogeography--South America.
| Biogeography--Tropics.
Classification: LCC QH84.5 .M68 2017 | DDC 577.2/2--dc23
LC record available at https://lccn.loc.gov/2016045862

Visit the Taylor & Francis Web site at
http://www.taylorandfrancis.com

and the CRC Press Web site at
http://www.crcpress.com

*I dedicate this book to Chico Mendes (1944–1988),
guardian angel of the Neotropical forests.*

Contents

List of Figures

Preface

The Neotropics comprise the tropical areas of the Americas, from Mexico to Argentina. Hundreds of biogeographic studies of plant and animal taxa from this region have been published for more than 150 years, producing a basic knowledge that is synthesized into regionalization schemes. In this book, I address two central questions of evolutionary biogeography: which areas are recognized within the Neotropical region and how did their biotas evolve? In the last few decades, molecular phylogenetics and parametric model-based biogeography have allowed the postulation of complex biogeographic scenarios for particular taxa, and the search for biotic patterns has been somewhat neglected. However, I feel that biogeographic regionalizations based on the distributional patterns of plant and animal taxa are still relevant in the twenty-first century because they constitute the background knowledge of systematic, ecological, evolutionary, and other kinds of studies.

The biogeographic regionalization of the Neotropical region presented herein consists of three subregions, two transition zones, seven dominions, and 53 provinces. For each unit, I provide the valid name according to the International Code of Area Nomenclature (ICAN) followed by a list of citations and synonyms, a brief characterization, and some endemic and characteristic taxa. In order to deal with biotic evolution, I refer to the identification of biotas through areas of endemism and generalized tracks, their relationships based on track and cladistic biogeographic analyses, and, when possible, the cenocrons or biotic subsets that have been identified within them. This attempt of synthesis is based on a vast bibliography that I have compiled for more than two decades. I feel grateful to many authors who have provided insights on the regionalization and evolution of the Neotropical region. Particularly inspirational for my work were Jorge Artigas, Ángel L. Cabrera, Joel Cracraft, Léon Croizat, Philip Darlington, Gonzalo Halffter, Michael Heads, René Jeannel, Guillermo Kuschel, Emilio Maury, Ernst Mayr, Paul Müller, Eduardo Rapoport, Osvaldo Reig, Raúl Ringuelet, Donn E. Rosen, Jerzy Rzedowski, Jay Savage, George G. Simpson, Arne Takhtajan, Alfred R. Wallace, and Abraham Willink.

I thank Lone Aagesen, Tania Escalante, Cecilia Ezcurra, Livia León-Paniagua, Isolda Luna, Silvio Nihei, Sergio Roig-Juñent, and Luis Sánchez-González for providing useful comments on a preliminary version of the manuscript. Many friends and colleagues provided useful discussions, helped with the bibliography, provided data, and shared their ideas with me: Lone Aagesen, Dalton de Sousa Amorim, Marcelo Arana, Fabrizio Cecca, Jorge Crisci, Guadalupe del Río, Malte C. Ebach, Amparo Echeverry, Tania Escalante, Celene Espadas, David Espinosa, Cecilia Ezcurra, Ignacio Ferro, Ismael Ferrusquía-Villafranca, Gustavo Flores, Oscar Flores-Villela, Jorge Fontenla, Irene Goyenechea-Mayer, Gonzalo Halffter, Michael

Heads, Liliana Katinas, Analía Lanteri, Livia León-Paniagua, Jonathan Liria, Jorge Llorente-Bousquets, Peter Löwenberg-Neto, Isolda Luna-Vega, Juan Márquez-Luna, Andrés Moreira-Muñoz, Adolfo Navarro-Sigüenza, Paula Posadas, Alejandra Ribichich, Gerardo Rodríguez-Tapia, Sergio Roig-Juñent, Adriana Ruggiero, Luis Sánchez-González, Claudia Szumik, Estrella Urtubey, and Mario Zunino.

Juan J. Morrone
Mexico City

Author

Juan J. Morrone is full professor of biogeography, systematics, and comparative biology at the Facultad de Ciencias, Universidad Nacional Autónoma de México (UNAM), Mexico. He works on phylogenetic systematics of weevils (Coleoptera: Curculionidae) and evolutionary biogeography and regionalization of the Neotropical and Andean regions.

He joined the Museo de Zoología "Alfonso L. Herrera" of the Facultad de Ciencias, Universidad Nacional Autónoma de México (UNAM), Mexico, in 1998, after working for some years at the Museo de La Plata, Universidad Nacional de La Plata (UNLP), Argentina, where he obtained his PhD degree. He is a Member of the Academia Mexicana de Ciencias, Fellow of the Willi Hennig Society, and Research Associate of the American Museum of Natural History and the Buffalo Museum of Science. He has authored 270 scientific papers and has authored, or edited, 29 books on evolutionary biogeography, phylogenetic systematics, biogeographic regionalization, biodiversity conservation, and evolution.

CHAPTER 1

Theoretical Background

Evolutionary biogeography integrates distributional, phylogenetic, molecular, and paleontological data in order to discover biogeographic patterns exhibited by plant and animal taxa, and assess the historical changes that have shaped biotic assembly (Morrone, 2009). Biogeographic regionalizations are hierarchical classifications categorizing geographic areas in terms of their endemic taxa and their relationships. They represent the syntheses of different evolutionary biogeographic analyses and, at the same time, constitute the background knowledge of other studies (e.g., systematic, ecological, or evolutionary).

EVOLUTIONARY BIOGEOGRAPHY

Evolutionary biogeography is the integrative study of distributional, phylogenetic, molecular, and paleontological data, aimed to discover biogeographic patterns and assess the historical changes that have shaped them (Morrone, 2009). It follows a stepwise approach. First, areas of endemism or generalized tracks are identified and considered as hypotheses about biotic identity based on the distributional congruence exhibited by different plant and animal taxa. Second, cladistic biogeographic analyses test these hypotheses, based on the available phylogenetic evidence on the taxa analyzed. Third, biogeographic regionalization is achieved based on cladistic biogeographic hypotheses. Fourth, the molecular dating of divergences between lineages and fossil data allows the identification of cenocrons, which represent subsets of taxa within a biota, identified by their common origin and evolutionary history. Cenocrons incorporate a temporal dimension that implies a time frame of the dispersal of sets of taxa into the biota. Finally, after the biotas and cenocrons have been identified, one may construct a geobiotic scenario by accounting biological and non-biological data to explain the episodes of vicariance/biotic divergence and dispersal/biotic convergence that have shaped the evolution of the biotas analyzed.

The dispersal–vicariance model followed herein assumes that the relationship between earth history and life is more complex than what is assumed in simpler models, because biotic history is reticulate. It treats vicariance as the default explanation for general biogeographic patterns and dispersal as the process that shapes the distribution of cenocrons or individual species (Morrone, 2015a). Both vicariance and dispersal are integrated in order to understand biotic assembly, incorporating the dating of the lineages and the identification of the cenocrons.

STEPS OF EVOLUTIONARY BIOGEOGRAPHY

The five steps of evolutionary biogeography (Figure 1.1) are as follows.

Step 1—Identification of biotas. Biotas are sets of spatiotemporally integrated plant and animal taxa that coexist in given areas. The identification of these sets of taxa constitutes the first step of an evolutionary biogeographic analysis. There are different ways to represent biotas graphically; the most common are generalized tracks and areas of endemism.

Generalized tracks indicate the preexistence of ancestral biotas that became fragmented by geological or tectonic events (Craw et al., 1999). They result from the significant superposition of different individual tracks, which operationally correspond to line graphs connecting the different localities or distributional areas of species or supraspecific taxa according to their geographic proximity (Morrone, 2015a). In the areas where two or more generalized tracks superimpose, nodes are identified. They are usually interpreted as tectonic and biotic convergence zones, where different biotas contact. Areas of endemism are areas of nonrandom distributional congruence among different taxa. They are identified by plotting the distributional ranges of different species or supraspecific taxa on a map and finding the areas of congruence between them.

There are different methods used to identify generalized tracks or areas of endemism, reviewed by Morrone (2009). The most commonly applied method for obtaining generalized tracks is parsimony analysis of endemicity (Echeverry and Morrone, 2010; Morrone, 2014a, 2015a). It includes the following steps (Figure 1.2).

1. Choose a set of biogeographic units across the study area, for example, localities, predefined areas of endemism, areas defined by physiographic criteria, or grid cells.
2. Construct individual tracks for species and/or supraspecific taxa, connecting the occurrence localities by a minimal spanning tree, either manually or using any of the programs available (e.g., Rojas-Parra, 2007; Liria, 2008; Echeverría-Londoño and Miranda-Esquivel, 2011).
3. Construct a data matrix, where rows represent the biogeographic units analyzed and columns represent the individual tracks. Each entry is coded as either "1" or "0," depending on whether each track is present or absent in the unit. A "?" code may be included in the case of doubtful occurrence in some geographic unit. A hypothetical unit coded as all zeros is added to the matrix in order to root the resulting cladogram(s).
4. Analyze the matrix with a parsimony algorithm, applying a phylogenetic software (e.g., Swofford, 2003; Goloboff et al., 2008). If more than one cladogram is found, calculate a strict consensus cladogram.

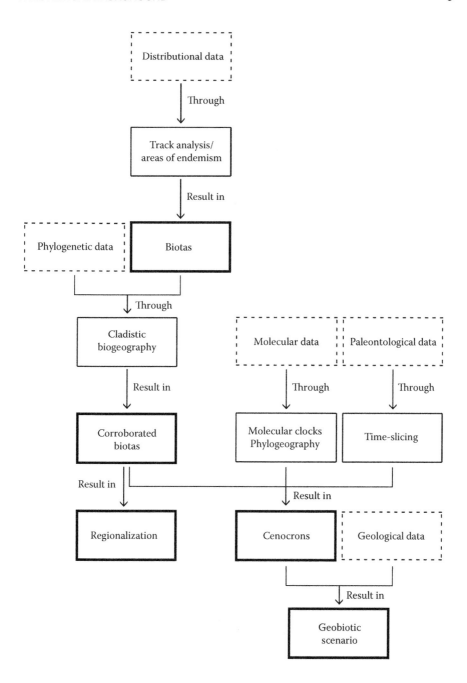

Figure 1.1 Flow chart showing the steps of evolutionary biogeography.

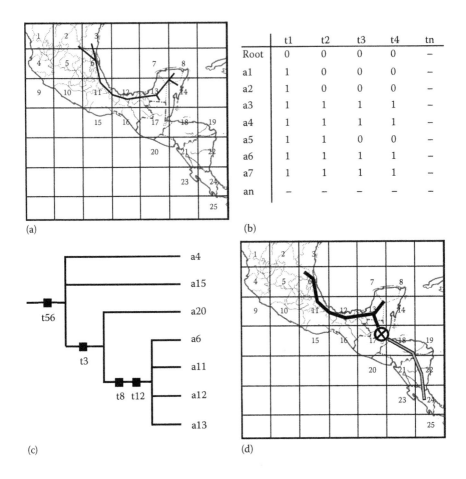

Figure 1.2 Steps of parsimony analysis of endemicity used to identify generalized tracks. (a) Map with an individual track represented; (b) data matrix; (c) cladogram obtained; and (d) map with two generalized tracks and one node.

5. Identify generalized tracks in the resulting cladogram based on the monophyletic groups of units defined by at least two individual tracks.

6. Remove from the data matrix the synapomorphic individual tracks that support the clades previously obtained and repeat steps 4–6 until no more individual tracks support any clade.

7. If some areas result in the overlap of two or more generalized tracks, identify them as nodes.

8. Represent the identified generalized tracks and nodes on a map.

Step 2—Testing relationships among biotas. Cladistic biogeography is an approach based on the correspondence between the phylogenetic relationships of different plant and animal taxa and the relationships between the areas that they inhabit. It uses information on the phylogenetic relationships between the taxa and their geographic distribution to postulate hypotheses on the relationships between the areas. When different taxa show the same pattern, such congruence is evidence of a common biotic history, shaped by vicariance. A cladistic biogeographic analysis begins by constructing taxon-area cladograms, from the taxonomic cladograms of two or more different taxa, by replacing their terminal taxa with the areas they inhabit. Then, resolved area cladograms are obtained from the taxon-area cladograms (when demanded by the method applied). Finally, a general area cladogram is constructed, based on the information contained in the resolved area cladograms. General area cladograms based on the information from the different resolved area cladograms represent hypotheses on the biogeographic history of the taxa analyzed and the areas where they are distributed.

There are different methods for obtaining general area cladograms, reviewed by Morrone (2009), including parsimony analysis of paralogy-free subtrees (Morrone, 2009, 2014c). It includes the following steps (Figure 1.3):

1. Identify paralogy-free subtrees in the taxon-area cladograms, by eliminating areas duplicated or redundant in the descendants of a node, so that geographic paralogy is eliminated or reduced significantly and data are associated only with informative nodes.
2. Identify the components on the paralogy-free subtrees obtained.
3. Compile a data matrix, scoring with "1" the presence of a component in an area and "0" its absence. A hypothetical unit coded as all zeros is added to the matrix in order to root the resulting cladogram(s).
4. Analyze the matrix with a parsimony algorithm by applying phylogenetic software (e.g., Swofford, 2003; Goloboff et al., 2008) to identify the general area cladogram.

Step 3—Regionalization. As the geographic distributions of taxa have limits and these limits are repeated for different taxa, they allow the recognition of biotas. Once they have been identified, they may be ordered hierarchically by cladistic biogeography and used to provide a biogeographic regionalization. It implies the recognition of successively nested areas, for which classically the following five categories have been used: kingdom, region, dominion, province, and district.

In cases where it is difficult to determine the exact boundaries of two regions, some authors have identified transition zones (Ferro and Morrone, 2014). They represent events of biotic "hybridization," promoted by historical and ecological changes that allowed the mixture of different biotas. When a transition zone is recognized, it means implicitly that it belongs simultaneously to both regions.

Step 4—Identification of cenocrons. Cenocrons are sets of taxa that share the same biogeographic history, constituting identifiable subsets within a biotic component by their common biotic origin and evolutionary history. After identifying the biotas, time slicing, intraspecific phylogeography, and molecular divergence dating help establish when the cenocrons assembled in the identified biotas, incorporating a time perspective to the study of biotic evolution. Time slicing consists of using fossils or molecular divergence dating to divide the complete set of taxa analyzed into slices according to the time when they occurred (Cecca et al., 2011). Intraspecific phylogeography, the analysis of the processes governing the geographic distribution of

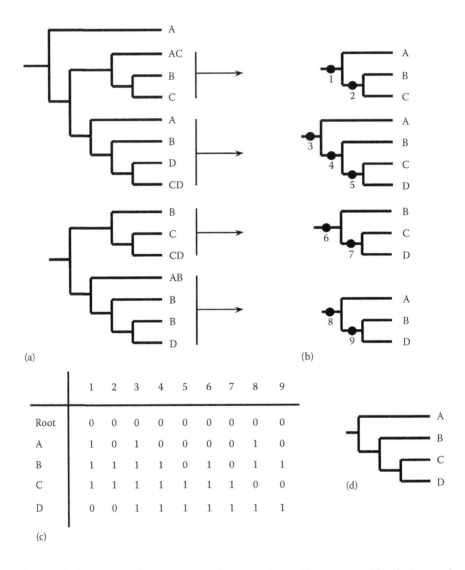

Figure 1.3 Steps of parsimony analysis of paralogy-free subtrees used to identify a general area cladogram. (a) Two taxon-area cladograms; (b) four paralogy-free subtrees derived from the area cladograms; (c) data matrix; and (d) general area clado-gram obtained.

genealogical lineages within and among closely related species, based on molecular data (Avise, 2000), may offer insight into when and how relatively recent cenocrons incorporated to a biota. Phylogenetic hypotheses based on molecular data may be used to calculate molecular divergence dates under the assumption that the rate of molecular evolution is approximately constant over time in all lineages (Zuckerland and Pauling, 1965) and considering that the "ticks," which correspond to mutations, do not occur at regular intervals, but rather at random points in time. Time can be measured in arbitrary units and then calibrated in millions of years by reference to the fossil record or geological data (Magallón, 2004), giving minimum estimates of the age of a clade, and may be used to elucidate the relative minimum age of the cenocron to which it belongs.

Step 5—Construction of a geobiotic scenario. Once we have identified the biotas and cenocrons, we may construct a geobiotic scenario. By accounting biological and nonbiological data, we can integrate a plausible scenario to help explain the episodes of vicariance/biotic divergence and dispersal/biotic convergence that have shaped biotic evolution. Geographic features may be classified in terms of their impact on dispersal and vicariance. The most important are barriers (e.g., geographic features that hinder dispersal) and corridors (geographic features that facilitate dispersal). When dealing with long-term changes in broad biotic patterns, continental drift may be a relevant factor (Cox and Moore, 1998). Distributional patterns are affected directly by the splitting and collision of land masses, but also new mountains, oceans, or land barriers change the climatic patterns upon the land masses.

BIOGEOGRAPHIC REGIONALIZATION

A biogeographic regionalization is a hierarchical system that categorizes geographic areas in terms of their endemic taxa. During the last two centuries, several biogeographic regionalizations have been proposed for the world and for particular areas. Biogeographic regionalization is historically rooted in the nineteenth century, with classificatory and nomenclatural procedures similar to those of systematics. De Candolle (1820) proposed that phytogeographic kingdoms should harbor a minimum of endemic species and genera so as to be accepted. Sclater (1858) postulated that zoogeographic regions should be based on endemic families, and Drude (1884) suggested the same for phytogeographic kingdoms. Wallace (1876) discussed some principles that should be applied for obtaining natural regionalizations.

In the last few decades, there have been different proposals for producing biogeographic regionalizations in a more objective way, although there is little agreement on the use of different methods available. Some authors consider that to produce natural biogeographic regionalizations, successively nested endemism should be the basic criterion used (Escalante, 2009), as has been the case traditionally. After recognizing natural areas based on endemic taxa (generalized tracks or areas of endemism), the areas are arranged hierarchically (general area cladograms) and given names. To denote this biogeographic hierarchy, five basic levels may be used: kingdoms (also known as realms), regions, dominions, provinces, and districts; in some cases, subkingdoms, subregions, or subprovinces are also recognized (Ebach et al., 2008). In addition, transition zones are recognized, representing areas of overlap, with a gradient of replacement and partial segregation between biotas belonging to different regions (Ferro and Morrone, 2014).

NOMENCLATURAL CONVENTIONS

In the biogeographic regionalization of the Neotropical region, I have followed the nomenclatural conventions set out in the International Code of Area Nomenclature (ICAN; Ebach et al., 2008). ICAN provides a universal naming system to standardize area names used in biogeography and other disciplines, where names are grouped under more inclusive area names in order to represent a biogeographic hierarchy (kingdoms, regions, dominions, provinces, and districts).

The notion of priority for using the oldest available names instead of new names is followed (ICAN, Art. 2.8). Sclater (1858) was adopted as the date of the starting point of biogeographic nomenclature because it constitutes the first widely adopted world biogeographic regionalization. In some cases, widely used names were kept instead of older synonyms, applying a criterion analogous to the *nomen conservandum* convention of taxonomic nomenclature in order to provide a better stability (Morrone, 2014b).

FORMAT USED IN THE BOOK

For each biogeographic unit recognized, the valid name is provided, followed by a list of citations and synonyms arranged in chronological order and a brief characterization. Maps with the areas recognized are provided. For each province, some endemic and characteristic taxa are listed, and the main vegetational types are mentioned. For the different areas recognized, individual tracks representing the distribution of selected taxa are mapped. Whenever they have been identified, brief references to subprovinces and districts are provided. In order to deal with the biotic evolution of the Neotropical region, I refer to the identification of biotas (as either areas of endemism or generalized tracks), their relationships based on a track, and cladistic biogeographic analyses and, when possible, the cenocrons or biotic subsets that have been identified within the biotas.

CHAPTER 2

Historical Background

The Neotropical region corresponds to the tropics of the New World, from Mexico to Argentina. The early history of its biogeographic regionalization began with Sclater (1858), Wallace (1876), and Engler (1879). During the nineteenth and twentieth centuries, several authors have provided maps recognizing subregions, dominions, and provinces in the Neotropical region or areas within it (Morrone, 2010b).

The first formal definition of the Neotropical region was provided by Sclater (1858). He divided the world into six zoogeographic regions, based on bird taxa (Passeriformes): Palearctic, Ethiopian, Indian, Australian, Nearctic, and Neotropical. The Neotropical region, as defined by Sclater, included the West Indies, southern Mexico, Central America, the whole South America, and the Galapagos and Falkland Islands. Sclater (1858) provided a list of genera of the Neotropical region. Two decades later, Wallace (1876) accepted Sclater's scheme and applied it to other vertebrate taxa, promoting the use of this regionalization as an organizing principle of biogeographic inquiry (Whittaker et al., 2013). According to the Sclater-Wallace system, the Neotropical region (Figure 2.1) comprises South America and Central America and reaches as far north as central Mexico, where it limits with the Nearctic region. The Sclater-Wallace system was followed by many zoogeographers (e.g., Murray, 1866; Huxley, 1868; Kirby, 1872; Allen, 1892; Sclater, 1894; Heilprin, 1897; Bartholomew et al., 1911; Mello-Leitão, 1937; Darlington, 1957; Morain, 1984; Fleming, 1987). It is considered the "standard" system, especially for authors analyzing the distribution of vertebrate taxa (Cox, 2001).

Engler (1879, 1882) provided a phytogeographic regionalization of the world, recognizing the Holarctic, South American, Paleotropical, and Old Oceanic kingdoms. The South American phytogeographic kingdom is equivalent to the Neotropical zoogeographic region of Sclater and Wallace, but in South America it is restricted to the tropical areas. The southern temperate part of South America was assigned by Engler to the Old Oceanic kingdom, together with New Zealand's South Island, the subantarctic islands, most of Australia, and South Africa. This kingdom was later renamed to Austral kingdom (Engler, 1899). Engler's system was widely adopted by phytogeographers, although some authors preferred to exclude South Africa and Australia from the Antarctic or Holantarctic kingdom, leaving only southern South America and New Zealand (Good, 1947; Mattick, 1964; Cabrera and Willink, 1973;

Neotropical region

Figure 2.1 Map of the Neotropical region and its subregions. (From Wallace, A. R., *The geographical distribution of animals. Vol. I and II*. Harper and Brothers, New York, 1876.)

Takhtajan, 1986). Others further separated South America and New Zealand into different kingdoms (Drude, 1884; Diels, 1908).

Several studies based on plant and invertebrate taxa suggested a more restrictive definition of the Neotropical region. For example, some authors excluded the southern portion and the Andean area of South America from the Neotropical region, because of their closest links with other Austral areas, mainly Australia, Tasmania, New Guinea, New Zealand, and South Africa (Blyth, 1871; Engler, 1882; Drude, 1884; Gill, 1885; Allen, 1892; Lydekker, 1896; Diels, 1908; Jeannel, 1938, 1942; Good, 1947; Monrós, 1958; Rapoport, 1968; Cabrera and Willink, 1973; Amorim and Tozoni, 1994; Morrone, 2002, 2006; Moreira-Muñoz, 2007). Kuschel (1964) discussed the geographic distribution of plant and animal taxa from the southern continents, and based on the phylogenetic relationships of several taxa and the possibility of Antarctica having behaved in the past as a land bridge, proposed an Austral region, to which he assigned the Patagonian (southern South America), South African, and Australian subregions. In this restricted sense, the Neotropical region corresponds to the tropics of South America, Central America, southern Mexico, and the West Indies, excluding the Andean area and southern South America, which belong to the Andean region (Austral kingdom), and northern Mexico, which belongs to the Nearctic region (Holarctic kingdom). Additionally, two transition zones have been recognized: the Mexican transition zone in the overlap between the Nearctic and Neotropical regions, and the South American transition zone between the Neotropical and Andean regions (Morrone, 2006, 2015b).

Some recent quantitative analyses have recovered the Neotropical region as the Sclater-Wallace system (Kreft and Jetz, 2010; Holt et al., 2013a; Rueda et al., 2013); however, they are based only on vertebrate taxa, usually at the species level. The analysis of Procheş and Ramdhani (2012) showed a more restricted Neotropical region, excluding southern South America. Studies considering other taxa (invertebrates and plants) and the information provided by supraspecific taxa (see discussion by Szumik and Goloboff, 2015) are strongly needed for an appropriate evaluation of this issue.

BIOGEOGRAPHIC PLACEMENT OF THE NEOTROPICAL REGION

Sclater (1858) grouped the Nearctic and Neotropical regions into the Neogean creation and the remaining regions in the Paleogean creation. This scheme follows basically the division into the New and Old World, which does not represent a natural regionalization.

Newbigin (1950) recognized five regions for the world: Northern Lands (Holarctic), Mediterranean region, Northern Paleotropical Desert, Intertropical region, and Austral region (southern parts of the southern continents). The Intertropical region comprises all the tropical areas of the world, including the Neotropics, Africa, and southern and southeast Asia.

Rapoport (1968) recognized three biogeographic regions for the world. The Holarctic region comprises the areas of the Northern Hemisphere, including the Nearctic and Palearctic regions of previous authors. The Holotropical region comprises the Neotropical region, Africa, Southeast Asia, and the Pacific islands. The Holantarctic region comprises southern South America, South Africa, Australia, New Guinea, New Zealand, and Antarctica.

I (Morrone, 2015b) provided a consensus scheme, dividing the world into the Holarctic, Holotropical, and Austral kingdoms and nine regions (Figure 2.2). The

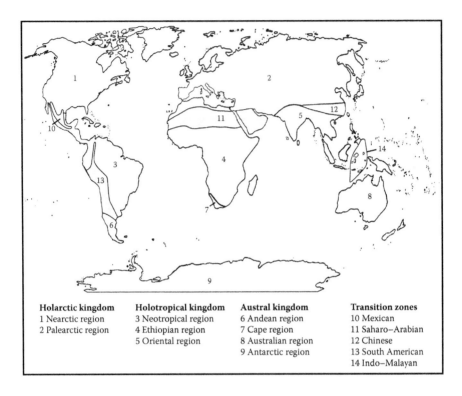

Holarctic kingdom	Holotropical kingdom	Austral kingdom	Transition zones
1 Nearctic region	3 Neotropical region	6 Andean region	10 Mexican
2 Palearctic region	4 Ethiopian region	7 Cape region	11 Saharo–Arabian
	5 Oriental region	8 Australian region	12 Chinese
		9 Antarctic region	13 South American
			14 Indo–Malayan

Figure 2.2 Map of the world biogeographic regionalization, indicating transition zones. (Modified from Morrone, J. J., *Journal of Zoological Systematics and Evolutionary Research*, 53: 249–257, 2015.)

Neotropical region belongs to the Holotropical kingdom, which corresponds to the tropical areas of the world, approximately between 30° south latitude and 30° north latitude. From a paleogeographic viewpoint, the Holotropical kingdom corresponds to the western portion of the Gondwanaland paleocontinent (Crisci et al., 1993) or tropical Gondwanaland (Amorim et al., 2009). Additionally, the Neotropical region has two transition zones: the Mexican transition zone in the boundary with the Nearctic region (Holarctic kingdom) and the South American transition zone in the boundary with the Andean region (Austral kingdom). This means that the Neotropical region exhibits biotic connections with the other tropical regions belonging to the Holotropical kingdom, mostly shaped by vicariance, and also connections with regions of the Holarctic and Austral kingdoms, mostly due to dispersal (Morrone, 2015b).

EARLY REGIONALIZATION OF THE NEOTROPICAL REGION

There have been several proposals recognizing subregions, dominions, provinces, and other units within the Neotropical region (see reviews by Fernández and Cuezzo, 1997; Sánchez Osés and Pérez-Hernández, 1998, 2005; Morrone, 2010b,

2014b). Some of these proposals have dealt with the Neotropical region as a whole, whereas others have been restricted to North America, to South America, to the Antilles, or even to single countries. Particularly interesting are the regionalizations of Mexico and Argentina, because they involve the northern and southern boundaries, respectively, of the Neotropical region.

Wallace (1876) provided the first regionalization of the Neotropical region, recognizing four subregions within it: Mexican (southern Mexico and Central America), Antillean (West Indies), Brazilian (tropical South America), and Chilean (southern or temperate South America). Wallace considered that the Chilean subregion was transitional to the Australian region and that the Mexican subregion was transitional to the Nearctic region. These subregions were followed by several authors during the last decades of the nineteenth century and the first decades of the twentieth century (Heilprin, 1887; Lydakker, 1896; Bartholomew et al., 1911; Mello-Leitão, 1937; Hershkovitz, 1969).

Mello-Leitão (1937), based on arachnids, modified slightly Wallace's scheme, recognizing five subregions: Mexican (Mexico and Central America), Antillean (Antilles, except Trinidad and Tobago), Brazilian (South America west of the Andes, from Colombia and Guyana to Bahía Blanca, Argentina), Andean–Patagonian (Andean cordillera, from Colombia to Chile), and Western Insular (Galapagos and Juan Fernández Islands). Mello-Leitão (1937) considered that the Mexican subregion had a mixture of Nearctic and Neotropical faunas. Within the Brazilian subregion, he recognized five provinces, which he named Bororô, Hyléa, Gê, Guaraní, and Tupí.

Cabrera and Yepes (1940) proposed a geographic scheme of South America, based on the distribution of mammals. They considered that Sclater's Guianan–Brazilian and Patagonian subregions were characterized by the proportion of species belonging to different mammal orders. The Guianan–Brazilian subregion corresponded to the tropics of South America, predominating the lowlands with forests and savannas. The highest elevations were situated in the northwestern part of the subregion, in the Ecuadorian and Colombian Andes. Its southern boundary followed an oblique line that goes from northwest to southeast, from northern Peru to central Argentina. Characteristic mammals correspond basically to the orders Marsupialia, Chiroptera, Primates, and Xenarthra. The Patagonian subregion corresponded to the rest of the continent, extending in most of Peru, Bolivia, Argentina, and all of Chile. Sclater (1858) and Wallace (1876) had previously named it Chilean subregion, but Cabrera and Yepes (1940) considered more appropriate to name it Patagonian. They cited characteristic species of Cervidae, Camelidae, and Rodentia. Additionally, Cabrera and Yepes (1940) defined 11 smaller divisions, which they named districts (considered by themselves as equivalent to provinces): Sabana, Amazonian, Tropical, Subtropical, Tupí, Pampasic, Patagonian, Subandean, Chilean, Andean, and Incasic (Figure 2.3). These districts were based on physiographic criteria and the presence of some mammal species.

Smith (1941) considered that Mexico was of great interest, because the Nearctic and Neotropical regions mix in the country, making it particularly difficult to establish a clear boundary between them. Considering that available distributional data on animal taxa were insufficient for zoogeographic studies, Smith (1941) proposed that

1 Sabana district
2 Amazonian district
3 Tropical district
4 Subtropical district
5 Tupí district
6 Pampasic district
7 Patagonian district
8 Subandean district
9 Chilean district
10 Andean district
11 Incasic district

Figure 2.3   Map of the zoogeographic regionalization of South America. (Modified from Cabrera, A. and J. Yepes, *Mamíferos sud-americanos (vida, costumbres y descripción)*. Historia Natural Ediar, Buenos Aires, 1940.)

the distribution of the lizard genus *Sceloporus* (Phrynosomatidae) was appropriate for these studies, because it was rich in species and widespread in the country. He considered that the provinces situated approximately south of the Tropic of Cancer belonged to the Neotropical region (Mexican subregion) and that the provinces situated north of it belonged to the Nearctic region. The Neotropical provinces are the Chiapas Highlands, Tapachultecan, Petén, Yucatecan, Veracruzan, Tehuanan, and Lower Balsas Basin, and the Nearctic provinces are the Sandiegan, Arizonian, Apachian, Chihuahuan, Peninsular, Tamaulipan, Sinaloan, Durangoan, Austral–Eastern, Cabo de Baja California, Austral–Central, Hidalgan, Austral–Western, Upper Balsas Basin, Guerreran, and Oaxacan Highlands (Figure 2.4).

Ringuelet (1961) published an essay on several issues of the zoogeography of Argentina, highlighting the relevance of considering both ecological and historical criteria when regionalizing an area. Ringuelet (1961) considered that the primary divisions within a subregion were the dominions, which he considered equivalent to Mello-Leitão's provinces and Cabrera and Yepes' districts. He postulated that the fauna of Argentina included 10 "lineages," which may be considered cenocrons: primitive Austral–American, Gondwanic, Notogeic, Brazilian, Afro-Brazilian, Palearctic, Nearctic, Pacific, intrusive (freshwater), and autochthonous. In contrast to previous proposals, Ringuelet (1961) recognized three subregions: Guianan–Brazilian, Andean–Patagonian, and Araucanian, the latter corresponding to the *Nothofagus* forests of southern South America. The Guianan–Brazilian subregion comprised the Subtropical and Pampasic dominions, whereas the Andean–Patagonian subregion comprised the Andean, Central, and Patagonian dominions (Figure 2.5). Ringuelet (1961) considered that each of these dominions harbored a particular fauna, with taxa belonging to different lineages but ecologically similar. Ringuelet (1961) also postulated that the Guianan–Brazilian fauna extended southward in the past, reaching northern Patagonia, and that there was a secular dynamism in the boundaries between the Guianan–Brazilian and the Andean–Patagonian subregions.

Ryan (1963) defined biotic provinces in Central America, following Dice (1943), who had considered particular ecological associations, vegetation types, flora, fauna, climate, physiography, and soil. Ryan (1963) considered that his delineation of biotic provinces based only on terrestrial mammals was preliminary and that future analyses of other taxa would allow to revise it. He mapped the distributions of the Central American mammal species (only a few ones were excluded from the study), used similarity indices, and established isolines on a map. Based on this analysis, Ryan (1963) identified 10 biotic provinces: Yucatán, Chiapas–Guatemalan Highlands, Escuintla–Usulután, Lempira–Tegucigalpa, Mosquito, Chinandega, Nicaraguan Montane, Puntarenas–Chiriquí, Guatuso–Talamanca, and Chocó–Darién (Figure 2.6).

West (1964) provided a detailed regionalization of Mexico and Central America. He considered a basic division into three large areas: the extratropical dry lands in northern Mexico and the Baja California peninsula, the tropical highlands in the mountains of the Sierras Madre, and the tropical lowlands from the Pacific and Mexican Gulf coasts of Mexico and most of Central America. Within these large areas, he recognized several smaller units (Figure 2.7). The extratropical land included the Mesa del Norte, Sonora and northern Sinaloa, Baja California, and Tamaulipas subhumid lowlands.

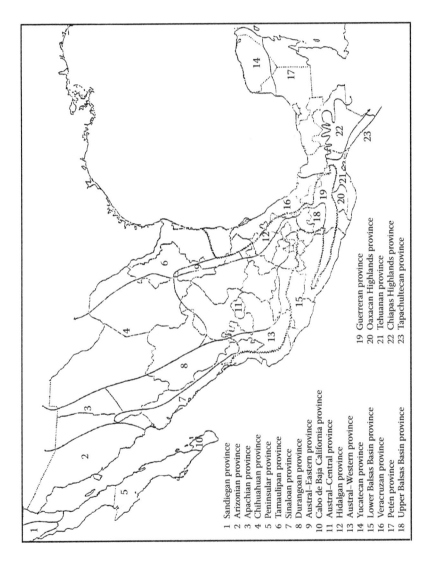

Figure 2.4 Map of the zoogeographic regionalization of Mexico. (Modified from Smith, H., *Anales de la Escuela Nacional de Ciencias Biológicas*, 2: 103–110, 1941.)

1 Sandiegan province
2 Arizonian province
3 Apachian province
4 Chihuahuan province
5 Peninsular province
6 Tamaulipan province
7 Sinaloan province
8 Durangoan province
9 Austral–Eastern province
10 Cabo de Baja California province
11 Austral–Central province
12 Hidalgan province
13 Austral–Western province
14 Yucatecan province
15 Lower Balsas Basin province
16 Veracruzan province
17 Petén province
18 Upper Balsas Basin province
19 Guerreran province
20 Oaxacan Highlands province
21 Tehuanan province
22 Chiapas Highlands province
23 Tapachultecan province

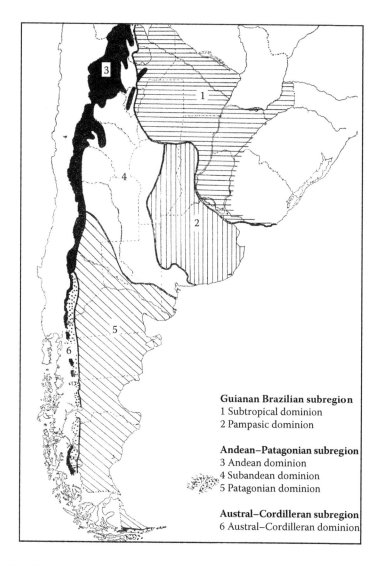

Guianan Brazilian subregion
1 Subtropical dominion
2 Pampasic dominion

Andean–Patagonian subregion
3 Andean dominion
4 Subandean dominion
5 Patagonian dominion

Austral–Cordilleran subregion
6 Austral–Cordilleran dominion

Figure 2.5 Map of the zoogeographic regionalization of Argentina. (Modified from Ringuelet, R. A., *Physis* (Buenos Aires), 22: 151–170, 1961.)

The tropical highlands included the Sierra Madre Occidental, Sierra Madre Oriental, Mesa Central, Sierra and Mesa del Sur, and the Highlands of Costa Rica and western Panama. Within the tropical lowlands, West (1964) included the Caribbean–Gulf lowlands (Petén–Yucatán rainforest, southern Veracruz–Tabasco rainforest, and Mosquito coast, among others), Pacific lowlands (savannas of central Panama, Azuero rainforest, and coastal lowlands of southwestern Mexico, among others), and dry interior tropical basins (Balsas–Tepalcatepec basin and Valley of Chiapas).

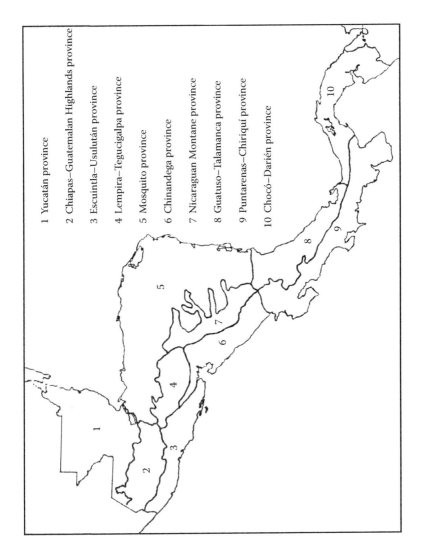

Figure 2.6 Map of the zoogeographic regionalization of Central America Mexico according to Ryan (1963). (Modified from Rapoport, E. H., "Algunos problemas biogeográficos del Nuevo Mundo con especial referencia a la región Neotropical." In: Delamare Debouteville, C. and E. H. Rapoport (eds.), *Biologie de l'Amerique Australe. Vol. 4.* CNRS, Paris, pp. 55–110, 1968.)

1 Yucatán province

2 Chiapas–Guatemalan Highlands province

3 Escuintla–Usulután province

4 Lempira–Tegucigalpa province

5 Mosquito province

6 Chinandega province

7 Nicaraguan Montane province

8 Guatuso–Talamanca province

9 Puntarenas–Chiriquí province

10 Chocó–Darién province

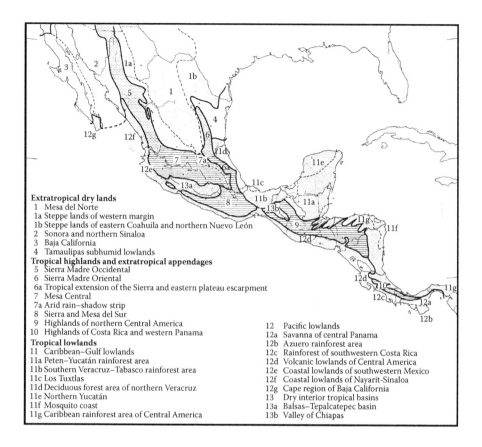

The following labels appear within the map:

Extratropical dry lands
1 Mesa del Norte
1a Steppe lands of western margin
1b Steppe lands of eastern Coahuila and northern Nuevo León
2 Sonora and northern Sinaloa
3 Baja California
4 Tamaulipas subhumid lowlands
Tropical highlands and extratropical appendages
5 Sierra Madre Occidental
6 Sierra Madre Oriental
6a Tropical extension of the Sierra and eastern plateau escarpment
7 Mesa Central
7a Arid rain–shadow strip
8 Sierra and Mesa del Sur
9 Highlands of northern Central America
10 Highlands of Costa Rica and western Panama
Tropical lowlands
11 Caribbean–Gulf lowlands
11a Peten–Yucatán rainforest area
11b Southern Veracruz–Tabasco rainforest area
11c Los Tuxtlas
11d Deciduous forest area of northern Veracruz
11e Northern Yucatán
11f Mosquito coast
11g Caribbean rainforest area of Central America

12 Pacific lowlands
12a Savanna of central Panama
12b Azuero rainforest area
12c Rainforest of southwestern Costa Rica
12d Volcanic lowlands of Central America
12e Coastal lowlands of southwestern Mexico
12f Coastal lowlands of Nayarit-Sinaloa
12g Cape region of Baja California
13 Dry interior tropical basins
13a Balsas–Tepalcatepec basin
13b Valley of Chiapas

Figure 2.7 Map of the zoogeographic regionalization of Mexico. (Modified from West, R. C., "The natural regions of Middle America." In: West, R. C. (ed.), *Handbook of Middle American Indians. Vol. 1.* University of Texas Press, Austin, pp. 363–383, 1964.)

Rapoport (1968) provided an insightful review of the biogeography of the New World, with particular emphasis on the Neotropical region. He divided it into the Central American (or Mexican), Antillean, Guianan–Brazilian, Andean–Patagonian, and Araucanian subregions, which he characterized in terms of their fauna. Rapoport (1968) considered that the Andean–Patagonian subregion was transitional between the Guianan–Brazilian and Araucanian subregions. He analyzed, in particular, the boundary between the Guianan–Brazilian and Andean–Patagonian subregions, which he named the *subtropical line* (Figure 2.8), discussing the delimitation of previous authors (see also Ruggiero and Ezcurra, 2003). Finally, Rapoport discussed the relationships of the Neotropical region with the tropics of Africa and southeastern Asia, considering the possibility of establishing a Holotropical region to group them.

Fittkau (1969) analyzed Mello-Leitão's (1937, 1943) and Cabrera and Yepes' (1940) zoogeographic divisions of South America, finding that although they had been based on different animals (arachnids and mammals, respectively), the areas

Figure 2.8 Map showing the boundaries between the Guianan–Brazilian and Andean–Patagonian subregions. (Modified from Rapoport, E. H., "Algunos problemas bio-geográficos del Nuevo Mundo con especial referencia a la región Neotropical." In: Delamare Debouteville, C. and E. H. Rapoport (eds.), *Biologie de l'Amerique Australe. Vol. 4*. CNRS, Paris, pp. 55–110, 1968.)

delimited were rather similar. Fittkau (1969) accepted the Guianan–Brazilian and Patagonian subregions but considered that their boundaries were difficult to place because of the existence of transition zones between them. He presented a system of 13 zoogeographic provinces (Figure 2.9), which are similar to Cabrera and Yepes' districts: Central America, Caquetio, Hyléa, Bororô, Carirí, Tupí, Guaraní, Incasic, Pampasic, Patagonian, Subandean, Chilean, and Andean.

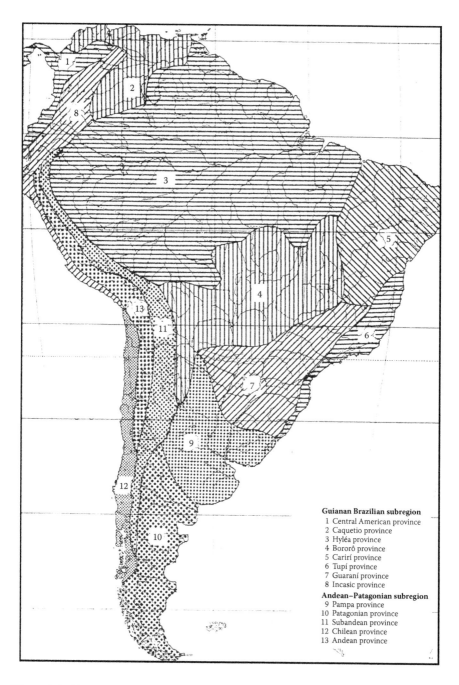

Figure 2.9 Map of the zoogeographic regionalization of South America. (Modified from Fittkau, E. J., "The fauna of South America." In: Fittkau, E., J. J. Illies, H. Klinge, G. H. Schwabe, and H. Sioli (eds.), *Biogeography and ecology in South America. Vol. 2*. Junk, The Hague, pp. 624–650, 1969.)

Kuschel (1969) analyzed the biogeography of South America, recognizing the Brazilian and Patagonian subregions (Figure 2.10), based on the distribution of Coleoptera. The Brazilian subregion occupied all the tropical areas east of the Andes, reaching the south in Antofagasta, Chile, and including also the Galapagos Islands. The Patagonian subregion occupied the Andes above 3000–3500 m and Patagonia, corresponding to the Chilean subregion of Wallace (1876). Kuschel (1969)

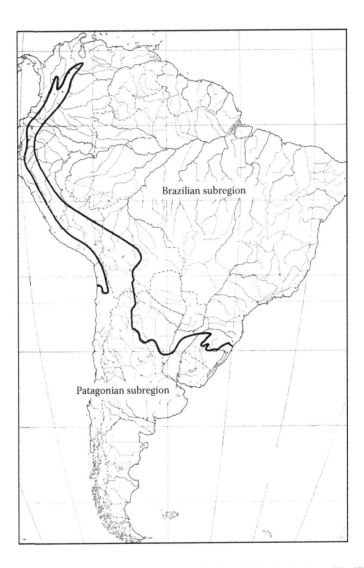

Figure 2.10 Map of the zoogeographic regionalization of South America. (Modified from Kuschel, G., "Biogeography and ecology of South American Coleoptera." In: Fittkau, E., J. J. Illies, H. Klinge, G. H. Schwabe, and H. Sioli (eds.), *Biogeography and ecology in South America. Vol. 2*. Junk, The Hague, pp. 709–722, 1969.)

also compared the ecology of the beetles of both subregions, referring especially to the ground, forest, litter, soil, and aquatic faunas.

Cabrera (1971) provided a phytogeographic regionalization of Argentina, which was based on his previous studies (Cabrera, 1951, 1953, 1957, 1958) where he used extensive floristic data. Cabrera (1971) recognized 13 provinces (Figure 2.11), which he grouped into two regions and five dominions. The Neotropical region comprises three dominions: Amazonian (Yungas and Paraná provinces), Chacoan (Chacoan, Espinal, Prepuna, Monte, and Pampean provinces), and Andean–Patagonian (High Andean, Puna, and Patagonian provinces). The Antarctic region includes

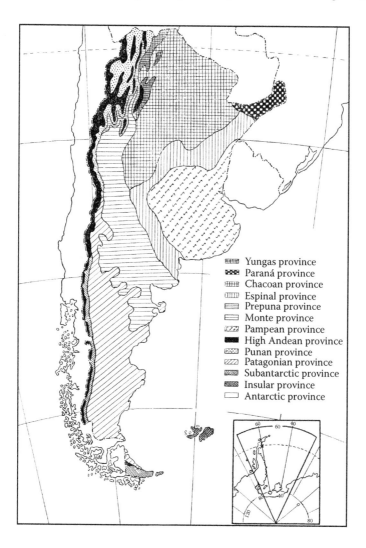

Figure 2.11 Map of the phytogeographic regionalization of Argentina. (Modified from Cabrera, A. L., *Boletín de la Sociedad Argentina de Botánica*, 14: 1–42, 1971.)

two dominions: Subantarctic (Subantarctic and Insular provinces) and Antarctic (Antarctic province). For each phytogeographic province, Cabrera (1971) provided a general description and a characterization of the vegetation, listed the endemic or characteristic species, and, in many cases, described districts. Ribichich (2002) analyzed this phytogeographic system, finding some incompatibilities with Cabrera's hypotheses concerning the assembly of the flora of these phytogeographic units.

MODERN BIOGEOGRAPHIC REGIONALIZATION

After decades of separate phytogeographic and zoogeographic regionalizations, Cabrera and Willink (1973) proposed a biogeographic regionalization of Latin America based on plant and animal taxa. Within the Neotropical region, they recognized five dominions: Caribbean (Mexico, Central America, and the Antilles), Amazonian, Guianan, Chacoan, and Andean–Patagonian. For the Neotropical region from North and Central America, Cabrera and Willink (1973) recognized five provinces, which they classified into two dominions (Figure 2.12). The Caribbean dominion includes the Mountain Mesoamerican, Mexican Xerophyllous, Caribbean, and Guajira provinces; the Amazonian dominion includes the Pacific province. For South America, Cabrera and Willink (1973) recognized 23 provinces for the Neotropical region (Figure 2.13), which they classified into five dominions. These are the Caribbean dominion (Guajira

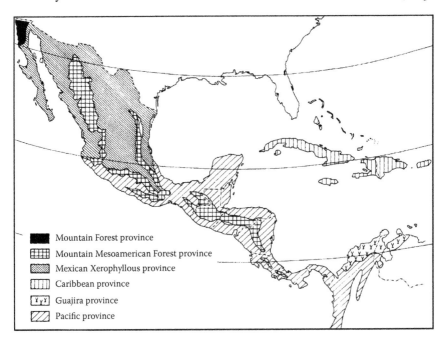

Figure 2.12 Map of the biogeographic regionalization of North and Central America. (Modified from Cabrera, A. L. and A. Willink, *Biogeografía de América Latina*. Monografía 13, Serie de Biología, OEA, Washington, D.C., 1973.)

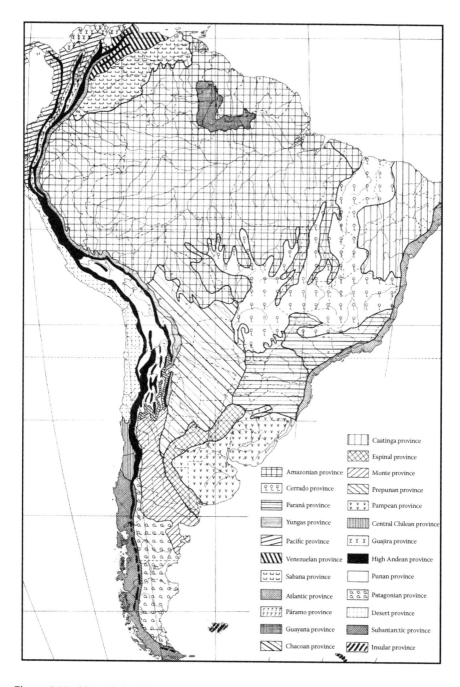

Figure 2.13 Map of the biogeographic regionalization of South America. (Modified from Cabrera, A. L. and A. Willink, *Biogeografía de América Latina*. Monografía 13, Serie de Biología, OEA, Washington, D.C., 1973.)

and Galapagos Islands provinces), Amazonian dominion (Amazonian, Pacific, Yungas, Venezuelan, Cerrado, Paraná, Sabana, Atlantic, and Páramo provinces), Guianan dominion (Guyana province), Chacoan dominion (Caatinga, Chacoan, Espinal, Prepunan, Monte, and Pampean provinces), and Andean–Patagonian dominion (High Andean, Punan, Desert, Central Chilean, and Patagonian provinces). Cabrera and Willink's (1973) Neotropical region did not include the southernmost area of South America, which was assigned to the Antarctic region. This scheme has been widely adopted and found to be useful for characterizing and naming geographic areas of many plant and animal taxa.

Müller (1973) analyzed the geographic distribution of Neotropical vertebrate taxa. He identified 40 dispersal centers (Figure 2.14): Central American Rainforest, Central American Montane Forest, Yucatán, Central American Pacific, Cocos Island, Costa Rican, Talamanca Páramo, Barranquilla, Santa Marta, Sierra Nevada, Magdalena, Cauca, Colombian Montane Forest, Colombian Pacific, North Andean, Catatumbo, Venezuelan Coastal Forest, Venezuelan Montane Forest, Caribbean, Roraima, Pantepui, Guyanan, Pará, Ucayali, Amazon, Yungas, Puna, Marañón, Andean Pacific, Galapagos, Caatinga, Campo Cerrado, Serra do Mar, Paraná, Uruguayan, Chaco, Monte, Pampa, Patagonian, and *Nothofagus*. For each center, Müller (1973) mapped the distributional areas of several endemic species. Additionally, for some centers, he recognized nested subcenters: the Talamanca Montane Forest and Guatemalan Montane Forests subcenters within the Central American Montane Forest center; the Mosquito and Chiriquí subcenters within the Costa Rican center; the Peruvian Andes and the Bogotá subcenters within the North Andean center; the Maracaibo and Venezuela subcenters within the Caribbean center; the Ucayali and Napo subcenters within the Amazon center; the Ecuadorian and Peruvian subcenters within the Andean Pacific center; and the Pernambuco and Bahia subcenters within the Serra do Mar center. Some of these centers and subcenters are coincident with areas recognized by other authors, whereas others represent smaller nested units.

Ringuelet (1975) undertook a global analysis of South American freshwater fishes. After discussing several previous proposals, Ringuelet (1975) proposed 20 provinces (Figure 2.15), which he grouped into two subregions and seven dominions. The Brazilian subregion comprises seven dominions: Orinoco–Venezuelan (Maracaibo, Caribbean Coast, Orinoquia, and Trinidad provinces), Pacific or Transandean (Northern Pacific and Guayas provinces), Magdalenan (Magdalena province), Andean (North Andean, Titicaca, and Cuyan Subandean provinces), Guianan–Amazonian (Guianan and Amazonian provinces), Paraná (Alto Paraguay, Alto Paraná, and Parano–Platense provinces), and Eastern Brazil (Northeastern Brazil, San Francisco River, and coastal rivers of Southeastern Brazil provinces). The Austral subregion includes the Chilean and Patagonian provinces.

Udvardy (1975) introduced a unified regionalization of the world, intended for biogeographic and conservation purposes, where he recognized the Palearctic, Nearctic, Africotropical, Indomalayan, Oceanian, Australian, Antarctic, and Neotropical kingdoms. Within the Neotropical kingdom, Udvardy (1975) recognized 47 provinces (Figure 2.16): Campechean, Panamanian, Colombian Coastal, Guianan, Amazonian, Madeiran, Serra do Mar, Brazilian Rainforest, Brazilian Planalto, Valdivian Forest, Chilean *Nothofagus*, Everglades, Sinaloan, Guerreran, Yucatecan, Central American,

1 Central American Rainforest province
2 Central American Montane Forest province
3 Yucatán province
4 Central American Pacific province
5 Coco province
6 Costa Rican province
7 Talamanca Paramo province
8 Baranquilla province
9 Santa Marta province
10 Sierra Nevada province
11 Madalena province
12 Cauca province
13 Colombian Montane Forest province
14 Colombian Pacific province
15 North Adean province
16 Catatumbo province
17 Venezuelan Coastal Forest province
18 Venezuelan Montane Forest province
19 Carribean province
20 Roraima province
21 Pantepui province
22 Guyanan province
23 Para province province
24 Ucayali province
25 Amazon province

26 Yungas province
27 Puna province
28 Marañon province
29 Andean Pacific province
30 Galápagos province
31 Caatinga province
32 Campo Cerrado province
33 Serra do Mar province
34 Parana province
35 Uruguayan province
36 Chaco province
37 Monte province
38 Pampa province
39 Patagonian province
40 *Nothofagus* province

Figure 2.14 Map showing the dispersal centers of Neotropical vertebrates. (Modified from Müller, P., *The dispersal centers of terrestrial vertebrates in the Neotropical realm: A study in the evolution of the Neotropical biota and its native landscapes.* Junk, The Hague, 1973.)

Venezuelan Dry Forest, Venezuelan Deciduous Forest, Ecuadorian Dry Forest, Caatinga, Gran Chaco, Chilean Araucaria Forest, Chilean Sclerophyll, Pacific Desert, Monte, Patagonian, Llanos, Campos Limpos, Babacu, Campos Cerrados, Argentinean Pampas, Uruguayan Pampas, Northern Andean, Colombian Montane, Yungas, Puna, Southern Andean, Bahamas–Bermudan, Cuban, Greater Antillean, Lesser Antillean, Revillagigedo Island, Cocos Island, Galapagos Islands, Fernando de Noronha Island, South Trindade Island, and Lake Titicaca.

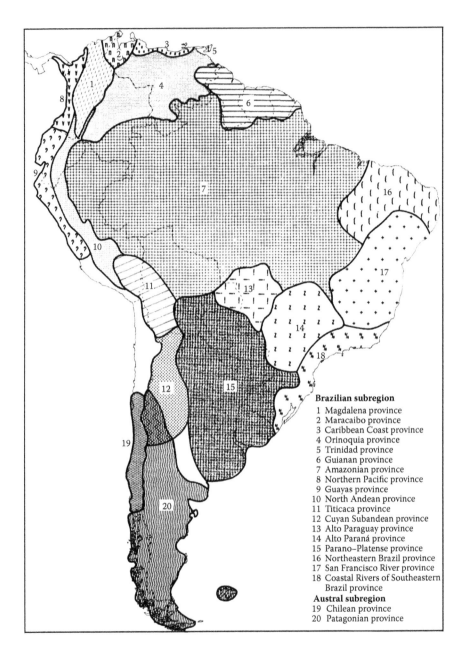

Brazilian subregion

1 Magdalena province
2 Maracaibo province
3 Caribbean Coast province
4 Orinoquia province
5 Trinidad province
6 Guianan province
7 Amazonian province
8 Northern Pacific province
9 Guayas province
10 North Andean province
11 Titicaca province
12 Cuyan Subandean province
13 Alto Paraguay province
14 Alto Paraná province
15 Parano–Platense province
16 Northeastern Brazil province
17 San Francisco River province
18 Coastal Rivers of Southeastern
    Brazil province
Austral subregion
19 Chilean province
20 Patagonian province

Figure 2.15 Map of the regionalization of South America. (Modified from Ringuelet, R. A., *Ecosur*, 2: 1–122, 1975.)

1 Campechean province
2 Panamanian province
3 Colombian Coastal province
4 Guianan province
5 Amazonian province
6 Madeiran province
7 Serra do Mar province
8 Brazilian Rainforest province
9 Brazilian Planalto province
10 Valdivian Forest province
11 Chilean Nothofagus province
12 Everglades province
13 Sinaloan province
14 Guerreran province
15 Yucatecan province
16 Central American province
17 Venezuelan Dry Forest province
18 Venezuelan Deciduous Forest province
19 Ecuadorian Dry Forest province
20 Caatinga province
21 Gran Chaco province
22 Chilean *Araucaria* Forest province
23 Chilean Sclerophyll province
24 Pacific Desert province
25 Monte province
26 Patagonian province
27 Llanos province
28 Campos Limpos province
29 Babacu province
30 Campos Cerrados province
31 Argentinean Pampas province
32 Uruguayan Pampas province
33 Northern Andean province
34 Colombian Montane province
35 Yungas province
36 Puna province
37 Southern Andean province
38 Bahama–Bermudan province

39 Cuban province
40 Greater Antillean province
41 Lesser Antillean province
42 Revillagigedo Island province
43 Cocos Island province
44 Galapagos Island province
45 Fernando de Noronha Island province
46 South Trinidade Island province
47 Lake Titicaca province

Figure 2.16   Map of the provinces of Latin America. (Modified from Udvardy, M. D. F., *A classifi-cation of the biogeographical provinces of the world.* International Union for Conservation of Nature and Natural Resources Occasional Paper 18, Morges, 1975.)

Rzedowski (1978) published a phytogeographic synthesis of Mexico, where he discussed the physiographic and climatic bases for phytogeography, the geographic origins of the Mexican flora, and the vegetation types of the country. In order to explain the origins of the Mexican flora, Rzedowski (1978) postulated the existence of six floral elements: meridional (taxa with Central and South American affinities), boreal (Nearctic), Antillean, eastern Asian, African, and endemic. He also proposed 17 phytogeographic provinces (Figure 2.17), which he classified into two kingdoms and four regions. The Holarctic kingdom includes the North American Pacific region (California and Guadalupe Island provinces) and the Mountain Mesoamerican region (Sierra Madre Occidental, Sierra Madre Oriental, Meridional Mountains, and Transisthmic Mountains provinces).

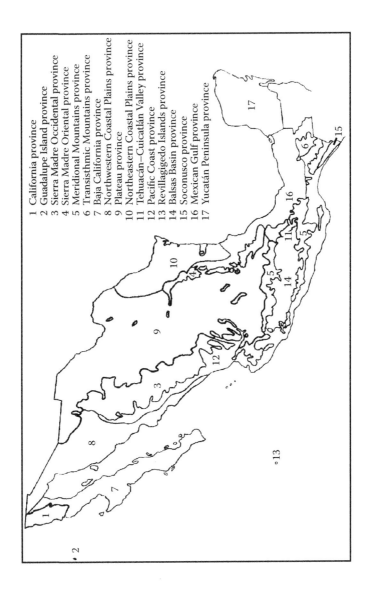

1 California province
2 Guadalupe Island province
3 Sierra Madre Occidental province
4 Sierra Madre Oriental province
5 Meridional Mountains province
6 Transisthmic Mountains province
7 Baja California province
8 Northwestern Coastal Plains province
9 Plateau province
10 Northeastern Coastal Plains province
11 Tehuacán–Cuicatlán Valley province
12 Pacific Coast province
13 Revillagigedo Islands province
14 Balsas Basin province
15 Soconusco province
16 Mexican Gulf province
17 Yucatán Peninsula province

Figure 2.17 Map of the phytogeographic regionalization of Mexico. (Modified from Rzedowski, J., *La vegetación de México*. Editorial Limusa, Mexico City, 1978.)

The Neotropical kingdom includes the Mexican Xerophytic region (Baja California, Northwestern Coastal Plains, Plateau, Northeastern Coastal Plains, and Tehuacán–Cuicatlán Valley provinces) and the Caribbean region (Pacific Coast, Revillagigedo Islands, Balsas Basin, Soconusco, Mexican Gulf Coast, and Yucatán Peninsula provinces). Rzedowski's (1978) scheme has been widely used by Mexican botanists.

Rivas-Martínez and Tovar (1983) proposed that the Andean highlands, the Pacific coast between 5° and 38°, and Patagonia constituted a biogeographic unit within the Neotropical kingdom, which they named Andean subkingdom. The other biogeographic units recognized by them for South America were the Caribbean–Amazonian subkingdom, the group of Chacoan regions, and the Caatinga and Subantarctic regions (Figure 2.18). Within the Andean subkingdom, the authors

1 Páramo region
2 Puna region
3 Pacific Desert region
4 Central Chilean region
5 Patagonian region
6 Caribbean–Amazonian subkingdom
7 Group of Chacoan regions
8 Caatinga region
9 Andean Subantarctic region

Figure 2.18   Map of the regionalization of South America. (Modified from Rivas-Martínez, S. and O. Tovar, *Collectanea Botanica* [Barcelona], 14: 515–521, 1983.)

recognized the Páramo, Puna, Pacific Desert, Central Chilean, and Patagonian regions, each characterized by different endemic plant genera.

Takhtajan (1986) provided a phytogeographic regionalization of the world. Within the Neotropical kingdom, he recognized seven regions: Caribbean (comprising the Baja California peninsula, Mexican lowlands, Revillagigedo Islands, southern Florida and the Florida Keys, the West Indies, Bahamas, Bermudas, Guatemala–Panama, northern Colombia and northern Venezuela), Venezuela and Surinam (Orinoco Basin, Venezuelan highlands, sandstone mountains of British Guiana, and Surinam), Amazon (lowlands of the Amazon basin and eastern coasts of Brazil), Central Brazilian (central Brazil, highlands of eastern Brazil, and Chaco), Pampas (Uruguay, southeastern Brazil, and eastern Argentina), Andean (Galapagos and Cocos Islands, Andes, Atacama desert, and Chilean sclerophyll zone), and Fernandezian (Juan Fernández and Desventuradas Islands).

Samek et al. (1988) provided a phytogeographic regionalization of the Caribbean, which they treated as a separate region. Within this region, Samek et al. (1988) recognized three subregions and 11 provinces (Figure 2.19). The Central American subregion includes the Revillagigedo Islands, Central American Pacific Coast, Mexican Gulf, and Guatemala–Panama provinces; the Antillean subregion comprises the Bermudas–Bahamas–Florida, Cuba, Hispaniola, Jamaica, Puerto Rico, and Lesser Antilles provinces; and the Colombia and Septentrional Venezuela subregion comprise the homonymous province. This regionalization integrates in a single unit areas from Mexico, Central America, the Antilles, and northwestern South America, as stated by other authors, for example, Good (1947), Takhtajan (1986), and Morrone (2001a,e, 2006).

Rivas-Martínez and Navarro (1994) recognized a Neotropical–Austroamerican kingdom. They recognized 49 provinces (Figure 2.20), which they classified into subkingdoms and regions. The Neotropical subkingdom includes the Caribbeo-Mexican region (Floridan, Cuban, and Antillean provinces), Colombian–Mesoamerican region (Mesoamerican, Colombian, Ecuatorian, and Galapagos Islands provinces), Venezuelan region (Septentrional Venezuelan, Llanos, and Tepuis provinces), Amazonian region (Loreto, Rio Negro, Madeira, Acre-Madre de Dios, Roraima–Trombetas, Xingu–Tapajós, Guayanas, and Amazon Delta provinces), Brazilian–Paraná region (Cerrado, Tocantins, Beni, Pantanal, Atlantic, Paraná, and Caatinga provinces), Andean region (Peruvian, Bolivian, Argentinean–Atacaman, Monte, Páramo, and Yunga provinces), Chacoan region (Septentrional Chaco, Meridional Chaco, and Andean Chaco provinces), and Peruvian Pacific Desert region (Peruvian Desert and Atacama Hyperdesert provinces). The Austroamerican subkingdom includes the Pampean region (Semitropical Pampean, Central Pampean, and Xerophytic Pampean provinces), Mesochilean–Patagonian region (Desert Mesochilean and Central Chilean provinces), Andean–Patagonian subregion (Mediterranean Andean, Septentrional Patagonian, and Meridional Patagonian provinces), and Valdivian–Magellanic region (Valdivian, Austroandean, Fuegian, Juan Fernández Islands, and Antarctic provinces). Rivas-Martínez et al. (2011) presented a revised version of their regionalization, with the same provinces but arranged in a different scheme.

Dinerstein et al. (1995) proposed a system of ecoregions for Latin America and the Caribbean. These 178 ecoregions were classified based on their major ecosystem types into tropical broadleaf forests (tropical moist broadleaf forests

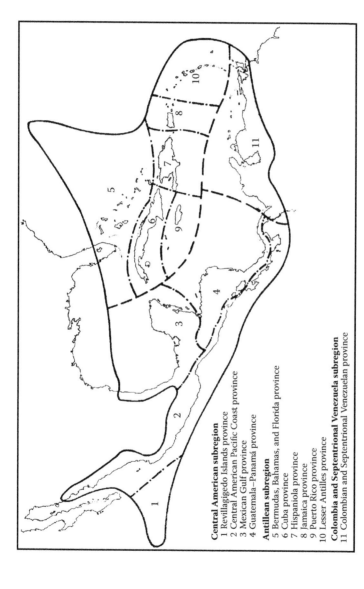

Central American subregion
1 Revillagigedo Islands province
2 Central American Pacific Coast province
3 Mexican Gulf province
4 Guatemala–Panamá province

Antillean subregion
5 Bermudas, Bahamas, and Florida province
6 Cuba province
7 Hispaniola province
8 Jamaica province
9 Puerto Rico province
10 Lesser Antilles province

Colombia and Septentrional Venezuela subregion
11 Colombian and Septentrional Venezuelan province

Figure 2.19 Map of the phytogeographic regionalization of the Caribbean. (Modified from Samek, V. et al., *Revista del Jardín Botánico Nacional* [La Habana], 9: 25–38, 1988.)

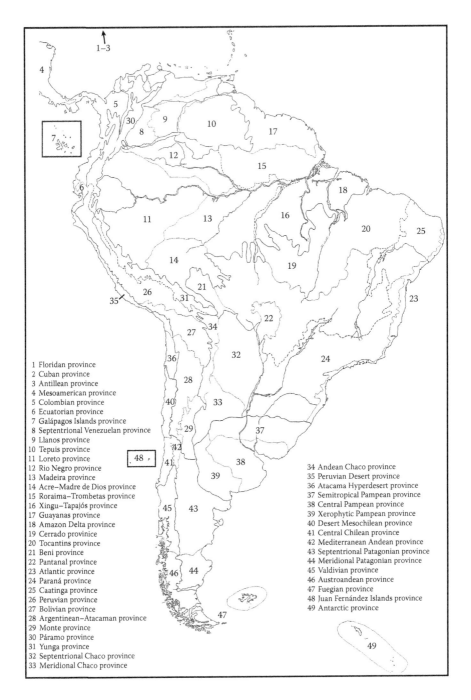

1 Floridan province
2 Cuban province
3 Antillean province
4 Mesoamerican province
5 Colombian province
6 Ecuatorian province
7 Galápagos Islands province
8 Septentrional Venezuelan province
9 Llanos province
10 Tepuis province
11 Loreto province
12 Rio Negro province
13 Madeira province
14 Acre–Madre de Dios province
15 Roraima–Trombetas province
16 Xingu–Tapajós province
17 Guayanas province
18 Amazon Delta province
19 Cerrado provinice
20 Tocantins province
21 Beni province
22 Pantanal province
23 Atlantic province
24 Paraná province
25 Caatinga province
26 Peruvian province
27 Bolivian province
28 Argentinean–Atacaman province
29 Monte province
30 Páramo province
31 Yunga province
32 Septentrional Chaco province
33 Meridional Chaco province

34 Andean Chaco province
35 Peruvian Desert province
36 Atacama Hyperdesert province
37 Semitropical Pampean province
38 Central Pampean province
39 Xerophytic Pampean province
40 Desert Mesochilean province
41 Central Chilean province
42 Mediterranean Andean province
43 Septentrional Patagonian province
44 Meridional Patagonian province
45 Valdivian province
46 Austroandean province
47 Fuegian province
48 Juan Fernández Islands province
49 Antarctic province

Figure 2.20 Map of the regionalization of Latin America. (Modified from Rivas-Martínez, S. and G. Navarro, *Mapa biogeográfico de Suramérica*. Published by the authors, Madrid, 1994.)

and tropical dry broadleaf forests), conifer/temperate broadleaf forests (temperate forests and tropical and subtropical coniferous forests), grasslands/savannas/shrublands (grasslands, savannas and shrublands, flooded grasslands, and montane grasslands), and xeric formations (Mediterranean scrub and *restingas*). Dinerstein et al. (1995) grouped the ecoregions into seven major bioregions (Figure 2.21):

1 Caribbean bioregion

2 Northern Andes bioregion

3 Orinoco bioregion

4 Amazonia bioregion

5 Central Andes bioregion

6 Eastern South America bioregion

7 Southern South America bioregion

Figure 2.21   Map showing the bioregions of Latin America and the Caribbean. (Modified from Dinerstein, E. et al., *A conservation assessment of the terrestrial ecoregions of Latin America and the Caribbean*. The World Bank, Washington, D.C., 1995.)

Caribbean, Northern Andes, Orinoco, Amazonia, Central Andes, Eastern South America, and Southern South America.

BIOGEOGRAPHIC REGIONALIZATION
IN THE TWENTY-FIRST CENTURY

I (Morrone, 2001a) synthesized some previous schemes and, mostly based on track analyses of plant and animal taxa, provided a biogeographic regionalization of Latin America and the Caribbean. This scheme divided the Neotropical region into four subregions: Caribbean, Amazonian, Chacoan, and Paraná. Then, the Mexican and South American transition zones were formalized and incorporated into the scheme (Morrone, 2004a, 2006). The resulting biogeographic classification of the Neotropical region is as follows: Mexican transition zone (Sierra Madre Occidental, Sierra Madre Oriental, Transmexican Volcanic Belt, Sierra Madre del Sur, and Chiapas provinces), Caribbean subregion (Mexican Pacific Coast, Mexican Gulf, Balsas Basin, Eastern Central America, Western Panamanian Isthmus, Yucatán Peninsula, Bahama, Cuba, Cayman Islands, Jamaica, Hispaniola, Puerto Rico, Lesser Antilles, Chocó, Maracaibo, Venezuelan Coast, Trinidad and Tobago, Magdalena, Venezuelan Llanos, Cauca, Galapagos Islands, Western Ecuador, Arid Ecuador, and Tumbes–Piura provinces), Amazonian subregion (Napo, Imerí, Guyana, Humid Guyana, Roraima, Amapá, Varzea, Ucayali, Madeira, Tapajós–Xingu, Pará, Pantanal, and Yungas provinces), Chacoan subregion (Caatinga, Cerrado, Chaco, and Pampa provinces), Paraná subregion (Brazilian Atlantic Forest, Paraná Forest, and *Araucaria angustifolia* Forest provinces), and South American transition zone (North Andean Páramo, Coastal Peruvian Desert, Puna, Atacama, Prepuna, and Monte provinces).

Espinosa et al. (2008) reviewed four regionalizations previously proposed for Mexico using different criteria, namely, morphotectonics (Ferrusquía-Villafranca, 1990), plant taxa (Rzedowski and Reyna-Trujillo, 1990), herpetofauna (Casas-Andreu and Reyna-Trujillo, 1990), and mammals (Ramírez-Pulido and Castro-Campillo, 1990), as well as some consensus systems (Morrone et al., 2002; Morrone, 2005). They recognized 20 Mexican provinces (Figure 2.22), classified into the Nearctic and Neotropical regions. The Nearctic region includes the California province and the Mexican transition zone (Sierra Madre Occidental, Sierra Madre Oriental, Transmexican Volcanic Belt, Sierra Madre del Sur, Soconusco, Chiapas Highlands, and Oaxaca provinces). The Neotropical region is divided into the Northern Arid Neotropics (Baja California, Cabo, Revillagigedo Islands, Northern Plateau, Southern Plateau, and Sonora provinces) and the Mesoamerican Subhumid and Humid Neotropics (Pacific Coast, Balsas Basin, Tamaulipas, Mexican Gulf, Petén, and Yucatán provinces).

During the last two decades, several cladistic biogeographic analyses have suggested that the previously recognized Amazonian and Caribbean subregions were not natural areas (Nihei and de Carvalho, 2007; Sigrist and de Carvalho, 2009; Pires and Marinoni, 2010; Echeverry and Morrone, 2013). I (Morrone, 2014c) undertook a cladistic biogeographic analysis using smaller units to test their naturalness, based

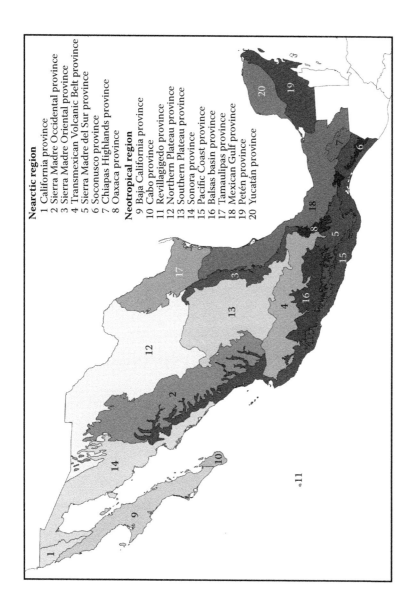

Nearctic region
1 California province
2 Sierra Madre Occidental province
3 Sierra Madre Oriental province
4 Transmexican Volcanic Belt province
5 Sierra Madre del Sur province
6 Soconusco province
7 Chiapas Highlands province
8 Oaxaca province

Neotropical region
9 Baja California province
10 Cabo province
11 Revillagigedo province
12 Northern Plateau province
13 Southern Plateau province
14 Sonora province
15 Pacific Coast province
16 Balsas basin province
17 Tamaulipas province
18 Mexican Gulf province
19 Petén province
20 Yucatán province

Figure 2.22 Map of the biogeographic regionalization of Mexico. (Modified from Espinosa Organista, D. et al., "El conocimiento biogeográfico de las especies y su regionalización natural." In: Sarukhán, J. (ed.), *Capital natural de México. Vol. I. Conocimiento actual de la biodiversidad.* Conabio, Mexico City, pp. 33–65, 2008.)

Mexican transition zone
1 Sierra Madre Occidental province
2 Sierra Madre Oriental province
3 Transmexican Volcanic Belt province
4 Sierra Madre del Sur province
5 Chiapas Highlands province

Antillean subregion
6 Bahama province
7 Cuba province
8 Cayman Islands province
9 Jamaica province
10 Hispaniola province
11 Puerto Rico province
12 Lesser Antilles province

Brazilian subregion
Mesoamerican dominion
13 Pacific Lowlands province
14 Balsas Basin province
15 Veracruzan province
16 Yucatán Peninsula province
17 Mosquito province

Pacific dominion
18 Guatuso–Talamanca province
19 Puntarenas–Chiriquí province
20 Chocó–Darién province
21 Guajira province
22 Venezuelan province
23 Trinidad province
24 Magdalena province
25 Sabana province
26 Cauca province
27 Galápagos Islands province
28 Western Ecuador province
29 Ecuadorian province

Boreal Brazilian dominion
30 Napo province
31 Imerí province
32 Pantepui province
33 Guianan Lowlands province
34 Roraima province
35 Pará province

South Brazilian dominion
36 Ucayali province
37 Madeira province
38 Rondônia province
39 Yungas province

Chacoan subregion
Southeastern Amazonian dominion
40 Xingu–Tapajós province
Chacoan dominion
41 Caatinga province
42 Cerrado province
43 Chaco province
44 Pampean province

Paraná dominion
45 Atlantic province
46 Paraná province
47 *Araucaria* Forest province

South American transition zone
48 Páramo province
49 Desert province
50 Puna province
51 Atacama province
52 Cuyan High Andean province
53 Monte province

Figure 2.23 Map of the biogeographic regionalization of Morrone (2014b).

on the taxon-area cladograms of 36 plant and animal taxa, which included insects (Orthoptera, Hemiptera, Coleoptera, Diptera, and Lepidoptera), spiders (Araneae), vertebrates (Aves and Mammalia), and plants. The main finding was that both subregions as previously considered by some authors were not natural units and had to be redefined. The resulting general area cladogram showed a first split separating the Antilles from the continent and a second one dividing the continental areas into a northwestern and a southeastern component. Within the northwestern component, the areas follow the sequence northern Amazonia, southwestern Amazonia, northwestern South America, and Mesoamerica. Within the southeastern component, the areas follow the sequence southeastern Amazonia, Chaco, and Paraná. Based on this analysis, I proposed a revised regionalization (Morrone, 2014b), which constitutes the basis of this book. This regionalization recognizes the Antillean, Brazilian, and Chacoan subregions, and the Mexican and South American transition zones, and, within them, seven dominions and 53 provinces (Figure 2.23).

CHAPTER 3

The Neotropical Region

The Neotropical region comprises the tropics of the New World in most of South America, Central America, southern and central Mexico, and the Antilles (Rapoport, 1968; Fittkau, 1969; Cabrera and Willink, 1973; Löwenberg-Neto, 2014; Morrone, 2014b, 2015b). This region was recognized originally by Sclater (1858) and its circumscription in zoogeography was consolidated by Wallace (1876). It has received several alternative names and has been also treated as a kingdom (Engler, 1882; Heilprin, 1887; Lydekker, 1896), subkingdom (Rivas-Martínez and Navarro, 1994), or subregion (Schmidt, 1954).

Its boundaries differ according to different authors. The Andes and southern South America were originally part of the region, but have been excluded by several authors, especially phytogeographers and zoogeographers working with invertebrates (e.g., Blyth, 1871; Engler, 1882; Lydekker, 1896; Good, 1947; Monrós, 1958; Kuschel, 1964; Cabrera and Willink, 1973; Amorim and Tozoni, 1994; Moreira-Muñoz, 2007). These areas are currently assigned to the Andean region of the Austral kingdom (Morrone, 2014b, 2015b). Southern Florida has been assigned to the Neotropical region in some regionalizations (e.g., Takhtajan, 1986; Samek et al., 1988; Rivas-Martínez and Navarro, 1994). Recent analyses, however, indicate that it belongs to the Nearctic region (Smith et al., 2004; Francisco-Ortega et al., 2007; Escalante et al., 2013).

From an ecoregional perspective, most of the Neotropical region corresponds to the Humid Tropical domain (Bailey, 1998). Within it, there are mostly rainforests and savannas in the lowlands, mountains with altitudinal zonation (e.g., in the Mexican and South American transition zones), and some areas assigned to the Dry domain (e.g., Caatinga and Monte).

NEOTROPICAL REGION

Neotropical region—Sclater, 1858: 143 (regional.); Murray, 1866: 297 (regional.); Kirby, 1872: 437 (regional.); Wallace, 1876: 78 (regional.); Sclater, 1894: 98 (regional.); Lydekker, 1896: 25 (regional.); Sclater and Sclater, 1899: 52 (regional.); Bartholomew et al., 1911: 9 (regional.); Newbigin, 1913: 221 (regional.); Mello-Leitão, 1937: 221 (regional.); Orfila, 1941: 85 (regional.); Lane, 1943: 409 (regional.); Vivó, 1943:

9 (regional.); Cabrera, 1951: 24 (regional.), 1953: 108 (regional.); Schmidt, 1954: 328 (regional.); Monrós, 1958: 143 (regional.); Halffter, 1964: 51 (biotic evol.), 1965: 2 (biotic evol.); Rapoport, 1968: 61 (biotic evol. and regional.); Fittkau, 1969: 624 (regional.); Hershkovitz, 1969: 3 (regional.); Cabrera, 1971: 5 (regional.); Cabrera and Willink, 1973: 32 (regional.); Ávila-Pires, 1974a: 134 (regional.); Cabrera, 1976: 1 (regional.); Bǎnǎrescu and Boşcaiu, 1978: 253 (regional.); Cadle, 1982: 1 (regional.); Pielou, 1992: 6 (regional.); Vuilleumier, 1993: 12 (biotic evol.); Amorim and Pires, 1996: 188 (regional.); Fernández and Cuezzo, 1997: 6 (regional.); Huber and Riina, 1997: 284 (glossary); Ortega and Arita, 1998: 772 (regional.); Morrone, 1999: 2 (regional.); Morrone et al., 1999: 510 (PAE); Zuloaga et al., 1999: 18 (regional.); Ippi and Flores, 2001: 50 (PAE); Marino et al., 2001: 115 (PAE); Morrone, 2001a: 25 (regional.), 2001d: 66 (regional.); Morrone, 2002: 150 (regional.); Morrone et al., 2002: 91 (regional.); Huber and Riina, 2003: 275 (glossary); Lücking, 2003: 43 (compat. anal.); MacDonald, 2003: 317 (text); Morrone, 2004a: 157 (track anal.), 2004b: 43 (regional.); Corona and Morrone, 2005: 37 (track anal.); Escalante et al., 2005: 202 (PAE); Martínez-Gordillo and Morrone, 2005: 23 (track anal.); Morrone, 2005: 238 (regional.); Procheş, 2005: 610 (cluster anal. and regional.); Sánchez Osés and Pérez-Hernández, 2005: 145 (regional.); Viloria, 2005: 448 (regional.); Morrone, 2006: 477 (regional.); Quijano-Abril et al., 2006: 1268 (track anal.); Espinosa Organista et al., 2008: 58 (regional.); López et al., 2008: 1564 (PAE); Löwenberg-Neto et al., 2008: 374 (macroecol.); Roig-Juñent et al., 2008: 24 (biotic evol.); Löwenberg-Neto and de Carvalho, 2009: 1751 (PAE); de Carvalho and Couri, 2010: 295 (track anal.); Morrone, 2010a: 34 (regional.); Urtubey et al., 2010: 505 (track anal. and clad. biogeogr.); Arana et al., 2011: 18 (track anal.); Löwenberg-Neto et al., 2011: 1942 (macroecol.); Pires and Marinoni, 2011: 8 (track anal.); Coulleri and Ferrucci, 2012: 105 (track anal.); Ferretti et al., 2012: 1 (track anal.); Mercado-Salas et al., 2012: 459 (track anal.); Procheş and Ramdhani, 2012: 263 (cluster anal. and regional.); Campos-Soldini et al., 2013: 16 (track anal.); Ferro, 2013: 323 (biotic evol.); Gutiérrez-Velázquez et al., 2013: 282 (PAE); Holt et al., 2013: 77 (cluster anal. and regional.); Puga-Jiménez et al., 2013: 1180 (PAE); Lamas et al., 2014: 955 (clad. biogeogr.); Morrone, 2014b: 26 (regional.), 2014c: 206 (clad. biogeogr.); Cione et al., 2015: 44 (biotic evol.); Klassa and Santos, 2015: 520 (endem. anal.); del Río et al., 2015: 1294 (track anal. and clad. biogeogr.); Morrone, 2015b: 85 (regional.).

Austro–Columbian region—Huxley, 1868: 315 (regional.).

Columbian region—Blyth, 1871: 428 (regional.).

South American kingdom—Engler, 1882: 345 (regional.).

Neotropical kingdom—Heilprin, 1887: 73 (regional.); Diels, 1908: 150 (regional.); Rizzini, 1963: 46 (regional.); Laubenfels, 1970: 34; Müller, 1973: 6 (regional.); Udvardy, 1975: 41 (regional.); Rzedowski, 1978: 104 (regional.); Rivas-Martínez and Tovar, 1983: 516 (regional.); Morain, 1984: 177 (text); Borhidi and Muñiz, 1986: 4 (regional.); Takhtajan, 1986: 250 (regional.); Muñiz, 1996: 283 (regional.); Huber and Riina, 1997: 233 (glossary); Brown et al., 1998: 31 (veget.); Olson et al., 2001: 934 (ecoreg.); Beierkuhnlein, 2007: 191 (text); Kreft and Jetz, 2010: 2044 (cluster anal. and regional.); Moreira-Muñoz, 2011: 136 (biotic evol. and regional.).

Tropical American region—Blanford, 1890: 49 (regional.); Rizzini, 1963: 46 (regional.), 1997: 622 (regional.).

American Tropical kingdom—Allen, 1892: 207 (regional.).

Neotropical area—Clarke, 1892: 381 (regional.).

Tropical region—Merriam, 1892: 33 (regional.).
Neogeic kingdom—Lydekker, 1896: 64 (regional.).
Neotropical subregion—Schmidt, 1954: 328 (regional.); Smith, 1983: 462 (cluster anal.
    and regional.); Morrone, 1996: 104 (regional.); Morrone and Coscarón, 1996: 1
    (PAE); Katinas et al., 1997: 112 (track anal.); Posadas et al., 1997: 2 (PAE).
Latin American region—Smith, 1983: 462 (cluster anal. and regional.).
Neotropical subkingdom—Rivas-Martínez and Navarro, 1994: map (regional.); Rivas-
    Martínez et al., 2011: 26 (regional.).
South American region—Cox, 2001: 519 (regional.).

Endemic and Characteristic Taxa

Taxa endemic to the Neotropical region include many genera and suprageneric groups. Neotropical plant families are Bromeliaceae, Cactaceae, and Heliconiaceae (Cabrera and Willink, 1973; Kress, 1990; Benzing, 2000). Additionally, there are many plant genera, e.g., *Astrocasia, Celianella, Chascotheca, Chonocentrum, Gonatogyne, Hyeronima, Jablonskia, Phyllanoa, Richeria,* and *Tacarcuna* among the Euphorbiaceae (Martínez-Gordillo and Morrone, 2005). Morrone (2001a,d) listed several insect taxa endemic to this region. Some generic examples include *Agra* (Carabidae; Erwin and Pogue, 1988), *Actinote, Agrias, Heliconius,* and *Prepona* (Nymphalidae; Brower, 1994; Viloria, 2005), *Nesiostrymon* and *Terra* (Lycaenidae; Viloria, 2005), *Nymphidium* and *Theope* (Riodinidae; Viloria, 2005), *Rhinacloa* (Miridae; Schuh and Schwartz, 1985; Figure 3.1), *Thecomyia* (Sciomyzidae; Pires and Marinoni, 2011), *Partamona* (Apidae; Camargo and Pedro, 2003), *Synoeca* (Menezes et al., 2015), and *Ptychoderes* (Anthribidae; Mermudes and Napp, 2006; Figure 3.2). Among Opiliones (Arachnida), families Agoristenidae, Biantidae, Cosmetidae, Cranaidae, Fissiphalliidae, Gerdesiidae, Gonyleptidae, Stygnidae, and Stygnopsidae are endemic to the Neotropics (Pinto-da-Rocha, 1997; Acosta, 2002; Kury, 2013; Bragagnolo et al., 2015). Haffer (1985) identified 22 bird families that are endemic or restricted to the Neotropics (e.g., hummingbirds [Trochilidae], toucans [Ramphastidae], ovenbirds [Furnariidae], antbirds [Thamnophilidae], and tanagers [Thraupidae]). Mammal taxa commonly considered as endemic of the Neotropical region include opossums (Didelphidae), sloths (Bradypodidae), anteaters (Myrmecophagidae), New World monkeys (Aotidae, Atelidae, Callitrichidae, Cebidae, and Pitheciidae), and Caviomorpha (Poux et al., 2006). Cione et al. (2015) listed the following endemic mammals: *Mazama americana, M. gouazoubira, Tayassu pecari, Leopardus pardalis, Puma yaguaroundi, Eira barbara, Lontra longicaudis, Procyon cancrivorus, Caluromys* spp., *Tapirus* spp., *Cebuella* spp., *Alouatta* spp., *Ateles* spp., *Cebus apella, Myoprocta* spp., *Proechimys* spp., *Coendou* spp., *Hydrochoerus hydrochaeris, Nectomys* spp., *Oecomys* spp., *Oryzomys* spp., *Sciurus aestuans, S. spadiceus, Priodontes maximus, Dasypus novemcinctus, Myrmecophaga tridactyla,* and *Tamandua tetradactyla.*

Several plant and animal taxa extend their distribution further north and south of the Neotropical region, corresponding their distributions to the Neotropical region in the broad sense, that is, including the Mexican and South American transition

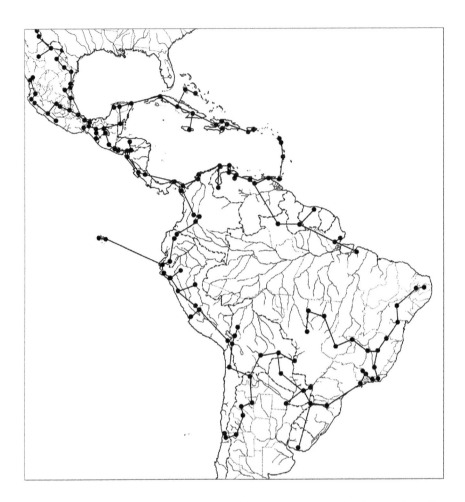

Figure 3.1   Map with the individual track of the Neotropical genus *Rhinacloa* (Miridae). (Data from Schuh, R. T. and M. D. Schwartz, *Bulletin of the American Museum of Natural History*, 179: 382–470, 1985.)

zones. There are other taxa that extend their distributions even to the Nearctic or the Andean regions. For example, the rodent superfamily Octodontoidea has some species in Patagonia and the Andes (Upham and Patterson, 2012; Ojeda et al., 2013), and some species of the assassin bug genus *Melanolestes* (Reduviidae) extend their distribution to the United States (Coscarón and Morrone, 1997).

Biotic Relationships

The relationships of the biogeographic regions of the world were represented by Morrone (2004b) in a general area cladogram (Figure 3.3). According to it, the Neotropical region belongs to the Holotropical kingdom, exhibiting biotic

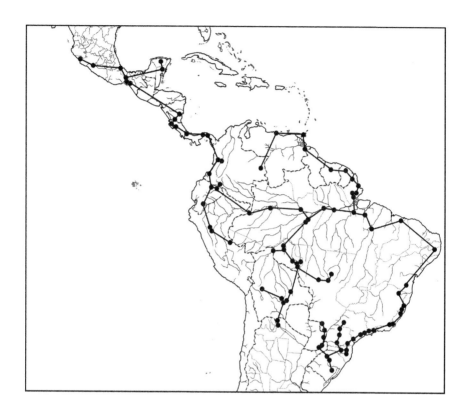

Figure 3.2   Map with the individual track of the Neotropical genus *Ptychoderes* (Anthribidae). (Data from Mermudes, J. R. and D. S. Napp, *Zootaxa*, 1182: 1–130, 2006.)

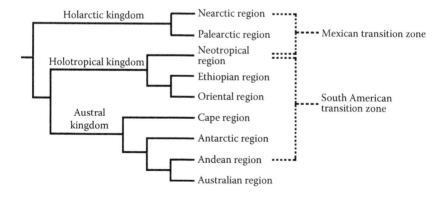

Figure 3.3   General area cladogram depicting the relationships of the biogeographic regions of the world. (Modified from Morrone, J. J., *Neotropica*, 42: 103–114, 1996.)

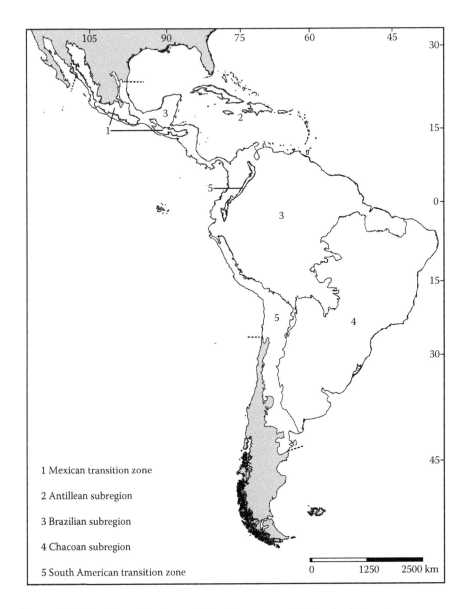

1 Mexican transition zone

2 Antillean subregion

3 Brazilian subregion

4 Chacoan subregion

5 South American transition zone

Figure 3.4   Map of the subregions and transition zones of the Neotropical region.

relationships with the Ethiopian and Oriental regions, as suggested by Newbigin (1950), Rapoport (1968), and Morrone (2002). The existence of pantropical taxa, evidencing these relationships, is hypothesized to have been caused by vicariance of a former eastern Gondwanaland (Crisci et al., 1993) or tropical Gondwanaland (Amorim et al., 2009). In addition to these relationships with other tropical areas, the Neotropical region exhibits biotic relationships with the Nearctic region through the

Mexican transition zone and with the Andean region through the South American transition zone (Morrone, 2004, 2006, 2014b).

Regionalization

The Neotropical region comprises the Antillean, Brazilian, and Chacoan subregions. In North America, the Neotropical region overlaps with the Nearctic region in the Mexican transition zone, whereas in South America it overlaps with the Andean region in the South American transition zone (Figure 3.4). These transition zones belong simultaneously to the Neotropical region and the Nearctic and Andean regions, respectively (Morrone, 2014b). Each of these subregions and transition zones is dealt herein in a different chapter.

CHAPTER 4

The Mexican Transition Zone

The Mexican transition zone is the area where the Neotropical and Nearctic regions overlap (Figure 4.1). In its broad sense, it comprises southwestern United States, Mexico, and most of Central America (Halffter, 1987; Zunino and Halffter, 1988; Gutiérrez-Velázquez et al., 2013). It is partially coincident with the areas named *Megamexico 3* (Rzedowski, 1991) and *biotic Mesoamerica* (Ríos-Muñoz, 2013). The Mexican transition zone in the strict sense, which is followed in this book, corresponds to the moderate to high elevation highlands of Mexico, Guatemala, Honduras, El Salvador, and Nicaragua (Morrone, 2006, 2010b, 2014b, 2015c; Espinosa Organista et al., 2008).

The earliest references to a transition zone between the Nearctic and Neotropical regions may be traced to Wallace (1876), Heilprin (1887), Drude (1890), and Dugès (1902). Hoffmann (1936) noted the presence of both Nearctic (septentrional) and Neotropical (meridional) insects in the Mexican fauna. Vivó (1943) identified a phytogeographic transition zone situated in the lowlands of Mesoamerica. Darlington (1957) formally named it Central American–Mexican transition zone, noting that it had received nothing like the attention given to Wallacea (Oriental–Australian transition zone) although it was probably equally important. Rapoport (1971) referred to the line separating the Nearctic and Neotropical regions as the *Anáhuac line*. Ortega and Arita (1998) analyzed the boundary between the Nearctic and Neotropical regions using distributional data of bats (Chiroptera). Cabrera and Willink (1973) considered that the flora and fauna of the mountain areas of Mexico were transitional between the Nearctic and Neotropical regions. Halffter (1987 and previous contributions) provided a general theory explaining the biotic assembly in the Mexican transition zone, based on distributional studies of Scarabaeid beetles.

MEXICAN TRANSITION ZONE

Mexican subregion (in part)—Wallace, 1876: 78 (regional.); Heilprin, 1887: 80 (regional.); Lydekker, 1896: 135 (regional.); Bartholomew et al., 1911: 9 (regional.); Mello-Leitão, 1937: 222 (regional.).
Mexican Highlands region—Engler, 1882: 345 (regional.).
Aztec province—Engler, 1882: 345 (regional.).

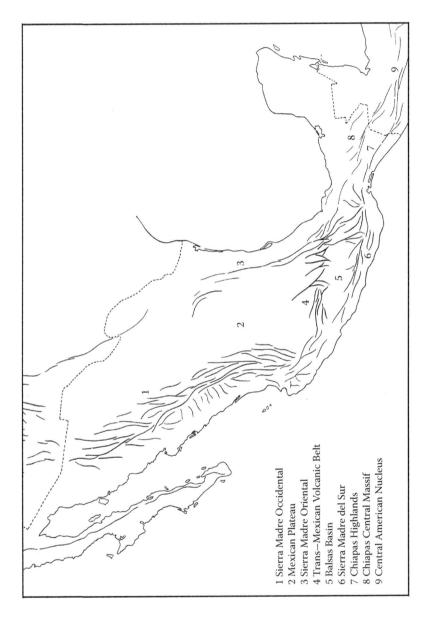

1 Sierra Madre Occidental
2 Mexican Plateau
3 Sierra Madre Oriental
4 Trans—Mexican Volcanic Belt
5 Balsas Basin
6 Sierra Madre del Sur
7 Chiapas Highlands
8 Chiapas Central Massif
9 Central American Nucleus

Figure 4.1   Map showing the main mountains of the Mexican transition zone. (Modified from Halffter, 1987.)

Guatemalan province—Engler, 1882: 345 (regional.).

Central American subarea (in part)—Clarke, 1892: 381 (regional.).

Central American subregion (in part)—Sclater and Sclater, 1899: 65 (regional.).

Central American–Mexican transition zone—Darlington, 1957: 456 (regional.).

Extratropical Highlands kingdom—West, 1964: 365 (regional.).

Tropical Highlands kingdom—West, 1964: 365 (regional.).

Mexican transition zone—Halffter, 1965: 4 (biotic evol.), 1974: 229 (biotic evol.), 1976: 13 (biotic evol.), 1978: 219 (biotic evol.); Castillo and Reyes-Castillo, 1984: 72 (biotic evol.); Halffter, 1987: 95 (biotic evol.); Kohlmann and Halffter, 1988: 112 (biotic evol.); Zunino and Halffter, 1988: 175 (biotic evol.); Kohlman and Halffter, 1990: 1 (biotic evol.); Llorente Bousquets, 1996: 50 (biotic evol.); Morón, 1996: 323 (biotic evol.); Ortega and Arita, 1998: 777 (regional.); Morrone and Márquez, 2001: 636 (track anal.); Márquez and Morrone, 2003: 23 (track anal.); Escalante et al., 2004: 327 (track anal.); Morrone, 2004a: 155 (track anal.); Corona and Morrone, 2005: 37 (track anal.); Morrone, 2005: 234 (regional.), 2006: 475 (regional.); Contreras-Medina et al., 2007a: 905 (clad. biogeogr.); Escalante et al., 2007a: 562 (clad. biogeogr.); Mariño-Pérez et al., 2007: 80 (track anal.); Halffter et al., 2008: 69 (biotic evol.); Escalante et al., 2009: 473 (endem. anal.); Morrone, 2010b: 355 (biotic evol.); Arana et al., 2011: 18 (track anal.); Coulleri and Ferrucci, 2012: 105 (track anal.); Mercado-Salas et al., 2012: 459 (track anal.); Campos-Soldini et al., 2013: 16 (track anal.); Gutiérrez-Velázquez et al., 2013: 282 (PAE); Miguez-Gutiérrez et al., 2013: 216 (clad. biogeogr.); Ruiz-Sánchez et al., 2013: 1337 (biotic evol.); Lamas et al., 2014: 955 (clad. biogeogr.); Morrone, 2014b: 27 (regional.), 2014c: 203 (clad. biogeogr.); Klassa and Santos, 2015: 520 (endem. anal.).

Mountain Mesoamerican province—Cabrera and Willink, 1973: 32 (regional.); Huber and Riina, 2003: 262 (glossary).

Mountain Mesoamerican region—Rzedowski, 1978: 101 (regional.); Luna Vega et al., 1999: 1301 (PAE).

Central America bioregion (in part)—Dinerstein et al., 1995: map 1 (ecoreg.).

Mexican Mountain's kingdom—Huber and Riina, 2003: 167 (glossary).

Mexican Mountain component—Morrone and Márquez, 2003: 219 (track anal.); Escalante et al., 2005: 199 (PAE).

Madrean Highlands area—Porzecanski and Cracraft, 2005: 266 (PAE).

Mexican Mountain transition zone—Espinosa Organista et al., 2008: 54 (regional.).

Endemic and Characteristic Taxa

There are several published studies that analyze the geographic distribution of different taxa in the Mexican transition zone. Rzedowski (1991) highlighted the richness and endemism of species of Mexican plants. He found that species richness is concentrated in the arid areas (considered herein to belong to the Nearctic region). For areas belonging to the Mexican transition zone, Rzedowski found a predominance of Nearctic taxa. Insects are among the best-studied taxa of the Mexican transition zone, including the classical contributions of Halffter (1964, 1965, 1974, 1976, 1978, 1987) on Scarabeid beetles. Castillo and Reyes-Castillo (1984) analyzed the species of *Petrejoides* (Coleoptera: Passalidae), finding different lineages characteristic of the Mexican transition zone. Kohlmann and Halffter (1988, 1990) analyzed the biotic evolution of the genera *Ateuchus* and *Canthon* (Coleoptera: Scarabaeidae), based on the phylogenetic relationships of their species. Liebherr (1991, 1994a,b) analyzed several taxa of Carabidae (Coleoptera), mapping distributional data of their species and classifying them according to Halffter's patterns. For example, species of the genus *Elliptoleus* (Figure 4.2) illustrate the Nearctic pattern, whereas those of the *Platynus degallieri* species group (Figure 4.3) illustrate the Mountain Mesoamerican pattern (Liebherr, 1994a). Vertebrate taxa include several genera of Rodentia (Mammalia), with Nearctic affinities (e.g., *Habromys* [Carleton et al., 2002; León-Paniagua et al., 2007], *Handleyomis* [Almendra et al., 2014], *Microtus* [Conroy et al., 2001], *Neotoma* [Edwards and Bradley, 2002a,b], *Peromyscus* [Harris et al., 2000; Tiemann-Boege et al., 2000], and *Reithrodontomys* [Arellano et al., 2005]). Among birds, the study of the *Chlorospingus ophthalmicus* species complex (Thraupidae) by García-Moreno and colleagues (2004) exemplifies the diversification of a taxon of Neotropical affinity.

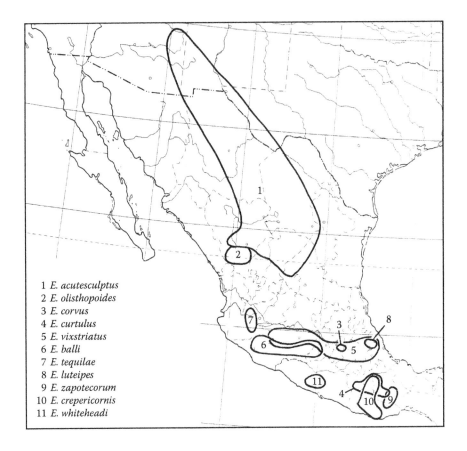

1 *E. acutesculptus*
2 *E. olisthopoides*
3 *E. corvus*
4 *E. curtulus*
5 *E. vixstriatus*
6 *E. balli*
7 *E. tequilae*
8 *E. luteipes*
9 *E. zapotecorum*
10 *E. crepericornis*
11 *E. whiteheadi*

Figure 4.2   Map with the distribution of the species of *Elliptoleus* in the Mexican transition zone. (Modified from Liebherr, 1994a.)

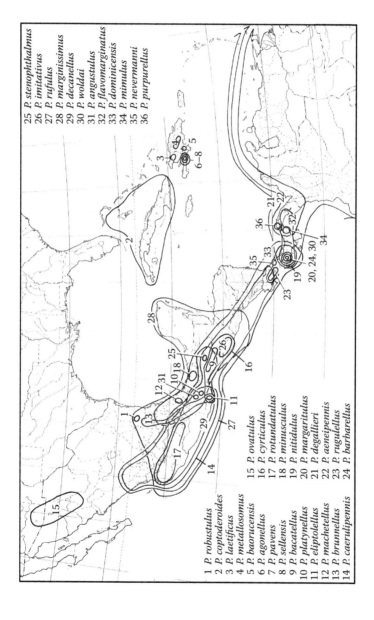

Figure 4.3 Map with the distribution of the species of the *Platynus degallieri* species group in the Mexican transition zone. (Modified from Liebherr, 1994a.)

Biotic Relationships

Based on its transitional nature, the biotic relationships of the Mexican transition zone lie with both the Nearctic and Neotropical regions. Rzedowski (1963, 1978, 1981) suggested that the flora of the cloud forests of the Mexican transition zone comprises basically three elements: Nearctic or temperate taxa, particularly represented by canopy trees; Neotropical or tropical taxa, particularly represented by herbs, epiphytes, and shrubs; and endemic taxa that are poorly represented at the generic level, but very significant if species are considered.

North and South America were separate continents during most of the Cenozoic. Their relatively recent biotic connection has been referred to as the Great American Biotic Interchange (GABI), especially when analyzing mammal taxa (Simpson, 1940, 1950; Stehli and Webb, 1985; Woodburne, 2010; Cione et al., 2015). Traditionally, it was assumed that the GABI began 3 mya, when a permanent Panama Isthmus developed, although it has been recognized that prior to its development the exchange of taxa was possible, with several phases occurring over 9 mya. The GABI has been one of the most important theories formulated to explain the biotic assembly in Mesoamerica and the Mexican transition zone, and has been discussed extensively in several biogeographic studies (for a recent review, see Cione et al., 2015). Woodburne (2010) reviewed the fossil evidence available and discussed the role of sea level changes and the Pliocene-Pleistocene vegetation changes in Central America and northern parts of South America, concluding that the GABI consisted of four major pulses that reflected the impact of glacial conditions in the Northern Hemisphere. GABI 1 (from 2.6 to 2.4 mya) resulted in a porcupine and a variety of Xenarthra dispersing to North America, and a group of carnivorans, equids, and a gomphotheriid proboscidean dispersing to South America. GABI 2 (at about 1.8 mya) entailed the dispersal of several Carnivora, Cetartiodactyla, one Perissodactyla, and two Proboscidea to South America, and a myrmecophagid Xenarthra to North America. GABI 3 included the biotic exchange of an opossum at 0.8 mya northward and a Carnivora and two Cetartiodactyla at 0.7 mya to South America. GABI 4 (at about 0.125 mya) reflected a diversity of carnivoran species, a sylvilagine rabbit, the genus *Equus*, and a glypdodont Xenarthra dispersing to South America. Woodburne (2010) found that more mammalian genera dispersed southward (32) than northward (17), especially in GABI 2, 3, and 4. The phylogenetic diversity was always greater in the southern taxa, with the northern taxa comprising mostly of Xenarthra, an erethizontid, two hydrochoerid rodents, and a marsupial.

Several authors have undertaken track analyses of the Mexican transition zone (Luna Vega et al., 1999; Morrone and Márquez, 2001; Escalante et al., 2004; Morrone and Gutiérrez, 2005; Andrés Hernández et al., 2006; Huidobro et al., 2006; Toledo et al., 2007; Espinosa Organista et al., 2008; Carcía-Marmolejo et al., 2008; Corona et al., 2009; Rosas et al., 2011). They have analyzed different taxa, namely, plants, insects, crustaceans, fish, and mammals. Some of the generalized tracks identified in these studies correspond to Halffter's patterns, but in other cases new patterns have resulted. Morrone and Márquez's (2001) track analysis of beetle (Coleoptera) taxa from the Mexican transition zone identified two generalized tracks. The northern generalized track comprises taxa distributed basically in mountain areas within the Sierra Madre Occidental, the Sierra Madre Oriental, the Trans–Mexican Volcanic Belt, the Balsas Basin, and the Sierra Madre del Sur. The southern generalized track comprises taxa distributed in the Sierra Madre de Chiapas and lowland areas in Chiapas, the Mexican Gulf, and the Mexican Pacific Coast, reaching the Panamanian Isthmus in the south. A detailed analysis by Espinosa Organista et al. (2008), based on plants, insects, and vertebrates, showed six patterns that are confined to specific mountain slopes (Figure 4.4):

Coastal mountain pattern: In the coastal slopes of the Sierra Madre Occidental, Sierra Madre Oriental, Sierra Madre del Sur, Chiapas Highlands, and the western and eastern extremes of the Trans–Mexican Volcanic Belt. Species supporting this track include *Cicindela fera* (Carabidae) and *Oryzomys melanotis* (Cricetidae). Some plant species extend their distribution to the mountains of Central America (e.g., *Blechnum schiedeanum, Campyloneurum ensifolium, Elaphoglossum muelleri*, and *Notholaena galeottii* [ferns], *Clethra alcoceri* [Clethraceae], and *Bletia purpurea* [Orchidaceae]).

Circum-Balsas River Basin subhumid mountain pattern: In the Trans–Mexican Volcanic Belt and Sierra Madre del Sur, predominantly in the slopes oriented to the Balsas River Basin. Species supporting this track include *Adiantum shepherdii* and *Asplenium muenchii* (ferns), *Juniperus flaccida* var. *poblana* (Cupressaceae), *Pinus pringlei* (Pinaceae), *Agave cupreata* (Asparagaceae), *Plectrohyla bistincta* (Hylidae), *Phrynosoma taurus* and *Sceloporus grammicus grammicus* (Phrynosomatidae), *Reithrodontomys fulvescens* (Cricetidae), *Campylorhynchus megalopterus* (Troglodytidae), and *Pipilo ocai* (Emberizidae). Some species extend their distribution to the dry slope of the Sierra de Juárez in Oaxaca.

Circum-Plateau semiarid and arid mountain pattern: In the mountains that surround the Mexican Plateau. Species supporting this track include *Phrynosoma orbiculare, Sceloporus grammicus disparilis*, and *S. scalaris* (Phrynosomatidae), *Crotalus molossus* and *C. pricei* (Viperidae), *Ambystoma tigrinum velascoi* (Ambystomatidae), *Peromyscus melanotis, P. difficilis*, and *Reithrodontomys megalotis* (Cricetidae), and *Pipilo fuscus* (Emberizidae). Some species are limited to specific sustrata, such as *Cheilanthes allosuroides* (Pteridaceae) on igneous rocks and *C. leucopoda* on limestones. Several species are shared with the Sonora province and southwestern United States (e.g., *Celtis lindheimeri* [Cannabaceae]); other species penetrate the Mexican Plateau (e.g., *Notholaena aschenborniana* [Pteridaceae]); and yet other species only surround the Plateau (e.g., *Pleopeltis polylepis* var. *erythrolepis* [fern], *Crotalus lepidus* [Viperidae], and *Peromyscus gratus* and *P. maniculatus* [Cricetidae]).

Mountain cloud pattern: Distributed discontinuously in the most humid areas of the Gulf of Mexico slope, which contain several species of Zamiaceae, tree ferns, Orchidaceae, and salamanders. Species supporting this track include *Cyathea bicrenata* (tree fern), *Clethra suaveolens* (Clethraceae), *Eleutherodactylus berkenbuschi* (Eleutherodactylidae), and *Sceloporus salvini* (Phrynosomatidae).

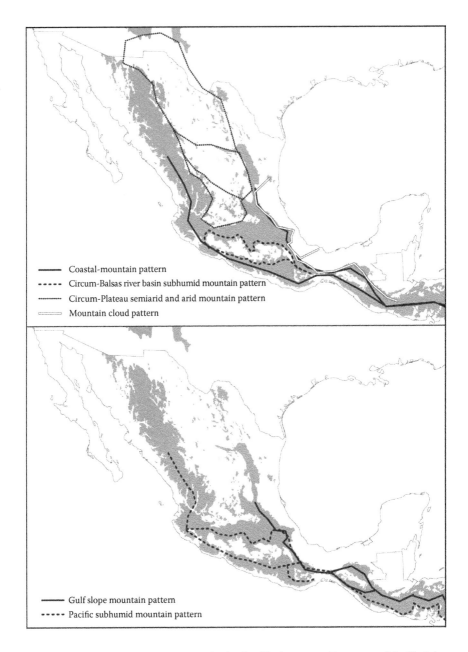

Figure 4.4 Maps with generalized tracks in the Mexican transition zone. (Modified from Espinosa Organista, D. et al., "El conocimiento biogeográfico de las especies y su regionalización natural." In: Sarukhán, J. (ed.), *Capital natural de México. Vol. I. Conocimiento actual de la biodiversidad*. Conabio, Mexico City, pp. 33–65, 2008.)

Gulf slope mountain pattern: In the Gulf of Mexico mountain slopes, in the transition between pine-oak, cloud, and rainforests. Species supporting this track include *Argyrochosma formosa, Anemia semihirsuta*, and *Polypodium puberulum* (ferns), *Pinus patula* (Pinaceae), *Taxus globosa* (Taxaceae), *Quercus lancifolia, Q. leiophylla*, and *Q. sapotiifolia* (Fagaceae), *Eleutherodactylus decoratus* (Eleutherodactylidae), *Ecnomiohyla miotympanum* (Hylidae), *Cryptotis mexicana* (Soricidae), and *Peromyscus furvus* (Cricetidae). Some species extend their distribution to Petén (Guatemala) and Belize, whereas others are restricted to the southern portion of the Gulf of Mexico as *Polypodium collinsii* (Polypodaceae).

Pacific subhumid mountain pattern: In the ecotone between pine-oak forests and tropical deciduous forests of the Pacific slope. Species supporting this track include *Anemia jaliscana, Cheilanthes aurantiaca*, and *Polypodium rzedowskianum* (ferns), *Pinus douglasiana* and *P. herrerae* (Pinaceae), *Agave rhodacantha* (Asparagaceae), *Quercus magnoliifolia, Q. splendens* and *Q. urbanii* (Fagacae), and *Cicindela aeneicollis* (Carabidae).

As expected from a transition zone, the track analyses have identified nodes (complex or "hybrid" areas) in the intersection of different generalized tracks. Nodes found in four of these studies (Escalante et al., 2004; Toledo et al., 2007; García-Marmolejo et al., 2008; Corona et al., 2009) are shown in Figure 4.5. The largest concentration of nodes lies in the Trans–Mexican Volcanic Belt, the Sierra Madre del Sur, and the southern parts of the Sierra Madre Occidental and Sierra Madre Oriental. Escalante et al. (2004) characterized the nodes found by them in terms of their climate, elevation, geology, and localization.

Several cladistic biogeographic studies have analyzed the Mexican transition zone (Liebherr, 1991, 1994a; Marshall and Liebherr, 2000; Flores-Villela and

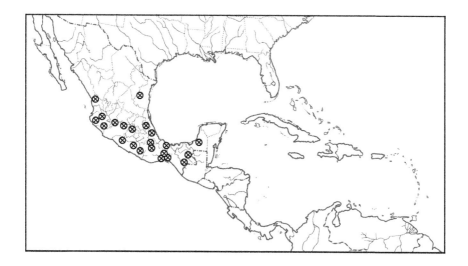

Figure 4.5 Map with nodes identified in the Mexican transition zone. (After Escalante et al., 2004; Toledo et al., 2007; García-Marmolejo et al., 2008; Corona et al., 2009.)

Goyenechea, 2001; Espinosa et al., 2006; Contreras-Medina et al., 2007a; Escalante et al., 2007; Flores-Villela and Martínez-Salazar, 2009; Miguez-Gutiérrez et al., 2013), showing little agreement in the general area cladograms obtained. Miguez-Gutiérrez et al. (2013) analyzed the relationships of the areas of endemism within the Mexican transition zone based on three different biogeographic regionalizations (Marshall and Liebherr, 2000; Flores-Villela and Goyenechea, 2001; Morrone, 2006). They constructed taxon-area cladograms for 10 genera of beetles, gymnosperms, lizards, and snakes, using the areas of the three regionalizations, and obtained general area cladograms using assumptions 0 and 1 (Figure 4.6). They found two groups of areas, one with Neotropical affinities and the other with

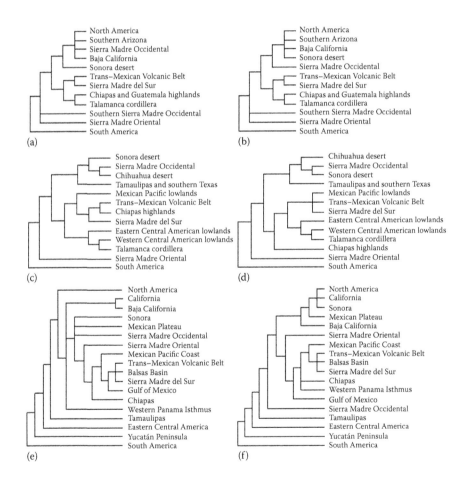

Figure 4.6   General area cladograms evaluated by Miguez-Gutiérrez et al. (2013). (a) Marshall and Liebherr's (2000) areas, assumption 0; (b) Marshall and Liebherr's (2000) areas, assumption 1; (c) Flores-Villela and Goyenechea's (2001) areas, assumption 0; (d) Flores-Villela and Goyenechea's (2001) areas, assumption 1; (e) Morrone's (2006) areas, assumption 0; (f) Morrone's (2006) areas, assumption 1.

Nearctic affinities. Some common patterns were the close relationships between the Sierra Madre del Sur and the Trans–Mexican Volcanic Belt, and between the Chiapas Highlands and the Talamanca ridge. Miguez-Gutiérrez and colleagues (2013) concluded that the most important vicariance events within the Mexican transition zone were the Trans–Mexican Volcanic Belt, which divided the areas with Nearctic affinities in the north and those with Neotropical affinities in the south, and the Tehuantepec Isthmus, which together with the Nicaraguan depression isolated Nuclear Central America.

Regionalization

The Mexican transition zone (Figure 4.7) comprises the Sierra Madre Occidental, Sierra Madre Oriental, Trans–Mexican Volcanic Belt, Sierra Madre del Sur, and Chiapas Highlands biogeographic provinces (Morrone, 2014b).

Cenocrons

In addition to the track and cladistic biogeographic analyses, there are contributions that attempted to identify cenocrons in the complex biota of the Mexican transition zone. Halffter formulated a general theory that explains how cenocrons, which evolved in different geographic areas, assembled in the transition zone (Halffter, 1964, 1965, 1974, 1976, 1978, 1987; Halffter et al., 1995; Gutiérrez-Velázquez et al., 2013; see also Morón, 1996). In order to infer these cenocrons, Halffter took into consideration the distribution of their closest relatives, species richness, the degree of species sympatry, the geologic history, and the diversity of habitats occupied by the species (Kohlmann and Halffter, 1990; Liebherr, 1991, 1994a; Morrone, 2015c).

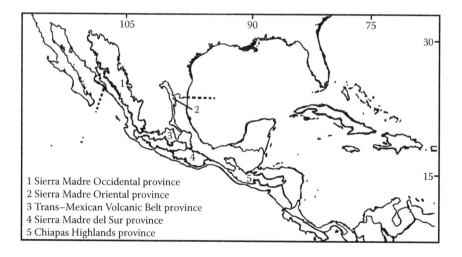

1 Sierra Madre Occidental province
2 Sierra Madre Oriental province
3 Trans–Mexican Volcanic Belt province
4 Sierra Madre del Sur province
5 Chiapas Highlands province

Figure 4.7 Map of the provinces of the Mexican transition zone.

Halffter (1974, 1978, 1987) classified Scarabaeid taxa distributed in the Mexican transition zone into five cenocrons:

Paleoamerican cenocron: It includes taxa distributed in the lowlands and mountains, with relict species belonging to groups widely ranged in the tropics of the Old World and species belonging to groups widely distributed in North America. They dispersed to the Mexican transition zone before the formation of the Mexican Plateau and the expansion of the deserts of western parts of North America.

Mexican Plateau cenocron: It includes taxa of ancient South American (Neotropical) origin, which are widely distributed in the Mexican Plateau and the highlands of Guatemala and Chiapas. They are rarely found in the mountains of the Mexican transition zone, where Nearctic and Paleoamerican taxa predominate.

Nearctic cenocron: It includes recent Holarctic and some Nearctic taxa, which are generally restricted to areas above 1,500 meters.

Mountain Mesoamerican cenocron: It includes groups distributed in the mountain forests of Central America that originally evolved in the Central American Nucleus (highlands of Chiapas, Guatemala, Honduras, El Salvador, and Nicaragua—north of Lake Nicaragua), and then dispersed northwest and southeast. They have ancient South American affinities and are distributed mainly in the mountains and cloud forests, although they penetrate occasionally into pine-oak forests.

Typical Neotropical cenocron: It includes taxa of recent origin distributed in the lowlands; they have not invaded the Mexican highlands and probably extended their ranges northward after the Pliocene.

Halffter and colleagues (1995) analyzed the cenocrons of three groups of beetles (Scarabaeinae, Geotrupinae, and Silphidae) along an altitudinal transect in the Mexican transition zone (State of Veracruz), from sea level up to an altitude of 4,282 meters. They found that the Paleoamerican, Nearctic, Mexican Plateau, Typical Neotropical, and Mountain Mesoamerican cenocrons corresponded to well-defined altitudinal zones, with an overlap of elements resulting in a transitional area. They concluded that the taxa assigned to the Paleoamerican cenocron did not have common ecological features, because of their ancient history in the Mexican transition zone and their adaptive plasticity (for an analysis of a taxon showing the heterogeneity of the Paleoamerican pattern, see Zunino and Halffter, 1988). Based on the analyses of Zunino and Halffter (1987), Halffter et al. (1995), and Halffter (pers. comm.), the Paleoamerican cenocron should be divided into different varieties. The Relict Paleoamerican species belong to the genera with wide distribution in the Old World, but represented in the Mexican transition zone by endemic species with very restricted distributions. The Mountain Paleoamerican species belong to lineages that were able to colonize the mountains of Mexico and, to a lesser extent, those of Central America, and that have undergone significant speciation driven by vicariance. The Mountain Mesoamerican variety corresponds to species that follow the Mesoamerican Mountain pattern, but have Old World affinities. The Mexican Plateau variety corresponds to Paleoamerican species distributed in the Mexican Plateau. The Tropical Paleoamerican species are found in tropical lowlands and moderate elevations, with a distribution very similar to those of the Typical Neotropical pattern, although their affinities are Paleoamerican (Old World). The altitudinal pattern analyzed by Halffter et al. (1995) parallels the latitudinal one. At lower elevations and southern latitudes, Typical Neotropical taxa are more abundant, whereas at higher elevations and northern latitudes, Nearctic and Paleoamerican taxa predominate. A similar pattern has been detected for bird taxa (Sánchez-González and Navarro-Sigüenza, 2009).

The cenocrons recognized by Halffter and collaborators dispersed into the Mexican transition zone in five successive stages (Morrone, 2015c; Figure 4.8). Initially, in the Jurassic–Cretaceous period, the Paleoamerican cenocron extended in Mexico. It represents the original (Holarctic) biota of the country; its taxa have evolved for a long period of time, and have been affected by the vicariance and biogeographic convergence events that shaped the complex biotic history of the country. From the Late Cretaceous to the Paleocene, the Plateau cenocron, which includes taxa widely distributed in the Mexican Plateau and rare in the mountains of the Mexican transition zone, dispersed from South America. In the Oligocene–Miocene, the Mountain Mesoamerican cenocron dispersed from the Central American Nucleus; it includes taxa with ancient South American affinities that evolved in the highlands of Chiapas, Guatemala, Honduras, El Salvador, and Nicaragua—north of Lake Nicaragua. In the Miocene–Pliocene, the Nearctic cenocron, which includes Holarctic and Nearctic taxa, dispersed from North America; its taxa are generally restricted to areas above 1,500 meters in temperate and cloud forests. Finally, in the Pliocene–Pleistocene, the Typical Neotropical cenocron dispersed from South America; it includes species widely distributed in tropical forests of the lowlands that did not invade the highlands.

SIERRA MADRE OCCIDENTAL PROVINCE

Sierra Madre Occidental province—Goldman and Moore, 1945: 253 (regional.); Moore, 1945: 218 (regional.); Stuart, 1964: 350 (regional.); Rzedowski, 1978: 102 (regional.); Casas-Andreu and Reyna-Trujillo, 1990: map (regional.); Ferrusquía-Villafranca, 1990: map (regional.); Ramírez-Pulido and Castro-Campillo, 1990: map (regional.); Rzedowski and Reyna-Trujillo, 1990: map (regional.); Anderson and O'Brien, 1996: 332 (biotic evol.); Ayala et al., 1996: 429 (regional.); Arriaga et al., 1997: 64 (regional.); Escalante et al., 1998: 285 (cluster anal. and regional.); Campbell, 1999: 114 (regional.); Luna Vega et al., 1999: 1301 (PAE); Morrone et al., 1999: 510 (PAE); Espinosa et al., 2000: 64 (PAE and regional.); Flores-Villela and Goyenechea, 2001: 174 (clad. biogeogr.); Morrone, 2001a: 34 (regional.), 2001e: 47 (regional.); Morrone and Márquez, 2001: 636 (track anal.); Morrone et al., 2002: 91 (regional.); Huber and Riina, 2003: 260 (glossary); Corona and Morrone, 2005: 38 (track anal.); Escalante et al., 2005: 202 (PAE); Morrone, 2005: 234 (regional.); Andrés Hernández et al., 2006: 901 (track anal.); Morrone, 2006: 476 (regional.); Contreras-Medina et al., 2007b: 408 (PAE); Mariño-Pérez et al., 2007: 80 (track anal.); Espinosa Organista et al., 2008: 56 (regional.); Morrone and Márquez, 2008: 19 (track anal.); León-Paniagua and Morrone, 2009: 1942 (clad. biogeogr.); Morrone, 2010b: 358 (biotic evol.); Domínguez-Domínguez et al., 2011: 2 (biotic evol.); Coulleri and Ferrucci, 2012: 105 (track anal.); Mercado-Salas et al., 2012: 459 (track anal.); Kobelkowsky-Vidrio et al., 2014: 2088 (PAE); Morrone, 2014b: 27 (regional.).

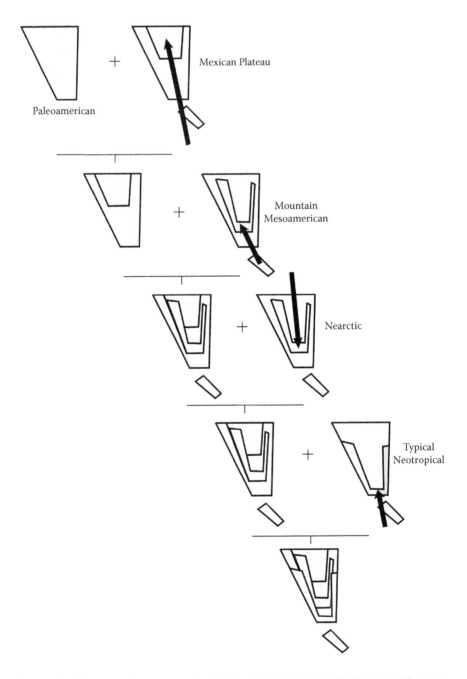

Figure 4.8 Diagrammatic representation of the development of the Mexican transition zone. (Modified from Morrone, 2015c.)

Sierra Madre Occidental region—West, 1964: 368 (regional.).
Sierra Madre Occidental Pine-Oak Forests ecoregion—Dinerstein et al., 1995: 97 (ecoreg.).
Madrean province—Brown et al., 1998: 30 (veget.).
Sierra Madre Occidental area—Katinas et al., 2004: 166 (PAE); Flores-Villela and
   Martínez-Salazar, 2009: 820 (clad. biogeogr.).

Definition

The Sierra Madre Occidental province is situated in western Mexico (states of
Chihuahua, Durango, Jalisco, Nayarit, Sinaloa, Sonora, and Zacatecas) at elevations
between 200 and 3,000 meters, with most of the area above 2,000 meters (Rzedowski,
1978; Morrone, 2001e, 2006, 2014b; Kobelkowsky-Vidrio et al., 2014). The Sierra
Madre Occidental connects the Rocky Mountains in the United States through its
portion known as the Madrean Sky Islands (southern Arizona and New Mexico) to
the Trans–Mexican Volcanic Belt, while separating the Sonoran and Chihuahuan
deserts (Kobelkowsky-Vidrio et al., 2014). Marshall and Liebherr (2000) combined
most of the Sierra Madre Occidental with the Mexican Plateau in a single biogeographic unit, based on their shared taxa.

Endemic and Characteristic Taxa

Morrone (2014b) provided a list of endemic and characteristic taxa. Some examples include *Selaginella mutica* var. *mutica* (Selaginellaceae), *Juniperus deppeana*
var. *pachyphlaea* (Juniperaceae), *Pinus engelmanii* (Pinaceae; Figure 4.9), *Quercus*

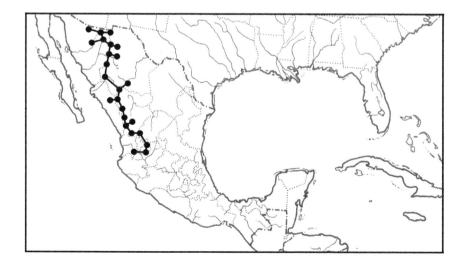

Figure 4.9   Map with the individual track of *Pinus engelmanni* (Pinaceae) in the Sierra Madre
Occidental province.

radiata, Q. tarahumara, and *Q. undata* (Fagaceae), *Elliptoleus olisthopoides* (Carabidae), *Coscinocephalus cribrifrons, Hologymnetis argenteola, Homoiosternus beckeri, Onthophagus brevifrons,* and *Strategus cessus* (Scarabaeidae), *Cyanocorax dickeyi* (Corvidae), *Peromyscus spicilegus* and *Reithrodontomys zacatecae* (Cricetidae), *Campostoma ornatum* (Cyprinidae), *Rhynchopsitta pachyrhyncha* (Psittacidae), *Glaucomys volans madrensis* and *Sciurus nayaritensis* (Sciuridae), and *Euptilotis neoxenus* (Trogonidae).

Vegetation

Temperate forests are predominant, especially pine and pine-oak forests (Rzedowski, 1978; Dinerstein et al., 1995). The montane forests of the Sierra Madre Occidental province harbor some of the world's most extensive subtropical coniferous forests (Dinerstein et al., 1995). Dominant plants include species of *Abies, Pinus,* and the southernmost present-day distribution of *Picea* (Graham, 2004). Endemic plant genera include *Arnicastrum, Pionocarpus, Pippenalia, Stenocarpha,* and *Trichocoryne* (Rzedowski, 1978).

Biotic Relationships

According to a parsimony analysis of endemicity (PAE) based on plant, insect, and bird taxa (Morrone et al., 1999) and another based on mammals (Escalante et al., 2005), the Sierra Madre Occidental is closely related to other areas in the Nearctic region. A track analysis based on species of Coleoptera (Morrone and Márquez, 2001) showed that this province is related to the Balsas Basin, Sierra Madre del Sur, Sierra Madre Oriental, and Trans–Mexican Volcanic Belt provinces. A cladistic biogeographic analysis, based on the Carabid (Coleoptera) genera *Elliptoleus* and *Calathus* (Liebherr, 1991), showed a close relationship between the Sierra Madre Occidental and the northern part of the Sierra Madre Oriental.

Regionalization

Smith (1941) and Moore (1945) have delimited nested units within the Sierra Madre Occidental province. I (Morrone, 2014b) accepted Moore's (1945) districts, but noting that Smith's (1941) names have nomenclatural priority. The Apachian district comprises the northern portion of the Sierra Madre Occidental province in the states of Chihuahua, Sonora, and part of Sinaloa; endemic taxa include four species of Sciuridae: *Peromyscus polius, Spermophilus madrensis, Tamias dorsalis carminis,* and *T. durangae* (Ceballos and Oliva, 2005). The Durangoan district comprises the southern portion of the Sierra Madre Occidental province, in the states of Jalisco, Nayarit, Sinaloa, Sonora, Zacatecas, and parts of Sinaloa and Chihuahua; endemic taxa include *Nelsonia neotomodon* (Cricetidae), *Tamias bulleri* (Sciuridae), and *Sorex emarginatus* (Soricidae; Ceballos and Oliva, 2005; Gámez et al., 2012).

Kobelkowsky-Vidrio et al. (2014) undertook a PAE of bird species, finding a separation along the highest elevations of the Sierra Madre Occidental between a southwestern area (Pacific slope) and a northeastern area (eastern slope). They found that this boundary coincides with the limit between tropical and temperate vegetation, and that bird species from the Pacific slope have Neotropical affinities, whereas those from the eastern slope have Nearctic affinities. They concluded that the area situated toward the Pacific slope could be regarded as part of the Mexican transition zone, whereas the area on the eastern slope corresponds to the Nearctic region. These results conflict with the already recognized districts, so more taxa should be analyzed to solve this issue.

Cenocrons

In order to analyze the relative contribution of geologic and environmental changes that have shaped the biotic evolution of the Sierra Madre Occidental, Domínguez-Domínguez et al. (2011) undertook the phylogeographic analysis of the fish *Campostoma ornatum* (Cyprinidae). They considered two alternative scenarios: the Sierra Madre Occidental acted as a biogeographic barrier, where the phylogeographic pattern would result from the mixture of western and eastern lowland lineages isolated during the Pleistocene interglacial stages; or the Sierra Madre Occidental was a center of diversification, where deep genetic divergences would have been shaped by geologic processes from the Paleogene. Analysis of specimens from different river basins by Domínguez-Domínguez et al. (2011) recovered two main clades: a northern clade, including the Atlantic drainage of the Conchos, the interior drainages of Santa Clara, Casas Grandes, and the western Pacific drainages of the Yaqui, Fuerte, Sonora, and Mayo rivers; and a southern clade, including the interior drainage of the Nazas and Aguanaval rivers and the Piaxtla basin. The morphological analysis showed the same groups. Based on the absence of sharp differentiation between the western (Pacific) and eastern (Atlantic) drainages, together with the lack of an altitudinal gradient of haplotidic diversity, the authors discarded a recent (Quaternary) origin for the observed patterns and inferred that older tecto-volcanic processes were crucial for the evolutionary history of the species. The northern clade has apparently followed episodes of taxon pulses, whereas the southern clade suffered a vicariant event in the Pliocene. The complex biogeographic history of *Campostoma ornatum*, with repeated events of vicariance and dispersal, could be related to the tecto-volcanic activity that occurred in the Sierra Madre Occidental since the Early Pliocene.

SIERRA MADRE ORIENTAL PROVINCE

Sierra Madre Oriental province—Goldman and Moore, 1945: 356 (regional.); Moore, 1945: 218 (regional.); Stuart, 1964: 350 (regional.); Rzedowski, 1978: 103 (regional.); Casas-Andreu and Reyna-Trujillo, 1990: map (regional.); Ferrusquía-Villafranca, 1990: map (regional.); Ramírez-Pulido and Castro-Campillo, 1990: map (regional.);

Rzedowski and Reyna-Trujillo, 1990: map (regional.); Anderson and O'Brien, 1996: 332 (biotic evol.); Ayala et al., 1996: 429 (regional.); Arriaga et al., 1997: 64 (regional.); Escalante et al., 1998: 285 (cluster anal. and regional.); Campbell, 1999: 114 (regional.); Luna Vega et al., 1999: 1301 (PAE); Morrone et al., 1999: 510 (PAE); Espinosa et al., 2000: 64 (PAE and regional.); Flores-Villela and Goyenechea, 2001: 174 (clad. biogeogr.); Morrone, 2001a: 35 (regional.), 2001e: 48 (regional.); Morrone and Márquez, 2001: 636 (track anal.); Morrone et al., 2002: 92 (regional.); Huber and Riina, 2003: 260 (glossary); Espinosa et al., 2004: 487 (regional.); Corona and Morrone, 2005: 38 (track anal.); Escalante et al., 2005: 202 (PAE); Morrone, 2005: 235 (regional.); Andrés Hernández et al., 2006: 902 (track anal.); Morrone, 2006: 476 (regional.); Contreras-Medina et al., 2007b: 408 (PAE); González-Zamora et al., 2007: 136 (track anal.); Mariño-Pérez et al., 2007: 80 (track anal.); Espinosa Organista et al., 2008: 56 (regional.); Morrone and Márquez, 2008: 19 (track anal.); Escalante et al., 2009: 473 (endem. anal.); León-Paniagua and Morrone, 2009: 1942 (clad. biogeogr.); Santa Anna del Conde Juárez et al., 2009: 374 (PAE); Morrone, 2010b: 358 (biotic evol.); Sanginés-Franco et al., 2011: 82 (PAE); Coulleri and Ferrucci, 2012: 105 (track anal.); Puga-Jiménez et al., 2013: 1180 (PAE); Morrone, 2014b: 29 (regional.).
Sierra Madre Oriental region—West, 1964: 368 (regional.).
Sierra Madre Oriental Pine–Oak Forests ecoregion—Dinerstein et al., 1995: 97 (ecoreg.).
Sierra Madre Oriental area—Marshall and Liebherr, 2000: 206 (clad. biogeogr.); Katinas et al., 2004: 166 (PAE); Flores-Villela and Martínez-Salazar, 2009: 820 (clad. biogeogr.).

Definition

The Sierra Madre Oriental province is situated in eastern Mexico (states of Coahuila, Durango, Guanajuato, Hidalgo, Nuevo León, Puebla, Querétaro, San Luis Potosí, Tamaulipas, Veracruz, and Zacatecas) at elevations above 1,500 meters (Rzedowski, 1978; Morrone, 2001e, 2014b; Luna et al., 2004; Sanginés-Franco et al., 2011). The Sierra Madre Oriental (Figure 4.10) is the second largest montane system in Mexico, with a length of 1,350 kilometers and altitudes averaging 1,500–2,000 meters, and in some places, reaching 3,000 meters. It runs northwest to southeast, beginning in the Sierra del Burro, close to the Bravo river in Tamaulipas, and ending in the Cofre de Perote (Veracruz), where it connects with the Trans–Mexican Volcanic Belt (González-Zamora et al., 2007).

Endemic and Characteristic Taxa

Morrone (2014b) provided a list of endemic and characteristic taxa. Some examples include *Cheilanthes decomposita* (Pteridaceae), *Agave inaequidens, A. horrida,* and *A. tenuifolia* (Asparagaceae), many species of Asteraceae and Cactaceae, *Priamides erostratus erostratinus* and *Pterourus palamedes leontis* (Papilionidae; Figure 4.11), *Odontotaenius zodiacus, Petrejoides laticornis, P. nebulosus, P. orizabae,* and *P. silvaticus* (Passalidae), *Typhlochactas* spp. (Superstitionidae), *Microtus quasiater, Peromyscus aztecus,* and *Reithrodontomys fulvescens* (Cricetidae), *Dendrortyx barbatus* (Odontophoridae), *Cryptotis obscura* and *C. parva berlandieri* (Soricidae), and *Glaucidium sanchezi* (Strigidae).

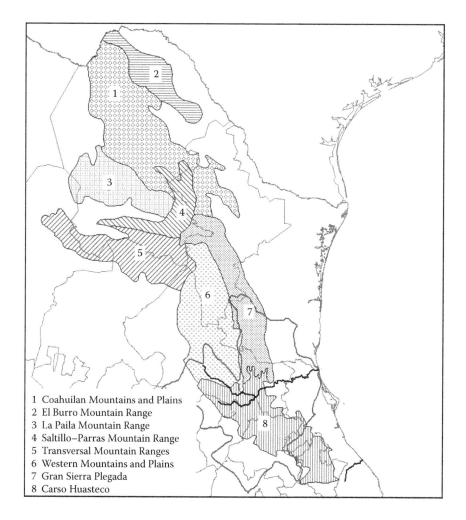

1 Coahuilan Mountains and Plains
2 El Burro Mountain Range
3 La Paila Mountain Range
4 Saltillo–Parras Mountain Range
5 Transversal Mountain Ranges
6 Western Mountains and Plains
7 Gran Sierra Plegada
8 Carso Huasteco

Figure 4.10   Map of the subprovinces of the Sierra Madre Oriental physiographic province. (Modified from Espinosa, D. et al., "Identidad biogeográfica de la Sierra Madre Oriental y posibles subdivisiones bióticas." In: Luna, I., J. J. Morrone, and D. Espinosa (eds.), *Biodiversidad de la Sierra Madre Oriental.* Las Prensas de Ciencias, UNAM, Mexico City, pp. 487–500, 2004.)

Vegetation

There are temperate forests, with oak forests being predominant; there are also pine forests (Rzedowski, 1978; Dinerstein et al., 1995; Sanginés-Franco et al., 2011). There are areas with xerophytic scrubland, mainly in the lowlands near the Mexican Plateau, where cacti are well represented (Santa Anna del Conde Juárez et al., 2009). The most important plant genera include *Greenmaniella*, *Loxothysanus*, and *Mathiasella* (Rzedoswki, 1978).

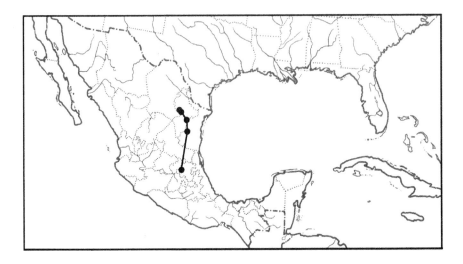

Figure 4.11 Map with the individual track of *Pterourus palamedes leontis* (Papilionidae) in the Sierra Madre Oriental province.

Biotic Relationships

According to a PAE based on plant, insect, and bird taxa (Morrone et al., 1999), the Sierra Madre Oriental province is closely related to the southern portion of the Mexican Plateau province. Another PAE based on mammals (Escalante et al., 2005) found that it is related to the Trans–Mexican Volcanic Belt province. A track analysis based on species of Coleoptera (Morrone and Márquez, 2001) showed that this province is related to the Sierra Madre Occidental, Sierra Madre del Sur, Balsas Basin, and Trans–Mexican Volcanic Belt provinces. Another track analysis (Sanginés-Franco et al., 2011), based on fern species, highlighted the complex biotic relationships of the Saltillo–Parras district.

A PAE based on bird taxa (Sánchez-González et al., 2008), a cladistic biogeographic analysis based on two Carabid genera (Liebherr, 1991) and the phylogeographic analysis of the furnariid bird *Lepidocolaptes affinis* (Arbeláez-Cortés et al., 2010) suggested that the Sierra Madre Oriental province may be a composite of two portions. The northern portion is related to the Sierra Madre Occidental and the southern is related to the Sierra Madre del Sur and the Trans–Mexican Volcanic Belt.

Regionalization

Espinosa et al. (2004, 2008) and Morrone (2014b) have recognized two subprovinces and four districts within the Sierra Madre Oriental province. The Austral–Oriental subprovince corresponds to the northern portion of the Sierra Madre Oriental province, north of the Moctezuma river (Espinosa et al., 2004); some endemic taxa include *Ageratina potosina*, *Flourensia monticola*, *Flyriella stanfordii*, *Grindelia greenmanii*, *Porophyllum filiforme*, *Rumfordia exauriculata*, *Senecio carnerensis*, *Stevia hintoniorum*, and *Verbesina daviesiae* (Asteraceae), *Sciurus alleni* (Sciuridae), and *Sorex milleri* (Soricidae; Espinosa et al., 2004; Ceballos and Oliva, 2005; González-Zamora et al., 2007). Within this subprovince, two districts have been recognized. The Saltillo–Parras district corresponds to the Pliegues Saltillo–Parras, Sierras Transversales, and the northern part of the Gran Sierra Plegada (Espinosa et al., 2004); some endemic taxa include *Pinus culminicola* (Pinaceae), *Acharagma roseana*, *Mammillaria grusonii*, and *Turbinicarpus booleanus* (Cactaceae), *Quercus sinuata breviloba* (Fagaceae), *Peromyscus hooperi* (Cricetidae), and *Myotis planiceps* (Vespertilionidae; Espinosa et al., 2004; Ceballos and Oliva, 2005; Santa Anna del Conde Juárez et al., 2009). The Potosí district corresponds to the Gran Sierra Plegada and parts of the Sierras and Llanuras Occidentales at Sierra de Potosí (Espinosa et al., 2004); some endemic taxa include *Pinus nelsonii* (Pinaceae), *Quercus rysophylla* (Fagaceae), *Guazuma ulmifolia* (Malvaceae), *Neotoma angustapalata*, and *Peromyscus ochraventer* (Cricetidae), and *Notiosorex villai* (Soricidae; Espinosa et al., 2004; Ceballos and Oliva, 2005).

The Hidalgo subprovince corresponds to the southern portion of the Sierra Madre Oriental province, south of the Moctezuma river, in the Carso Huasteco physiographic province (Espinosa et al., 2004); some endemic taxa include *Archibaccharis venturana*, *Chaptalia estribensis*, and *Verbesina coulteri* (Asteraceae), *Inga huastecana* (Fabaceae), and *Habromys simulatus*, *Megadontomys nelsoni*, *Peromyscus furvus*, and *Reithrodontomys sumichrasti* (Cricetidae; Espinosa et al., 2004; Ceballos and Oliva, 2005; Ramírez-Pulido et al., 2005; González-Zamora et al., 2007). Within this subprovince, two districts have been recognized. The Sierra Gorda district corresponds to the western part of the subprovince, delimited by the rivers Moctezuma and Verde, which are tributaries of the Pánuco river (Espinosa et al., 2004); some endemic taxa include *Agave tenuifolia* (Asparagaceae), *Styrax argenteus* subsp. *parvifolius* (Styracaceae), *Sciurus oculatus oculatus* (Sciuridae), and *Euderma phyllote phyllote* (Vespertilionidae; Espinosa et al., 2004). The Zacualtipán district corresponds to the eastern part of the subprovince, including the Sierra Norte de Puebla and Sierra de Zacualtipán (Espinosa et al., 2004); some endemic taxa include *Campyloneurum angustifolium* (Polypodiaceae), *Pinus pseudostrobus apulcensis* (Pinaceae), *Fleischmannia pycnocephala* (Asteraceae), *Aporocactus flagiformis* (Cactaceae), and *Smilax mollis* (Smilacaceae; Espinosa et al., 2004).

TRANS–MEXICAN VOLCANIC BELT PROVINCE

Austral–Western province—Smith, 1941: 108 (regional.); Huber and Riina, 2003: 257 (glossary).

Transverse Volcanic province—Goldman and Moore, 1945: 356 (regional.); Moore, 1945: 218 (regional.); Stuart, 1964: 351 (regional.); Huber and Riina, 2003: 331 (glossary).

Mesa Central area (in part)—West, 1964: 368 (regional.).

Meridional Mountains province (in part)—Rzedowski, 1978: 103 (regional.); Rzedowski and Reyna-Trujillo, 1990: map (regional.); Luna Vega et al., 1999: 1301 (PAE); Huber and Riina, 2003: 260 (glossary).

Neovolcanic Axis province—Casas-Andreu and Reyna-Trujillo, 1990: map (regional.); Escalante et al., 1998: 285 (cluster anal. and regional.); Espinosa Organista et al., 2008: 56 (regional.).

Neovolcanic province—Ferrusquía-Villafranca, 1990: map (regional.).

Mexican Transvolcanic Pine-Oak Forests ecoregion—Dinerstein et al., 1995: 97 (ecoreg.).

Mexican Alpine Tundra ecoregion—Dinerstein et al., 1995: 101 (ecoreg.).

Transverse Volcanic Belt province—Anderson and O'Brien, 1996: 332 (biotic evol.); Ayala et al., 1996: 429 (regional.).

Volcanic Axis province—Arriaga et al., 1997: 64 (regional.); Morrone et al., 1999: 510 (PAE); Espinosa et al., 2000: 64 (PAE and regional.).

Transvolcanic province—Brown et al., 1998: 29 (veget.).

Transvolcanic Mountain area—Marshall and Liebherr, 2000: 206 (clad. biogeogr.).

Transvolcanic Axis area—Flores-Villela and Goyenechea, 2001: 174 (clad. biogeogr.).

Trans–Mexican Volcanic Belt province—Morrone, 2001a: 35 (regional.), 2001e: 48 (regional.); Morrone and Márquez, 2001: 636 (track anal.); Morrone et al., 2002: 93 (regional.); Corona and Morrone, 2005: 38 (track anal.); Escalante et al., 2005: 202 (PAE); Morrone, 2005: 236 (regional.); Andrés Hernández et al., 2006: 902 (track anal.); Morrone, 2006: 477 (regional.); Torres Miranda and Luna Vega, 2006: 849 (track anal.); Corona et al., 2007: 1008 (track anal.); Escalante et al., 2007b: 486 (regional.); Espinosa and Ocegueda, 2007: 6 (regional.); Ferrusquía-Villafranca, 2007: 8 (paleogeogr.); Mariño-Pérez et al., 2007: 80 (track anal.); Martínez-Aquino et al., 2007: 449 (track anal.); Navarro-Sigüenza et al., 2007: 462 (PAE); Morrone and Márquez, 2008: 19 (track anal.); Morrone, 2010b: 358 (biotic evol.); Barrera-Moreno et al., 2011: 15 (track anal.); Coulleri and Ferrucci, 2012: 105 (track anal.); Gámez et al., 2012: 259 (regional.); Mercado-Salas et al., 2012: 459 (track anal.); Puga-Jiménez et al., 2013: 1180 (PAE); Suárez-Mota et al., 2013: 94 (cluster anal. and regional.); Morrone, 2014b: 31 (regional.).

Transverse Volcanic Axis province—Huber and Riina, 2003: 262 (glossary).

Trans–Mexican Volcanic area—Katinas et al., 2004: 166 (PAE).

Transvolcanic area—Flores-Villela and Martínez-Salazar, 2009: 820 (clad. biogeogr.).

Definition

The Trans–Mexican Volcanic Belt province is situated in central Mexico (states of Aguascalientes, Mexico City, Guanajuato, Jalisco, Mexico, Michoacán, Oaxaca, Puebla, Tlaxcala, and Veracruz), at elevations above 1,800 meters (Morrone, 2001e, 2006, 2014b; Luna et al., 2007). It corresponds to a large Neogene volcanic arc, encompassing 160,000 square kilometers and a length of almost 1,000 kilometers, situated between 18°30' and 21°30' N. The Trans–Mexican Volcanic Belt was built upon Cretaceous and Cenozoic magmatic provinces and a heterogeneous basement made of tectonostratigraphic terranes of different age and lithology (Ferrari et al., 2012). It originated during the Miocene around central Mexico, and then extended eastward and westward (Ruiz-Sánchez and Specht, 2013).

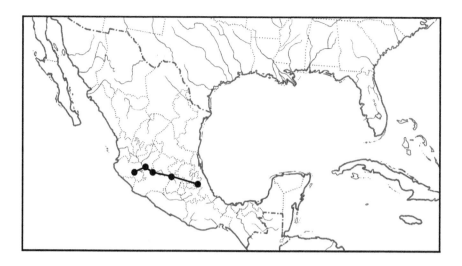

Figure 4.12 Map with the individual track of *Onthophagus hippopotamus* (Scarabaeidae) in the Trans–Mexican Volcanic Belt province.

Endemic and Characteristic Taxa

Morrone (2014b) provided a list of endemic and characteristic taxa. Some examples include *Montanoa frutescens* (Asteraceae), *Elliptoleus balli, E. corvus*, and *E. vixstriatus* (Carabidae), *Odontotaenius cuspidatus* (Passalidae), *Golofa globulicornis, Onthophagus fuscus canescens*, and *O. hippopotamus* (Scarabaeidae; Figure 4.12), *Nelsonia goldmani, Peromyscus gratus, Reithrodontomys chrysopsis*, and *R. zacatecae* (Cricetidae), *Cratogeomys tylorhinus* (Geomyidae), *Quiscalus palustres* (Icteridae), *Geothlypis speciosa* and *G. trichas chapalensis* (Parulidae), *Coturnicops noveboracensis goldmani* (Rallidae), and *Cryptotis alticola* and *Sorex orizabae* (Soricidae).

Vegetation

Temperate (pine-oak) forests are predominant; there are also *zacatonales* (alpine tundra) near the top of the volcanoes (Rzedowski, 1978; Dinerstein et al., 1995). The predominant plant genera are *Achaenipodium, Hintonella, Microspermum, Omiltemia, Peyritschia*, and *Silvia* (Rzedoswki, 1978).

Biotic Relationships

According to a PAE based on plant, insect, and bird taxa (Morrone et al., 1999), the Trans–Mexican Volcanic Belt province is closely related to the Pacific Lowlands, Balsas Basin, and Sierra Madre del Sur provinces. Another PAE based on mammals (Escalante et al., 2005) found that it is related to the Sierra Madre Oriental province.

A track analysis based on species of Coleoptera (Morrone and Márquez, 2001) showed that this province is related to the Sierra Madre Occidental, Sierra Madre del Sur, Sierra Madre Oriental, and Balsas Basin provinces. Corona et al. (2007) undertook a track analysis based on species of Coleoptera, finding three generalized tracks joining portions of this province with different provinces: one with the Sierra Madre Oriental, Chiapas Highlands, Veracruzan, and Sierra Madre del Sur provinces; a second track with the Sierra Madre Occidental, Sierra Madre del Sur, Balsas Basin, and Pacific Lowlands provinces; and a third with the Yucatán Peninsula, Chiapas Highlands, Sierra Madre Oriental, and Veracruzan provinces. They concluded that the Trans–Mexican Volcanic Belt might not represent a natural biogeographic unit, being transitional between the Nearctic and Neotropical regions (similar results were obtained by Sánchez-González et al., 2008). According to a cladistic biogeographic analysis based on insect, fish, reptile, and plant taxa (Marshall and Liebherr, 2000), the Trans–Mexican Volcanic Belt province is closely related to the Sierra Madre del Sur province.

Regionalization

Some units nested within this province identified by different authors (Moore, 1945; Rzedowski, 1978; Ferrusquía-Villafranca, 1990; Escalante et al., 2007b; Navarro-Sigüenza et al., 2007; Torres Miranda and Luna, 2007; Gámez et al., 2012; Suárez-Mota et al., 2013) were treated by Morrone (2014b) as two subprovinces and five districts. As a preliminary delimitation, I (Morrone, 2014b) considered Torres Miranda and Luna's (2007) districts (Figure 4.13), but noted that Moore's (1945) names have nomenclatural priority. The West subprovince corresponds to the western portion of the Trans–Mexican Volcanic Belt province. Endemic taxa include *Reithrodontomys goldmani* and *R. hirsutus* (Cricetidae), *Cratogeomys gymnurus* (Geomyidae), and *Liomys spectabilis* (Heteromyidae; Gámez et al., 2012). Within this subprovince, three districts have been recognized. The Jaliscan district corresponds to the northern part of the West subprovince in the state of Jalisco (Moore, 1945); endemic taxa include *Pappogeomys alcorni* (Geomyidae) and *Liomys spectabilis* (Heteromyidae; Barrera-Moreno et al., 2001; Moreno-Barrera et al., 2011; Gámez et al., 2012). The Otomí district corresponds to the central northern part of the West subprovince in the states of Jalisco, Guanajuato, Querétaro, Hidalgo, Michoacán, and Mexico (Moore, 1945); endemic taxa include *Habromys delicatulus* (Cricetidae) and *Zygogeomys trichopus* (Geomyidae; Ceballos and Oliva, 2005; Gámez et al., 2012). The Tarascan district corresponds to the central southern part of the West subprovince in the states of Jalisco, Michoacán, and Mexico (Moore, 1945); an endemic taxon is *Reithrodontomys sumichrasti nerterus* (Cricetidae; Ramírez-Pulido et al., 2005).

The East subprovince corresponds to the eastern portion of the Trans–Mexican Volcanic Belt province; some endemic taxa include *Neotoma nelsoni* and *Peromyscus bullatus* (Cricetidae), *Cratogeomys merriami* (Geomyidae), *Spermophilus perotensis* (Sciuridae) and *Sorex ventralis* (Soricidae; Ceballos and Oliva, 2005; Gámez et al., 2012). Within this subprovince, two districts have been recognized. The Aztec district

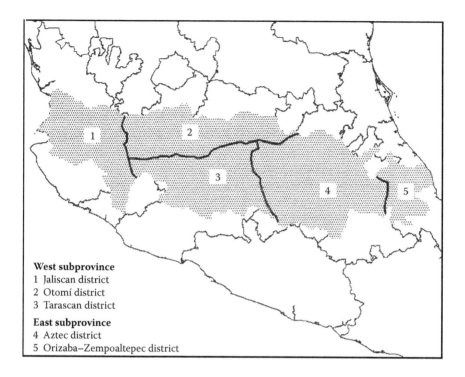

Figure 4.13   Map of the districts of the Trans–Mexican Volcanic Belt province. (Modified from Torres Miranda, A. and I. Luna, "Hacia una síntesis panbiogeográfica." In: Luna, I., J. J. Morrone and D. Espinosa (eds.), *Biodiversidad de la Faja Volcánica Transmexicana*. Las Prensas de Ciencias, UNAM, Mexico City, pp. 503–514, 2007.)

corresponds to the western part of the East subprovince in the states of Mexico, Hidalgo, Tlaxcala, Puebla, Morelos, and Guerrero (Moore, 1945); an endemic taxon is *Romerolagus* (Leporidae; Arriaga et al., 1997; Ceballos and Oliva, 2005; Escalante et al., 2005; Gámez et al., 2012). The Orizaba–Zempoaltepec district corresponds to southeastern Puebla, northern Oaxaca, and western Veracruz (Moore, 1945; Rzedowski, 1978); some endemic taxa include *Oaxacania* (Asteraceae), *Solisia* (Cactaceae), *Pringleochloa* (Poaceae), *Parallocorynus bicolor* (Belidae), *Habromys lepturus* and *Peromyscus mekisturus* (Cricetidae), and *Sorex macrodon* and *S. oreopolus* (Soricidae; Rzedowski, 1978; Ceballos and Oliva, 2005; Gámez et al., 2012; O'Brien and Tang, 2015).

Cenocrons

Ruiz-Sánchez and Specht (2013) undertook a phylogeographic analysis of *Nolina parviflora* (Asparagaceae) in order to evaluate the historical events that have influenced its distributional patterns in the Trans–Mexican Volcanic Belt. They combined a multi-locus phylogeographic analysis with phylogenetic-based estimates of divergence times, to place the genetic divergences within this species in a temporal context and relate them to abiotic factors. They found two well-supported clades: one including organisms from Jalisco and Zacatecas and another including the remaining populations. Divergence time estimates showed that both clades diverged from one another in the Late Oligocene to Late Pliocene, some subclades within them diverged from Mid- or Late Miocene to Pleistocene, and other subclades within them during the Pliocene and the Pleistocene. The authors concluded that the diversification of this species might have been driven by the volcanic episodes that shaped the development of the Trans–Mexican Volcanic Belt.

SIERRA MADRE DEL SUR PROVINCE

Sierra Madre del Sur province—Goldman and Moore, 1945: 358 (regional.); Moore, 1945: 218 (regional.); Stuart, 1964: 351 (regional.); Casas-Andreu and Reyna-Trujillo, 1990: map (regional.); Ramírez-Pulido and Castro-Campillo, 1990: map (regional.); Anderson and O'Brien, 1996: 332 (biotic evol.); Ayala et al., 1996: 429 (regional.); Arriaga et al., 1997: 65 (regional.); Campbell, 1999: 116 (regional.); Morrone et al., 1999: 510 (PAE); Espinosa et al., 2000: 64 (PAE and regional.); Morrone, 2001a: 39 (regional.), 2001e: 49 (regional.); Morrone and Márquez, 2001: 637 (track anal.); Morrone et al., 2002: 95 (regional.); Huber and Riina, 2003: 263 (glossary); Corona and Morrone, 2005: 38 (track anal.); Escalante et al., 2005: 202 (PAE); Morrone, 2005: 237 (regional.); Andrés Hernández et al., 2006: 902 (track anal.); Morrone, 2006: 477 (regional.); Mariño-Pérez et al., 2007: 80 (track anal.); Espinosa Organista et al., 2008: 57 (regional.); Morrone and Márquez, 2008: 20 (track anal.); Escalante et al., 2009: 473 (endem. anal.); León-Paniagua and Morrone, 2009: 1942 (clad. biogeogr.); Blancas-Calva et al., 2010: 562; Morrone, 2010b: 358 (biotic evol.); Escalante et al., 2011: 32 (track anal.); Coulleri and Ferrucci, 2012: 105 (track anal.); Mercado-Salas et al., 2012: 459 (track anal.); Puga-Jiménez et al., 2013: 1180 (PAE); Morrone, 2014b: 33 (regional.); Santiago-Alvarado et al., 2016: 431 (regional.).
Sierra and Mesa del Sur region—West, 1964: 368 (regional.).
Meridional Mountains province (in part)—Rzedowski, 1978: 103 (regional.); Rzedowski and Reyna-Trujillo, 1990: map (regional.); Luna Vega et al., 1999: 1301 (PAE); Huber and Riina, 2003: 260 (glossary).
Sierra Madre del Sur Pine-Oak Forests ecoregion—Dinerstein et al., 1995: 97 (ecoreg.).
Sierra Madre del Sur Highlands area—Flores-Villela and Goyenechea, 2001: 174 (clad. biogeogr.).
Sierra Madre del Sur area—Marshall and Liebherr, 2000: 206 (clad. biogeogr.); Katinas et al., 2004: 166 (PAE).
Sierra Madre del Sur ecoregion—Abell et al., 2008: 408 (ecoreg.).
Highlands of Southern Mexico area—Flores-Villela and Martínez-Salazar, 2009: 820 (clad. biogeogr.).

Definition

The Sierra Madre del Sur province comprises south central Mexico, between southern Michoacán, Guerrero, Oaxaca, and part of Puebla, at elevations above 1,000 meters (Morrone, 2001e, 2014b; Luna-Vega et al., 2016).

Endemic and Characteristic Taxa

Morrone (2014b) provided a list of endemic and characteristic taxa. Some examples include *Montanoa grandiflora*, *M. mollissima*, and *M. tomentosa* subsp. *microcephala* (Asteraceae), *Bursera aloexylon* (Burseraceae; Figure 4.14), *Naupactus stupidus*, *Pantomorus longulus*, *Phacepholis brevipes*, and *P. globicollis* (Curculionidae), *Cotinis ibarrai*, *Onthophagus bassarisus*, and *O. semiopacus* (Scarabaeidae), *Abronia mixteca* and *A. oaxacae* (Anguidae), *Cyanolica mirabilis* (Corvidae), *Megadontomys thomasi*, *Peromyscus megalops*, and *Reithrodontomys mexicanus* (Cricetidae), *Cryptotis goldmani* (Soricidae), and *Amazilia wagneri*, *Eupherusa cyanophrys*, *Lampornis margaritae*, and *Lophornis brachylopha* (Trochilidae).

Vegetation

Temperate forests, particularly pine-oak forests, are predominant; there are also xerophytic scrublands with cacti (Rzedowski, 1978; Dinerstein et al., 1995). The montane forests of the Sierra Madre del Sur province represent some of the world's most diverse and complex subtropical mixed hardwood-conifer forests (Dinerstein et al., 1995). Dominant plant genera include *Achaenipodium*, *Hintonella*, *Microspermum*, *Omiltemia*, *Peyritschia*, and *Silvia* (Rzedoswki, 1978).

Biotic Relationships

According to a PAE based on plant, insect, and bird taxa (Morrone et al., 1999), the Sierra Madre del Sur province is closely related to the Pacific Lowlands,

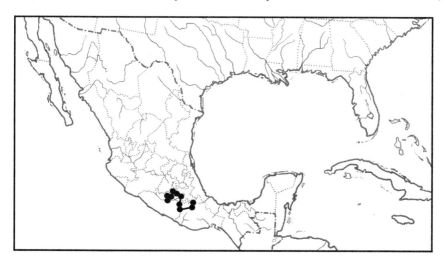

Figure 4.14 Map with the individual track of *Bursera aloexylon* (Burseraceae) in the Sierra Madre del Sur province.

Trans–Mexican Volcanic Belt, and Balsas Basin provinces. Another PAE based on mammals (Escalante et al., 2005) found that it is related to the Chiapas Highlands province and other areas assigned to the Neotropical region. A track analysis based on species of Coleoptera (Morrone and Márquez, 2001) showed that this province is related to the Sierra Madre Occidental, Balsas Basin, Sierra Madre Oriental, and Trans–Mexican Volcanic Belt provinces. A cladistic biogeographic analysis based on insect, fish, reptile, and plant taxa (Marshall and Liebherr, 2000) postulated that this province is related to the Trans–Mexican Volcanic Belt province.

Regionalization

Some nested units that have been identified within this province (Smith, 1941; Ferrusquía-Villafranca, 1990; Arriaga et al., 1997; Escalante et al., 1998) were treated by Morrone (2014b) as six districts. As a preliminary delimitation of these districts, I (Morrone, 2014b) combined the schemes of Ferrusquía-Villafranca (1990) and Escalante et al. (1998). The Guerrero district corresponds to the western part of the Sierra Madre del Sur province, basically in the state of Guerrero (Escalante et al., 1998); endemic taxa include *Peromyscus bakeri* (Cricetidae) and *Sylvilagus insonus* (Leporidae; Ceballos and Oliva, 2005). The Central Valleys district corresponds to the Central Valleys area, in the state of Oaxaca (Ferrusquía-Villafranca, 1990); an endemic taxon is *Melozone albicollis* (Emberizidae; Sánchez-González, pers. comm.). The Isthmian district corresponds to the portion of the Sierra Madre del Sur province that corresponds to the Isthmus of Tehuantepec (Ferrusquía-Villafranca, 1990); an endemic taxon is *Peromyscus melanurus* (Cricetidae; Ceballos and Oliva, 2005). The Nudo de Zempoaltépetl district corresponds to the Nudo Zempoaltépetl or Sierra Mixteca, in the states of Puebla and Oaxaca (Escalante et al., 1998); endemic taxa include *Microtus umbrosus* (Cricetidae), *Aimophila notosticta* (Emberizidae), and *Amazilia wagneri* (Trochilidae; García-Trejo and Navarro, 2004; Ceballos and Oliva, 2005). The Oaxacan Highland district corresponds to the highlands of the state of Oaxaca (Smith, 1941; Arriaga et al., 1997); endemic taxa include *Amphyteryx longicaudatus* (Amphypterygidae), *Neocordulia batesii* (Cordulidae), and *Habromys chinanteco, Megadontomys cryophilus*, and *Microtus oaxacensis* (Cricetidae; González Soriano and Novelo Gutiérrez, 1996; Ceballos and Oliva, 2005). The Sierra de Miahuatlán district corresponds to the Sierra de Mihauatlán in the state of Oaxaca (Escalante et al., 1998); endemic species include *Cryptotis peregrina* and *C. phillipsii* (Soricidae) and *Eupherusa cyanophrus* (Trochilidae; Ceballos and Oliva, 2005; Woodman, 2005; Sánchez-González, pers. comm.).

Blancas-Calva et al. (2010) analyzed the relationships of 26 subbasins of the Sierra Madre del Sur province, based on distributional bird data, applying a PAE. The cladogram that was obtained suggests the existence of three groups of subbasins: those from arid environments, located in north-southeastern Oaxaca; those from subhumid environments; and those from humid and environmentally more complex habitats.

A recent regionalization by Santiago-Alvarado et al. (2016), based on endemic plant and animal species, recognized two subprovinces. The Western Sierra Madre del Sur subprovince corresponds to the Jaliscan–Guerreran province (Ferrusquía-Villafranca, 1990), herein treated within the Balsas Basin province. The Eastern Sierra Madre del Sur subprovince, which corresponds to the Sierra Madre del Sur in the strict sense, includes the Guerreran and Oaxacan districts.

CHIAPAS HIGHLANDS PROVINCE

Chiapas Highlands province—Smith, 1941: 109 (regional.); Goldman and Moore, 1945: 359 (regional.); Arriaga et al., 1997: 66 (regional.); Espinosa et al., 2000: 64 (PAE and regional.); Huber and Riina, 2003: 103 (glossary); Espinosa Organista et al., 2008: 58 (track anal. and regional.); Morrone, 2014b: 34 (regional.).

Chiapas province—Barrera, 1962: 101 (regional.); Ferrusquía-Villafranca, 1990: map (regional.); Ramírez-Pulido and Castro-Campillo, 1990: map (regional.); Morrone et al., 1999: 510 (PAE); Morrone, 2001a: 45 (regional.), 2001e: 52 (regional.); Morrone and Márquez, 2001: 637 (track anal.); Morrone et al., 2002: 99 (regional.); Escalante et al., 2003: 570 (regional.), 2005: 202 (PAE); Morrone, 2006: 478 (regional.); Quijano-Abril et al., 2006: 1269 (track anal.); Mariño-Pérez et al., 2007: 80 (track anal.); Morrone and Márquez, 2008: 20 (track anal.); Escalante et al., 2009: 473 (endem. anal.); Morrone, 2010b: 358 (biotic evol.); Escalante et al., 2011: 32 (track anal.); Mercado-Salas et al., 2012: 459 (track anal.).

Chiapas–Guatemalan Highlands province—Ryan, 1963: 23 (regional.); Stuart, 1964: 357 (regional.); Miller, 1966: 783 (regional.); Campbell, 1999: 116 (regional.).

Honduran Upland province—Savage, 1966: 736 (biotic evol.).

Central American Montane Forest center—Müller, 1973: 14 (regional.).

Trans–Isthmic Mountains province—Rzedowski, 1978: 103 (regional.); Rzedowski and Reyna-Trujillo, 1990: map (regional.); Ayala et al., 1996: 429 (regional.); Luna Vega et al., 1999: 1301 (PAE); Huber and Riina, 2003: 260 (glossary).

Sierra Madre de Chiapas province—Casas-Andreu and Reyna-Trujillo, 1990: map (regional.); Anderson and O'Brien, 1996: 332 (biotic evol.).

Central American Montane Forests ecoregion—Dinerstein et al., 1995: 87 (ecoreg.).

Central American Pine–Oak Forests ecoregion—Dinerstein et al., 1995: 97 (ecoreg.).

Sierra Madre Moist Forests ecoregion—Dinerstein et al., 1995: 87 (ecoreg.).

Sierra Norte de Chiapas province—Escalante et al., 1998: 285 (cluster anal. and regional.).

Chiapan/Guatemalan Highland area—Marshall and Liebherr, 2000: 206 (clad. biogeogr.).

Highlands of Chiapas and Guatemala area—Flores-Villela and Goyenechea, 2001: 174 (clad. biogeogr.); Flores-Villela and Martínez-Salazar, 2009: 820 (clad. biogeogr.).

Highlands of Northern Central America region—Huber and Riina, 2003: 156 (glossary).

Chiapas–Fonseca ecoregion—Abell et al., 2008: 408 (ecoreg.).

Chortís block province (in part)—Townsend, 2014: 206 (paleogeogr.).

Definition

The Chiapas Highlands province comprises southern Mexico, Guatemala, Honduras, El Salvador, and Nicaragua; it basically corresponds to the Sierra Madre de Chiapas, from altitudes of 500 to 2,000 meters (Morrone, 2001e, 2006, 2014b). Marshall and Liebherr (2000) combined the Chiapas Highlands and Yucatán Peninsula provinces in a single province, based on their shared taxa.

Endemic and Characteristic Taxa

Morrone (2014b) provided a list of endemic and characteristic taxa. Some examples include *Juniperus comitana* (Cupressaceae), *Cecropia sylvicola* (Cecropiaceae), *Tetranema evolutum* (Scrophulariaceae), *Acropsopilio chomulae* (Caddidae), *Toonglasa indomita* (Lygaeidae), *Baronia brevicornis rufodiscalis* and *Priamides erostratus erostratus* (Papilionidae), *Bledius strenuus, Gansia andersoni*, and *Styngetus championi* (Staphylinidae), *Microtus guatemalensis, Oryzomys rhabdops, O. saturatior, Ototylomys phyllotis connectens, Peromyscus guatemalensis*, and *Tylomys bullaris* (Cricetidae), *Sceloporus malachitichus* (Phrynosomatidae; Figure 4.15), and *Bothriechis aurifer* and *B. bicolor* (Viperidae).

Vegetation

Different types of temperate forests (with species of *Pinus* and *Quercus* being dominant), savannas, and scrublands (Rzedowski, 1978; Dinerstein et al., 1995).

Biotic Relationships

According to Müller (1973), the Chiapas Highlands province is related to the Mosquito province. A PAE based on plant, insect, and bird taxa (Morrone et al., 1999) postulated that this province is closely related to the Veracruzan province.

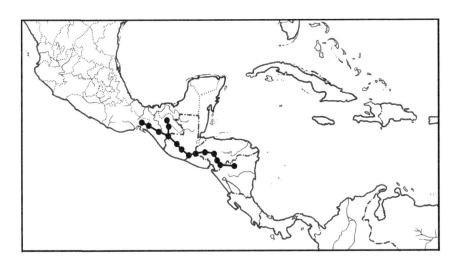

Figure 4.15   Map with the individual track of *Sceloporus malachitichus* (Phrynosomatidae) in the Chiapas Highlands province.

Another PAE based on mammals (Escalante et al., 2005) found that it is related to other areas from the Neotropical region. A track analysis based on species of Coleoptera (Morrone and Márquez, 2001) showed that this province is related to the Puntarenas–Chiriquí, Veracruzan, and Pacific Lowlands provinces.

Regionalization

Some nested units that have been identified within the Chiapas Highlands province (Ryan, 1963; Stuart, 1964; Müller, 1973; Rzedowski, 1978; Ferrusquía-Villafranca, 1990; Rzedowski and Reyna-Trujillo, 1990) were treated by Morrone (2014b) as six districts; their precise delimitation is not without doubt. The Sierra Madrean district corresponds to the western portion of the Chiapas Highlands province in eastern Oaxaca (Ferrusquía-Villafranca, 1990); some endemic taxa include *Habromys lophurus* and *Tylomys tumbalensis* (Cricetidae) and *Heteromys nelsoni* (Heteromyidae; Escalante et al., 2005). The Comitanian district corresponds to the southern portion of the Chiapas Highlands province in eastern Oaxaca and Chiapas (Ferrusquía-Villafranca, 1990). The Lacandonian district corresponds to the northern portion of the Chiapas Highlands province in Chiapas (Ferrusquía-Villafranca, 1990); an endemic taxon is *Cryptotis griseoventris* (Soricidae; Woodman, 2005). The Soconusco district is a narrow strip in the lower foothills of the Sierra Madre de Chiapas in Mexico and a part of Guatemala (Rzedowski, 1978); endemic taxa include the genera *Pinarophyllon* and *Plocaniophyllon* (Rubiaceae; Rzedowski, 1978). The Guatemalan Highland district corresponds to the highlands of Guatemala (Müller, 1973). The Nicaraguan Montane district corresponds to the highlands of Nicaragua (Ryan, 1963).

CHAPTER 5

The Antillean Subregion

The Antillean subregion comprises the Antilles or West Indies (Greater and Lesser Antilles) and the Bahamas (Morrone, 2004a, 2006, 2014b). The Greater Antilles include Cuba, Jamaica, Hispaniola, and Puerto Rico, and the Lesser Antilles include Grenada, The Grenadines, St. Vincent, Barbados, St. Lucia, Martinique, Dominica, Marie Galante, Guadeloupe, La Desirade, Montserrat, Antigua, Nevis, St. Kitts, Barbuda, St. Eustatius, Saba, St. Barthélemy, St. Martin, and Anguilla. The Antillean subregion was recognized originally by Wallace (1876). This subregion has been previously lumped with Central America (Clarke, 1892; Rivas-Martínez and Navarro, 1994; Holt et al., 2013) and with southern Mexico, Central America, and northwestern South America (Morrone, 1999, 2006).

ANTILLEAN SUBREGION

Antillean subregion—Wallace, 1876: 79 (regional.); Heilprin, 1887: 80 (regional.); Lydekker, 1896: 136 (regional.); Sclater and Sclater, 1899: 65 (regional.); Bartholomew et al., 1911: 9 (regional.); Mello-Leitão, 1937: 229 (regional.); Rapoport, 1968: 71 (biotic evol. and regional.); Bănărescu and Boşcaiu, 1978: 259 (regional.); Borhidi and Muñiz, 1986: 4 (regional.); Samek et al., 1988: 29 (regional.); Del Risco and Vandama, 1989: X.2.4 (regional.); Muñiz, 1996: 283 (regional.); Huber and Riina, 2003: 23 (glossary); Echeverry and Morrone, 2013: 1628 (track anal.); Morrone, 2014b: 35 (regional.), 2014c: 206 (clad. biogeogr.); Klassa and Santos, 2015: 520 (endem. anal.).

West Indian province—Engler, 1882: 345 (regional.).

Central American subarea (in part)—Clarke, 1892: 381 (regional.).

Antillean division—Merriam, 1892: 18 (regional.).

Caribbean province—Mello-Leitão, 1937: 246 (regional.); Cabrera and Willink, 1973: 38 (regional.); Brown et al., 1998: 32 (veget.).

Caribbean region (in part)—Good, 1947: 232 (regional.); Rzedowski, 1978: 107 (regional.); Takhtajan, 1986: 251 (regional.); Samek et al., 1988: 26 (regional.); Rangel et al., 1995d: 21 (veget.); Muñiz, 1996: 283 (regional.); Huber and Riina, 1997: 119 (glossary); A. Graham, 2003: 278 (paleofloras); Huber and Riina, 2003: 97 (glossary); Procheş and Ramdhani, 2012: 263 (cluster anal. and regional.).

West Indian subregion—Hershkovitz, 1969: 9 (regional.); Smith, 1983: 462 (cluster anal. and regional.); Sánchez Osés and Pérez-Hernández, 2005: 168 (regional.).

Caribbean dominion (in part)—Cabrera and Willink, 1973: 32 (regional.); Huber and Riina, 1997: 151 (glossary); Zuloaga et al., 1999: 18 (regional.); Huber and Riina, 2003: 124 (glossary).

Caribbean subregion (in part)—Rivas-Martínez and Navarro, 1994: map (regional.); Morrone, 1999: 2 (regional.); Morrone et al., 1999: 510 (PAE); Morrone, 2001a: 30 (regional.), 2001e: 46 (regional.); Corona and Morrone, 2005: 38 (track anal.); Morrone, 2005: 238 (regional.); Viloria, 2005: 449 (regional.); Quijano-Abril et al., 2006: 1268 (track anal.); Nihei and de Carvalho, 2007: 497 (clad. biogeogr.); Asiain et al., 2010: 178 (track anal.); Morrone, 2010a: 34 (regional.); Coulleri and Ferrucci, 2012: 105 (track anal.); Lamas et al., 2014: 955 (clad. biogeogr.).

Antillean province—Rivas-Martínez and Navarro, 1994: map (regional.); Cano et al., 2009: 528 (cluster anal. and regional.).

Caribbean area—Coscarón and Coscarón-Arias, 1995: 726 (areas of endem.).

Caribbean bioregion—Dinerstein et al., 1995: map (ecoreg.).

Antillean region—Huber and Riina, 2003: 23 (glossary).

Antillean dominion—Morrone, 2004a: 157 (track anal.); Corona and Morrone, 2005: 38 (track anal.); Morrone, 2006: 479 (regional.).

Caribbean component—Nihei and de Carvalho, 2004: 271 (clad. biogeogr.).

Caribbean Island hotspot—Smith et al., 2004: 112 (ecoreg.).

Greater Antilles area—Porzecanski and Cracraft, 2005: 266 (PAE).

Central–Eastern Antilles superprovince—Cano et al., 2009: 543 (cluster anal. and regional.).

Western Antilles superprovince—Cano et al., 2009: 543 (cluster anal. and regional.).

Caribbean–Mesoamerican region (in part)—Rivas-Martínez et al., 2011: 26 (regional.).

Panamanian region (in part)—Holt et al., 2013: 77 (cluster anal. and regional.).

Endemic and Characteristic Taxa

Taxa endemic to the Antillean subregion include many plant and animal genera and species. Antillean plants include several endemic taxa, with over 50 percent of the vascular plant species being endemic (Santiago-Valentin and Olmstead, 2004). Muñiz (1996) listed 25 plant genera that are endemic to the Antilles: *Borrichia* and *Gundlachia* (Asteraceae), *Calygonium, Tetrazygia,* and *Tetrazygiopsis* (Melastomataceae), *Cataesbaea, Ernodea, Neolagueria, Rondeletia, Scolosanthus,* and *Strumpfia* (Rubiaceae), *Coccothrinax* and *Thrinax* (Arecaceae), *Consolea* (Cactaceae), *Gesneria* and *Rhytidophyllum* (Gesneriaceae), *Metopium* (Anacardiaceae), *Oplonia* (Acanthaceae), *Oxandra* (Lauraceae), *Petitia* (Verbenaceae), *Reynosia* and *Sarcomphalus* (Rhamnaceae), *Rochefortia* (Borraginaeae), and *Wallenia* (Myrsinaceae). Francisco-Ortega et al. (2007) provided an updated list of 180 endemic genera, which belong to 47 families. Arthropods include several species of Formicidae (Cabrera and Willink, 1973) and Membracidae (Goldani et al., 2002). Among vertebrates, the genus *Cyclura* (Iguanidae; Figure 5.1) is diversified in the Antilles (Malone et al., 2000). The bird family Phaenicophilidae is endemic to the subregion (Barker et al., 2013). A summary of the mammal taxa of the Antilles (Dávalos, 2004) indicate that there are 38 endemic genera: five belong to

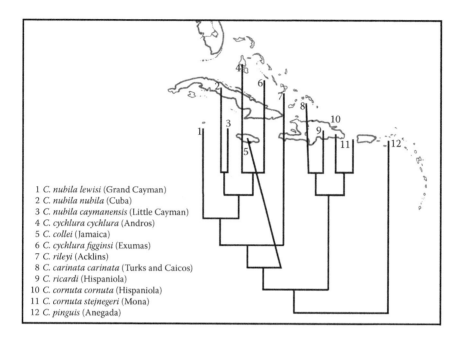

1 *C. nubila lewisi* (Grand Cayman)
2 *C. nubila nubila* (Cuba)
3 *C. nubila caymanensis* (Little Cayman)
4 *C. cychlura cychlura* (Andros)
5 *C. collei* (Jamaica)
6 *C. cychlura figginsi* (Exumas)
7 *C. rileyi* (Acklins)
8 *C. carinata carinata* (Turks and Caicos)
9 *C. ricardi* (Hispaniola)
10 *C. cornuta cornuta* (Hispaniola)
11 *C. cornuta stejnegeri* (Mona)
12 *C. pinguis* (Anegada)

Figure 5.1   Map with the distribution of the species of *Cyclura* (Iguanidae) in the Antilles, with their phylogenetic relationships indicated. (Modified from Malone C. L. et al., *Molecular Phylogenetics and Evolution*, 17: 269–279, 2000.)

Xenarthra (Megalonychidae), 18 to Rodentia (Muridae, Echimyidae, Capromyidae, and Heptaxodontidae), two to Lipotyphla (Nesophontidae and Solenodontidae), 10 to Chiroptera (Natalidae and Phyllostomidae), and three to Primates (Atelidae).

Biotic Relationships

The classical explanation of the origin of the Antillean biota, proposed before acceptance of continental drift and plate tectonics, postulated that the biota of the American continent colonized the islands by over-water dispersal (Wallace, 1876; Matthew, 1915; Simpson, 1953; Darlington, 1957). According to the biotic sources of the taxa, it has been usual to characterize the following elements: cosmopolitan (mainly pantropical and pan-Caribbean), West Indian (Greater Antilles, Lesser Antilles, or both), endemic and continental (Murphy and Lugo, 1995). As an alternative to this explanation, some authors (e.g., Allen, 1911; Barbour, 1914; Müller, 1973) postulated the existence of land bridges, formerly connecting the Antilles and the continent, basically considering the improbability of overwater dispersal.

Rosen (1976) provided the first dispersal–vicariance explanation of the Antillean biogeographic history, based on a track analysis of terrestrial, freshwater, and marine animal taxa, and the geotectonic evidence available (Figure 5.2). He postulated that in the Late Mesozoic–Early Cenozoic, the proto-Antilles were situated where Lower Central America is situated today. Then, they drifted eastward when a portion of the

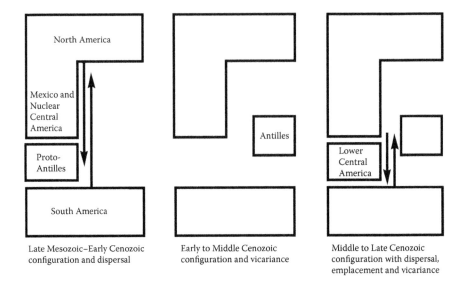

Figure 5.2 Dispersal–vicariance model for the Caribbean basin. (From Rosen, D. E., *Systematic Zoology*, 24: 431–464, 1976.)

Pacific seafloor began moving along two faults and originated the Antillean archipelago. Based on this model, Rosen (1976) concluded that dispersal is the simplest explanation for the generalized tracks identified and that vicariance predicted relationships that could be tested by phylogenetic hypotheses. Pregill (1981) rejected Rosen's model, considering that the existence of the proto-Antilles was contradicted by tectonic and fossil evidence and that vertebrates arrived to the Antilles by overwater dispersal from the Oligocene to the Quaternary. MacFadden (1981) considered that the historical biogeography of the Greater Antilles was very complex, and probably the result of both dispersal and vicariance events. Hedges (1982) considered that the proto-Antilles were supported by geological evidence, and that the Greater Antilles may have moved significantly to their present location, and that many different plate tectonic scenarios proposed suggested cautious biogeographic interpretations. He concluded that a vicariant origin for some (or all) of the Antillean biota was possible using the available plate tectonics reconstructions, but dispersal events may have also occurred.

Rauchenberger (1988) undertook a cladistic biogeographic analysis of 12 freshwater fish taxa of the Greater Antilles, namely gars of the genus *Atractosteus*, synbranchid eels of the genus *Ophisternon*, cichlids of the genus *Cichlasoma*, cyprindontiforms of the genera *Cyprinodon*, *Cubanichthys*, and *Rivulus* and poeciliids of the tribe Girardiini, the subgenera *Limia* and *Poecilia* of the genus *Poecilia*, and the *Gambusia nicaraguensis*, *G. puncticulata*, and *G. punctata* species groups. Based on their phylogenetic hypotheses, she obtained resolved area cladograms and a general area cladogram (Figure 5.3). Rauchenberger (1988) concluded that neither Cuba nor Hispaniola are natural areas of endemism (Cuba being divisible into its

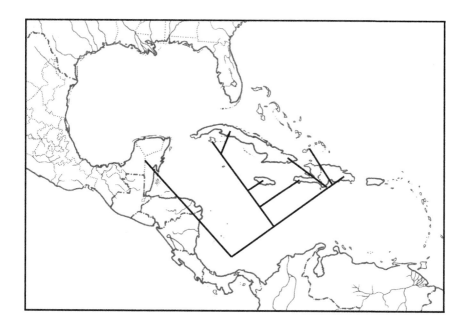

Figure 5.3   Map with the general area cladogram for freshwater fishes of the Greater Antilles. (From Rauchenberger, M., *Systematic Zoology*, 37: 356–365, 1988.)

western and eastern portions and Hispaniola into its southwestern and central portions); western Cuba, Isla de la Juventud, the Cayman Islands, Jamaica, and southwestern Hispaniola formed a monophyletic group; eastern Cuba was related to the Bahamas and central Hispaniola; and that these two island groups together form a group, which is most closely related to Nuclear Central America (Yucatán Peninsula, Guatemala, and Belize). Another cladistic biogeographic analysis (Crother and Guyer, 1996), based on arthropods, fish, mammals, and reptiles, led to several general area cladograms; the strict consensus cladogram (Figure 5.4) indicates a closer relationship between Hispaniola, Puerto Rico, Cuba, and Jamaica, with the other areas unresolved. The authors considered that it was reasonable to explain this pattern in terms of vicariance.

Williams (1989) reviewed the dispersal and vicariance hypotheses previously formulated. He considered that difficult colonizations were required by the dispersal hypotheses, and that massive early extinctions were required by the vicariance hypotheses. He postulated that the only way to make a decision between them was using the fossil record, in order to show either simultaneous arrivals of different taxa (for the vicariance hypotheses) or temporally spaced intervals (for the dispersal hypotheses). He considered that the scarce fossil record did not allow making such decision.

Iturralde-Vinent and MacPhee (1999; see also MacPhee and Iturralde-Vinent, 2005) analyzed the paleogeography of the Antilles (Figure 5.5), beginning with the opening of the Caribbean basin in the Middle Jurassic and running to the end of the

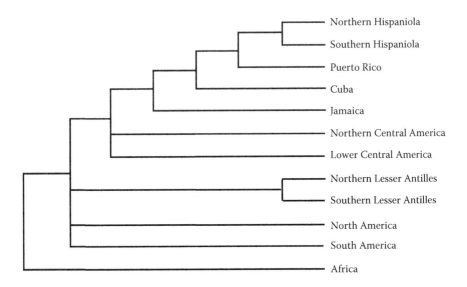

Figure 5.4   General area cladogram for several taxa of the Greater Antilles. (From Crother, B. I. and C. Guyer, *Herpetologica*, 52: 440–465, 1996.)

Middle Miocene. They postulated that while mammals and other terrestrial vertebrates might have occupied landmasses in the Antilles at any time, the existing Greater Antillean islands are no older than Middle Eocene; although earlier islands must have existed, it is not likely that they remained subaerial due to repeated transgressions, subsidence, and the K/T bolide impact and associated megatsunamis. Accordingly, they inferred that the Antillean mammal lineages of the existing fauna must all be younger than Middle Eocene, and according to the fossil record most mammal lineages entered the Greater Antilles around the Eocene–Oligocene transition. North and South America were physically connected as continental areas until the mid-Jurassic, and terrestrial connections between them since then can only have occurred via land bridges. In the Cretaceous, three major uplift events may have produced intercontinental land bridges involving the Cretaceous Antillean island arc: the Late Campanian/Early Maastrichtian uplift event is the one most likely to have resulted in a land bridge; the existing land bridge (Panamanian isthmus) was completed in the Pliocene; and evidence for a precursor bridge late in the Middle Miocene is ambiguous. Iturralde-Vinent and MacPhee (1999) hypothesized that during the Eocene–Oligocene transition, the developing northern Greater Antilles and northwestern South America were briefly connected by a landspan centered on the emergent Aves Ridge, named GAARlandia (GAAR = Greater Antilles + Aves Ridge). The massive uplift event that apparently permitted these connections was spent by 32 Ma; a general subsidence followed, ending the GAARlandia phase. The GAARlandia hypothesis involves both dispersal and continent–island vicariance, although the latter appears to be excludable for any time period since the mid-Jurassic. Even if vicariance occurred at that time, its relevance for understanding

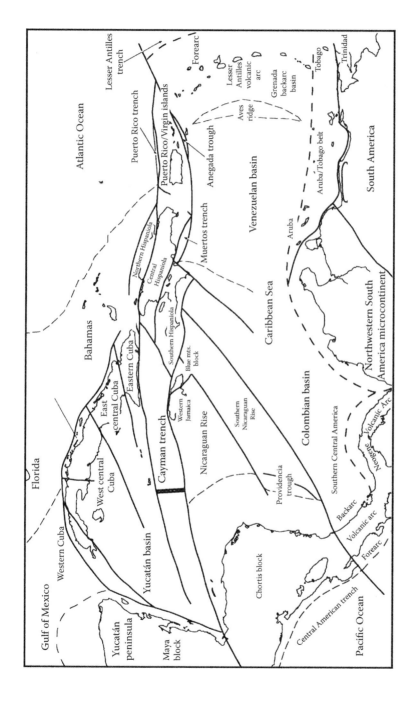

Figure 5.5 Map with the main geographic and tectonic features of the Caribbean basin. (From Iturralde-Vinent, M. A. and R. D. E. MacPhee, *Bulletin of the American Museum of Natural History*, 238: 1–95, 1999.)

the origin of the modern Antillean biota might be minimal. Over-water dispersal is inadequate as an explanation of observed distribution patterns of terrestrial faunas in the Greater Antilles, because prior to the Pliocene, regional paleoceanography was such that current flow patterns from major rivers would have delivered South American waifs to the Central American coast, not to the Antilles. Since at least three mammal lineages were already on one or more of the Greater Antilles by the Early Miocene, Hedges' inference as to the primacy of overwater dispersal appears to be at odds with the facts. By contrast, the landspan model is consistent with most aspects of Antillean land mammal biogeography as currently known.

A. Graham (2003) reviewed the known fossil floras of the Greater Antilles, southeastern United States, northeastern Mexico, Panama, and Colombia, in order to reconstruct the Cenozoic vegetation history. Based on the available tectonic models (Pindell and Barrett, 1990; Iturralde-Vinent and MacPhee, 1999), he discussed the possibility of past dispersal and vicariance events that shaped Antillean biogeography. He postulated that a Cretaceous volcanic island arc extended from the Mexico/Chortís block in the north to Ecuador in the south, gradually moved through the developing portal between North and South America and collided with the Bahamas platform in the Middle Eocene. Throughout this 70-million-year history, there was a very complex pattern of collision/separation and submergence/emergence that provided opportunities for vicariance and dispersal. Although both processes are viable options, A. Graham (2003) concluded that dispersal was a prominent means of diversification in the Antilles.

Vázquez-Miranda and colleagues (2007) analyzed the relationships of the Caribbean islands and nearby continental areas, applying a parsimony analysis of endemicity to bird distributional data obtained from published literature. They undertook three different analyses: species, genera, and species and genera combined. Based on the results of the combined analysis, they recognized four main areas (Figure 5.6): the Lesser Antilles (a paraphyletic group at the base of the cladogram), North America, South America (including the Dutch West Indies and Trinidad and Tobago), and the Greater Antilles as the sister area to Yucatán–Central America. Vázquez-Miranda and colleagues (2007) considered that their results supported Caribbean vicariance models and showed relative congruence with available phylogenetic data, although patterns appear to have been influenced also by dispersal.

Ali (2012) accepted that GAARlandia might explain the distribution of some taxa, but the great majority of the ancestors of terrestrial vertebrate species from the Greater Antilles arrived by overwater dispersal. As other authors have noted before, he considered that the fossil record did not exhibit a broad range of higher taxa, and at lower levels there are some broad adaptive radiations due to species exploiting a wide range of unoccupied niches (e.g., ground sloths, capromyid rodents, eleutherodactyline frogs, and anoline lizards). Additionally, given that since South America connected with Central and North America ca. 3 mya, there has been a massive biotic exchange; based on GAARlandia, a similar pattern should be expected for the Greater Antilles back in the mid-Cenozoic. Ali (2012) concluded that GAARlandia is a thought-provoking hypothesis, but more geological, biological, and, to a lesser

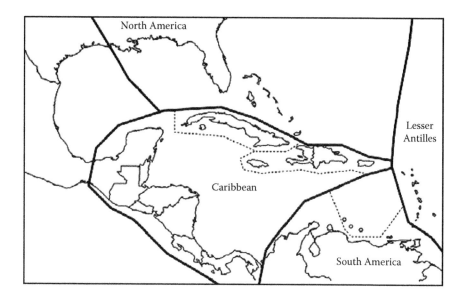

Figure 5.6 Map of the regionalization of the Caribbean basin islands and continental portions, based on a parsimony analysis of endemicity of bird distributions. (Modified from Vázquez-Miranda, H. et al., *Cladistics*, 23: 180–200, 2007.)

extent, paleoceanographical data are still required to accept its relevance for explaining Antillean biogeography.

Echeverry and Morrone (2013) conducted a cladistic biogeographic analysis to test the naturalness of the Caribbean subregion. They analyzed the biogeographic provinces of the subregion together with other provinces situated north and south of it (outgroup areas), obtaining 14 general area cladograms (Figure 5.7). They concluded that the analysis reflected a history of vicariance events and that the Caribbean subregion did not represent a natural biogeographic unit; that the Isthmus of Tehuantepec might not represent a conspicuous biogeographic barrier; and that the biogeographic relevance of the Isthmus of Panama exceeds the last 3 mya, which is the time it has connected North and South America.

It can be concluded that there are three main models of Antillean biogeography (Vázquez-Miranda et al., 2007; Echeverry, 2011; Matos-Maravi et al., 2014): (1) Rosen's (1976) proto-Antillean vicariance model, based on marine, freshwater, and terrestrial taxa (see also Rosen, 1985; Page and Lydeard, 1994; Crother and Guyer, 1996; Crawford and Smith, 2005); (2) Hedges' (1996a,b, 2001) overwater dispersal model, based on immunological distance data of vertebrates and supported by some molecular clocks (Hedges et al., 1992, 2001; Maxson, 1992); and (3) Iturralde-Vinent and MacPhee's (1999) land bridge theory of GAARlandia, based on the distributions of fossil and living mammals. Echeverry (2011) concluded that although the three models are coincident in treating vicariance and dispersal as alternative or complementary processes, only in Rosen's model both are explicitly alternative,

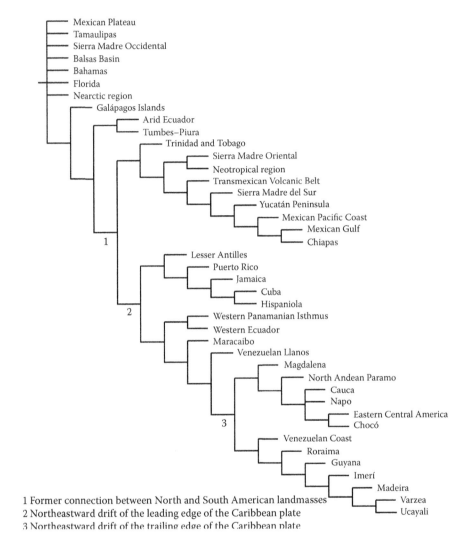

1 Former connection between North and South American landmasses
2 Northeastward drift of the leading edge of the Caribbean plate
3 Northeastward drift of the trailing edge of the Caribbean plate

Figure 5.7 Consensus of the general area cladograms. (From Echeverry, A. and J. J. Morrone, *Journal of Biogeography*, 40: 1619–1637, 2013.)

whereas in Hedges' and Iturralde-Vinent and MacPhee's models evidence that may be contrary to their interpretations is subestimated or ignored, in order to favor dispersal as the most important mechanism to explain the biotic origin of the Antilles.

I (Morrone, 2014c) obtained a general area cladogram showing the Neotropical areas in the strict sense (excluding the Mexican and South American transition zones) constitute a monophyletic biogeographic unit. Within it, a first dichotomy separates the Antilles and in a second dichotomy the continental areas are arranged into a northwestern and a southeastern component (Figure 5.8). Based on this general area

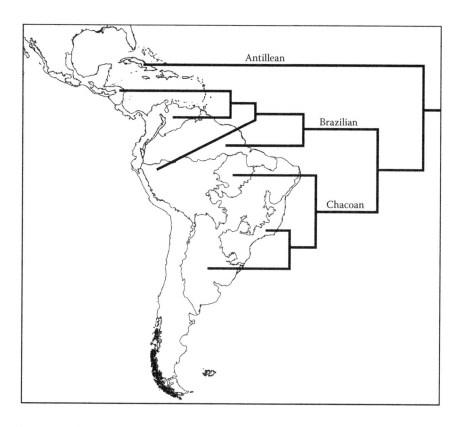

Figure 5.8 General area cladogram depicting the relationships of the biogeographic subregions and dominions of the Neotropical region. (From Morrone, J. J., *Cladistics*, 30: 202–214, 2014c.)

cladogram, a subdivision of the Neotropical region into three subregions (Antillean, Brazilian, and Chacoan) was proposed (Morrone, 2014b). The main events implied by this general area cladogram can be associated with known tectonic and geological information available. The oldest event corresponds to the former connection between North and South America during the Lower Jurassic to Lower Cretaceous (190–148 mya). This connection allowed the geodispersal of the Neotropical biota to North America, and would explain the presence of ancient Neotropical elements in the Mexican Transition Zone (and even the Nearctic region) in the early Cenozoic (Gentry, 1982; Calvillo-Canadell and Cevallos-Ferris, 2005; Morrone, 2005; Gutiérrez-García and Vázquez-Domínguez, 2013). The vicariance between the Antilles and the rest of the Neotropical region can be associated to the severance of the temporal connection represented by the leading edge of the Caribbean plate during its north-eastward drift, from the Eastern Pacific to the Western Atlantic (Briggs, 1994; Echeverry and Morrone, 2013; Lamas et al., 2014). The leading edge of the Caribbean plate reconnected temporarily North and South America 125–100 mya. This connection began to be severed in the Late Cretaceous (80 mya) and finished in the Miocene–Middle Pliocene (Pitman et al., 1993). The areas of the continental Neotropics share extensive hydrological connections and there is evidence that they constituted a superbasin that has persisted in relative isolation since at least the Late Cretaceous (Albert et al., 2011).

Regionalization

The Antillean subregion (Figure 5.9) comprises the Bahama, Cuban, Cayman Islands, Jamaica, Hispaniola, Puerto Rico, and Lesser Antilles provinces (Morrone, 2014b).

Cenocrons

There are several studies that have attempted to identify cenocrons within the Antilles. Gentry (1982) provided a general review of the Neotropical plant diversity, making a basic distinction into two major tropical elements: Laurasian-derived taxa and Gondwanan-derived taxa. Laurasian plants include primarily montane, higher altitude herbs and sometimes canopy trees mixed within the Gondwanan elements (e.g., Magnoliaceae, Pinaceae, Aquifoliaceae, Cyrillaceae, Rosaceae, and Ranunculaceae). They are not particularly species rich, and Gentry (1982) suggested that they were of recent arrival (Neogene–Quaternary). Additionally, there are some old Laurasian-derived groups that have diversified in lowland dry areas (e.g., Aristolochiaceae, Vitaceae, Rhamnaceae, Boraginaceae, and Buxaceae). According to Gentry (1982), Laurasian taxa are better represented in the Antilles than in other parts of the Neotropical region, where some of the taxa of the dry lowlands may be relicts of taxa that reached the proto-Greater Antilles during the Late Cretaceous–Paleogene. Gondwanan-derived plants include Amazonian- and Andean-centered taxa, the latter more abundant in low and mid-elevations of the Andes. The Amazonian-centered taxa include most of the Neotropical lowland canopy trees and

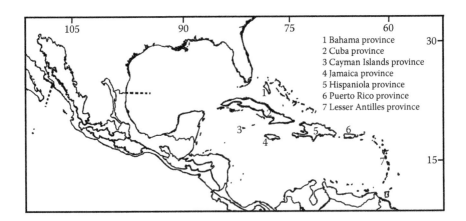

Figure 5.9 Map of the provinces of the Antillean subregion.

lianas and generally have few species per genus, whereas Andean-centered taxa are predominantly epiphytic or shrubby. Both are poorly represented in the Antilles.

Dávalos (2004) reviewed the fossil record and hypotheses of divergence based on molecular clocks of the Antillean mammals to assess the fit between proposed geological models and their diversification. She concluded that the colonization of the islands by mammals from South America, between Paleocene and Middle Miocene, accounted for the distribution of the majority of the lineages studied. Megalonychinae (Xenarthra), Hystricognatha (Rodentia), and Primates arrived during this window of colonization, fitting the pattern of divergence from the continent implied by the GAARlandia hypothesis. On the other hand, Choloepodinae (Xenarthra), Muridae (Rodentia), Lipotyphla, Mormoopidae, and Natalidae (Chiroptera) showed patterns of divergence from the mainland that are inconsistent with the GAARlandia hypothesis and seem to require taxon-specific explanations.

Ricklefs and Bermingham (2008) concluded that dating of island clades using molecular methods indicated the preeminence of overwater dispersal of most of the taxa inhabiting the West Indies. Additionally, they considered that direct connections with Mesoamerica in the Paleogene and subsequent land bridges or stepping stone islands linking the Antilles with Central and South America might have also facilitated dispersal. Based on the analysis of species-area relationships within different islands, the authors suggested that endemic radiations and extinctions have had a strong role in shaping their distributional patterns. Alonso et al. (2011) analyzed the phylogenetic relationships of 10 toad species of the genus *Peltophryne* from Cuba, Hispaniola, and Puerto Rico. They postulated that their common ancestor reached these islands ca. 33 mya, exactly when the GAARlandia land bridge is thought to have existed. As amphibians are largely (but not exclusively) salt-water intolerant, they could not have swum or rafted to the Greater Antilles, favoring instead the GAARlandia hypothesis.

BAHAMA PROVINCE

Bahamas–Bermudan province (in part)—Udvardy, 1975: 42 (regional.).
Bermuda, Bahama, and Southern Florida province (in part)—Samek et al., 1988: 30 (regional.).
Cuba–Western Bahamas subprovince (in part)—de la Cruz, 1989: XI.1.4 (regional.).
South Florida–Bahamas–Bermuda province (in part)—Muñiz, 1991: 284 (regional.).
Bahama province—Morrone, 2001a: 49 (regional.), 2001e: 53 (regional.), 2006: 479 (regional.), 2014b: 36 (regional.).
Bahama Archipelago ecoregion—Abell et al., 2008: 408 (ecoreg.).

Definition

The Bahama province comprises the archipelago of the Bahamas. It includes the islands of Abaco–Grand Bahama, Andros–Bimini, Cat, Crooked–Mayaguana, Exumas, Inaguas, Long–Ragged Island Range, Mona, New Providence–Eleutheras,

San Salvador–Rum Cay, St. Eustatius, St. Kitts, St. Lucia, St. Martin, St. Vincent, and Turks and Caicos (Morrone, 2001e, 2006, 2014b; Trejo-Torres and Ackerman, 2001).

Endemic and Characteristic Taxa

I (Morrone, 2014b) provided a list of endemic and characteristic taxa. Some examples include *Sabal bermudana* (Arecaceae), *Encyclia caicensis, E. inaguensis,* and *Tolumnia sasseri* (Orchidaceae), *Decuanellus bahamensis* (Curculionidae), *Enicocephalus insularis* (Enicocephalidae), *Epicrates chrysogaster* and *E. monensis* (Boidae), *Chironius vincenti, Clelia errabunda,* and *Mastigodryas bruesi* (Colubridae), *Icterus northropi* (Icteridae), *Sitta pusilla insularis* (Sittidae), and *Bothrops caribbaeus* and *B. lanceolatus* (Viperidae).

Vegetation

There are seasonally dry tropical forests and temperate (pine) forests (Dinerstein et al., 1995).

Biotic Relationships

Trejo-Torres and Ackerman (2001) found that the Bahama archipelago was related biotically to Isla de la Juventud and Anegada, which are geographically and geologically part of the Greater and Lesser Antilles, respectively. Isla de la Juventud (Cuban province) shows a strong affinity with northwestern Bahamas, whereas Anegada (Lesser Antilles province) is close to the rest of the Bahamas. A track analysis based on butterfly species (Fontenla, 2003) and a cladistic biogeographic analysis based on fish taxa (Rauchenberger, 1988) supported a close relationship of the Cuban, Hispaniola, and Bahama provinces.

Regionalization

Dinerstein et al. (1995) recognized two ecoregions, which were treated by me (Morrone, 2014b) as the Bahamian Dry Forests and Bahamian Pine Forests districts.

CUBAN PROVINCE

Cuban province—Udvardy, 1975: 42 (regional.); Rivas-Martínez and Navarro, 1994: map (regional.); Morrone, 2014b: 36 (regional.).

Cuba province—Samek et al., 1988: 30 (regional.); Del Risco and Vandama, 1989: X.2.4 (regional.); Muñiz, 1991: 284 (regional.); Morrone, 2001a: 49 (regional.), 2001e: 54 (regional.); Morrone and Márquez, 2001: 637 (track anal.); Huber and Riina, 2003: 258 (glossary); Morrone, 2006: 479 (regional.); Quijano-Abril et al., 2006: 1270 (track anal.).

Cuba–Western Bahamas subprovince (in part)—de la Cruz, 1989: XI.1.4 (regional.).

Cuban Cactus Scrub ecoregion—Dinerstein et al., 1995: 103 (ecoreg.).

Cuban Dry Forests ecoregion—Dinerstein et al., 1995: 93 (ecoreg.).
Cuban Moist Forests ecoregion—Dinerstein et al., 1995: 86 (ecoreg.).
Cuban Pine Forests ecoregion—Dinerstein et al., 1995: 96 (ecoreg.).
Cuban Wetlands ecoregion—Dinerstein et al., 1995: 100 (ecoreg.).
Cuban district—Huber and Riina, 2003: 115 (glossary).
Cuba–Cayman Islands ecoregion (in part)—Abell et al., 2008: 408 (ecoreg.).
Cuba subprovince—Cano et al., 2009: 543 (cluster anal. and regional.).

Definition

The Cuban province comprises the islands of Cuba, Isla de la Juventud, and several surrounding archipelagos in the Caribbean Sea (Morrone, 2001e, 2006, 2014b).

Endemic and Characteristic Taxa

I (Morrone, 2014b) provided a list of endemic and characteristic taxa. Some examples include *Dasytropis* and *Sapphoa* (Acanthaceae), *Leucocroton microphyllus* (Euphorbiaceae), *Behaimia, Pictetia angustifolia, P. marginata*, and *Poitea gracilis* (Fabaceae), *Cuphea lobelioides, Ginoria ginoirioides*, and *G. thomasiana* (Lythraceae), *Epilampra* spp. (Blaberidae), *Apogonalia imitatrix* (Cicadellidae), *Calisto aquilum* and *C. torrei* (Nymphalidae), *Polycentropus nigriceps* (Polycentropidae; Figure 5.10), *Capromys pilorides, Mesocapromys melanurus, Mysateles garridoi*, and *M. prehensilis* (Capromyidae), *Starnoenas cyanocephala* (Columbidae), *Torreornis inexpectata* (Emberizidae), *Trachemys decussata angusta*

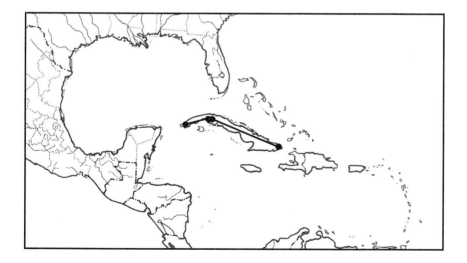

Figure 5.10 Map with the individual track of *Polycentropus nigriceps* (Polycentropidae) in the Cuban province.

and *T. decussata decussata* (Emydidae), *Xiphidiopicus percussus* (Picidae), and *Solenodon cubanus* (Solenodontidae).

Vegetation

The Cuban province includes a great diversity of vegetation types, namely, tropical humid forests, seasonally dry tropical forests, xerophytic scrublands, temperate pine forests, savannas and grasslands, freshwater vegetation, and mangroves (Borhidi, 1996). Rainforests and pine forests of the Cuban province are exceptionally diverse, with numerous endemic and relict species and higher taxa (Dinerstein et al., 1995). Dominant plant species include *Agave tubulata, Bactris cubensis, Cnidioscolus platyandrus, Coccothrinax crinita, Colpothrinax wrightii, Cupania glabra, Ekmanianthe actinophylla, Gaussia princeps, Monisia iguanaca, Pimenta officicinalis, Pinus tropicalis, Spathelia brittonii*, and *Tabebuia calcicola* (Cabrera and Willink, 1973).

Biotic Relationships

A cladistic biogeographic analysis based on fish taxa (Rauchenberger, 1988) indicated complex relationships of this province: its eastern portion is related to the Bahamas and central Hispaniola, whereas its western portion is related to Jamaica, Cayman Islands, and western Hispaniola. According to a cladistic biogeographic analysis based on vertebrate, crustacean, and insect taxa (Crother and Guyer, 1996), this province is closely related to the Hispaniola and Puerto Rico provinces.

Regionalization

Several authors (León, 1946; Panfilov, 1970; Voronov, 1970; Samek, 1973; Borhidi and Muñiz, 1986; de la Cruz, 1989; Del Riso and Vandana, 1989; Muñiz, 1996; López Almiral, 2005; Fontenla Rizo and López Almiral, 2008; Cano et al., 2009) have recognized nested units, which were treated by Morrone (2014b) as three subprovinces and 23 districts. I (Morrone, 2014b) considered Borhidi and Muñiz's (1986) scheme as a preliminary regionalization, although the names of the districts usually have older synonyms (Figure 5.11). Lists of synonyms are provided by Morrone (2014b) and lists of endemic taxa by Borhidi and Muñiz (1986), Sodrestrom et al. (1988), Muñiz (1996), and Huber and Riina (2003). The Western subprovince corresponds to western Cuba where nine districts have been recognized: Cajálbana, Cordillera de los Órganos, Guanahacabibes, Guane–Guajaibón, Northern and Central Pinos Island, Sierra del Rosario, Southern Pinos Island, White Sand Savanna, and Zapata districts. The central–eastern subprovince corresponds to central-eastern Cuba. Within it, seven districts have been recognized: Cascajal, Coastal Trinidad, Guamuhaya Mountains, Lomas de Habana–Matanzas, Motembo, Serpentinic Axis, and Sierras Calizas del Norte districts. The southeastern subprovince corresponds to southeastern Cuba, where seven districts have been recognized: Coastal, Eastern Central Valley, Gran Piedra Mountain, Maisí–Guantánamo Meridional Coast, Pilonense, Sagua–Baracoa, and Sierra Maestra districts.

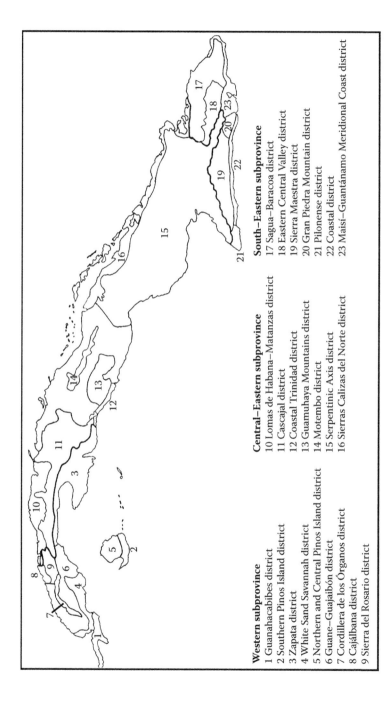

Figure 5.11 Map of the biogeographic districts of the Cuban province. (Modified from Borhidi, A. and O. Muñiz, *Acta Botanica Hungarica*, 32: 3–48, 1986.)

Western subprovince
1 Guanahacabibes district
2 Southern Pinos Island district
3 Zapata district
4 White Sand Savannah district
5 Northern and Central Pinos Island district
6 Guane–Guajaibón district
7 Cordillera de los Órganos district
8 Cajálbana district
9 Sierra del Rosario district

Central–Eastern subprovince
10 Lomas de Habana–Matanzas district
11 Cascajal district
12 Coastal Trinidad district
13 Guamuhaya Mountains district
14 Motembo district
15 Serpentinic Axis district
16 Sierras Calizas del Norte district

South–Eastern subprovince
17 Sagua–Baracoa district
18 Eastern Central Valley district
19 Sierra Maestra district
20 Gran Piedra Mountain district
21 Pilonense district
22 Coastal district
23 Maisí–Guantánamo Meridional Coast district

Cenocrons

A phytogeographic survey of Cuba (Borhidi, 1996) distinguished several elements within the Cuban flora, which may be considered as cenocrons: Caribbean (including Greater Antillean, Antillean, Antillean–Bahamian, Bahamian, Florida–Antillean, Florida–Antillean–Bahamian, Cuba–Florida–southwestern United States, North Caribbean, South Caribbean, and Pan-Caribbean species), Neotropical (including South American–Antillean, South American–Cuban, and Neotropical species), Pantropical, Extratropical (including North and Central American, American, Amphiatlantic, and cosmopolitan species), and adventive (Neotropical, Paleotropical, and Extratropical species). Acevedo-Rodríguez and Strong (2008), based on shared plant genera, suggested a close relationship between the Cuban and Hispaniola provinces. A track analysis based on butterfly species (Fontenla, 2003) supported a close relationship of the Cuban, Hispaniola, and Bahama provinces.

CAYMAN ISLANDS PROVINCE

Cayman Islands province—Morrone, 2001a: 50 (regional.), 2001e: 54 (regional.), 2006: 479 (regional.), 2014b: 42 (regional.).
Cuba–Cayman Islands ecoregion (in part)—Abell et al., 2008: 408 (ecoreg.).

Definition

The Cayman Islands province comprises the archipelago of the Grand Cayman, Little Cayman, and Cayman Brac islands (Morrone, 2001e, 2006, 2014b).

Endemic and Characteristic Taxa

Morrone (2014b) provided a list of endemic and characteristic taxa. Some examples include *Dendrophylax fawcettii* and *Myrmecophila albopurpurea* (Orchidaceae), *Centris caymanensis* (Apidae), *Setophaga vitellina* (Parulidae), and *Turdus ravidus* (Turdidae).

Vegetation

There are seasonally dry tropical forests and xerophytic scrublands (Dinerstein et al., 1995).

Biotic Relationships

The Cayman Islands have been included in the Jamaica province by Samek et al. (1988) and in the Cuban province by Borhidi (1996). Trejo-Torres and Ackerman (2001) showed their complex affinities with the Greater Antilles, the Yucatán Peninsula, Florida, and the Bahamas. A cladistic biogeographic analysis based on fish taxa (Rauchenberger, 1988) indicated a close relationship with the western portion of the Cuban province.

Regionalization

Dinerstein et al. (1995) recognized two ecoregions, which were treated by Morrone (2014b) as the Cayman Islands Dry Forests and the Cayman Islands Xeric Scrub districts.

JAMAICA PROVINCE

Greater Antilles province (in part)—Udvardy, 1975: 42 (regional.).
Jamaica province—Samek et al., 1988: 32 (regional.); Muñiz, 1991: 284 (regional.);
  Morrone, 2001a: 51 (regional.), 2001e: 54 (regional.), 2006: 479 (regional.), 2014b:
  42 (regional.).
Jamaica ecoregion—Abell et al., 2008: 408 (ecoreg.).
Jamaica subprovince—Cano et al., 2009: 543 (cluster anal. and regional.).

Definition

The Jamaica province corresponds to the island of Jamaica (Morrone, 2001e, 2006, 2014b).

Endemic and Characteristic Taxa

Morrone (2014b) provided a list of endemic and characteristic taxa. Some examples include *Bactris jamaicana* (Arecaceae), *Bursera aromatica* and *B. lunanii* (Burseraceae), *Mecranium purpurascens* (Melastomataceae), *Psorolyma sicardi* (Coccinellidae), *Strumigenys jamaicensis* (Formicidae), *Archaeoglenes pecki* (Tenebrionidae), *Geocapromys brownii* (Capromyidae), *Sphaerodactylus parkeri, S. richardsoni,* and *S. semasiops* (Gekkonidae), and *Amazona agilis* and *A. collaria* (Psittacidae).

Vegetation

There are tropical humid forests and seasonally dry tropical forests. Rainforests of the Jamaican province possess high levels of endemism in a wide range of taxa (Dinerstein et al., 1995).

Biotic Relationships

A cladistic biogeographic analysis based on fish taxa (Rauchenberger, 1988) indicated a close relationship with the Cayman Islands and the western portion of the Cuban province. Another cladistic biogeographic analysis based on vertebrate, crustacean, and insect taxa (Crother and Guyer, 1996) postulated that the Jamaica province is closely related to the Cuban, Hispaniola, and Puerto Rico provinces.

Regionalization

Dinerstein et al. (1995) recognized two ecoregions, which were treated by Morrone (2014b) as districts. The Jamaican Dry Forests district corresponds to coastal northern and southern Jamaica, whereas the Jamaican Moist Forests district corresponds to central Jamaica.

HISPANIOLA PROVINCE

Greater Antilles province (in part)—Udvardy, 1975: 42 (regional.).
Hispaniola province—Samek et al., 1988: 31 (regional.); Morrone, 2001a: 52 (regional.), 2001e: 55 (regional.); Huber and Riina, 2003: 259 (glossary); Corona and Morrone, 2005: 38 (track anal.); Morrone, 2006: 479 (regional.); Quijano-Abril et al., 2006: 1270 (track anal.); Cano et al., 2009: 543 (cluster anal. and regional.); Morrone, 2014b: 43 (regional.); Cano-Ortiz et al., 2016: 272 (cluster anal. and regional.).
Enriquillo Wetlands ecoregion—Dinerstein et al., 1995: 100 (ecoreg.).
Hispaniolan Dry Forests ecoregion—Dinerstein et al., 1995: 93 (ecoreg.).
Hispaniolan Moist Forests ecoregion—Dinerstein et al., 1995: 86 (ecoreg.).
Hispaniolan Pine Forests ecoregion—Dinerstein et al., 1995: 96 (ecoreg.).
Hispaniola ecoregion—Abell et al., 2008: 408 (ecoreg.).

Definition

The Hispaniola province comprises the island of Hispaniola, which corresponds to the Dominican Republic and Haiti (Morrone, 2001e, 2006, 2014b). This island is the result of the union-separation-union of different orographic units, formed in different geological times (Cano-Ortiz et al., 2016).

Endemic and Characteristic Taxa

Morrone (2014b) provided a list of endemic and characteristic taxa. Some examples include *Ekmaniopappus, Eupatorina, Fuertesia*, and *Salcedoa* (Asteraceae), *Garciadelia* (Euphorbiaceae), *Coeloneurum* (Solanaceae), *Amblytropidia* (Acrididae), *Antilliscaris darlingtoni* and *Ardistomis hispaniolensis* (Figure 5.12), *Barylaus puncticeps, Platynus biramosus, P. cristophe, P. jaegeri*, and *P. transcibao* (Carabidae), *Kuschelaxius discifer, Micromyrmex asclepia*, and *Sicoderus championi* (Curculionidae), *Hemiphileurus dispar, H. laeviceps, H. phratrius*, and *H. scutellaris* (Scarabaeidae), *Trachemys decorata* and *T. stejnegeri vicina* (Emydidae), and *Chamaelinorops* and *Cyclura ricordi* (Iguanidae). According to Cano-Ortiz et al. (2016), this province harbors 1,582 endemic species of plants.

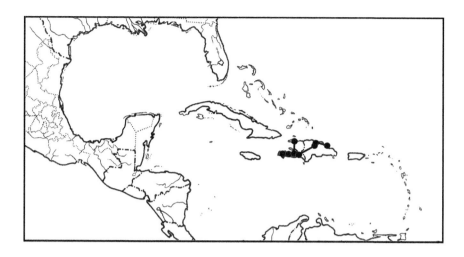

Figure 5.12   Map with the individual track of *Ardistomis hispaniolensis* (Carabidae) in the Hispaniola province.

Vegetation

There are tropical humid forests, seasonally dry tropical forests, temperate (montane and pine) forests, and mangroves. Pine forests of the Hispaniola province are exceptionally diverse with numerous endemic species (Dinerstein et al., 1995). Tropical humid forests are characterized by *Cecropia peltata*, *Manilkara nitida*, *Tetragastris balsamifera*, *Didymopanax morototoni*, *Guarea trichiloides*, and *Genipa americana*; seasonally dry tropical forests by *Coccoloba uvifera*, *Plumeria* spp., *Calophyllum antillanum*, *Bucida buceras*, *Curatella americana*, and *Byrsonina* spp.; mountain forests by *Didymopanax tremulans*, *Brunellia comocladifolia*, *Prunus occidentalis*, *Dendropanax arborea*, *Sloanea ilicifolia*, and *Weinmannia pinnata*; pine forests by *Pinus occidentalis*; and mangroves by *Rhizophora mangle*, *Laguncularia racemosa*, and *Avicennia nitida* (Cabrera and Willink, 1973).

Biotic Relationships

The genus *Barylaus* (Coleoptera: Carabidae), with one species from Hispaniola, is cladistically more closely related to Old World taxa rather than to other Neotropical taxa, emphasizing the isolation of this province, especially when highland taxa are considered (Liebherr, 1986). Acevedo-Rodríguez and Strong (2008), based on shared plant genera, suggested a close relationship between the Cuban and Hispaniola provinces. A track analysis based on butterfly species (Fontenla, 2003) supported a close relationship of the Cuban, Hispaniola, and Bahama provinces.

A cladistic biogeographic analysis based on fish taxa (Rauchenberger, 1988) showed complex relationships for this province: its central portion is related to the Bahamas and eastern Cuba, whereas its southwestern portion is related to Jamaica,

Cayman Island, and western Cuba. Another cladistic biogeographic analysis based on vertebrate, crustacean, and insect taxa (Crother and Guyer, 1996) postulated that this province is closely related to the Puerto Rico province.

Regionalization

Cano et al. (2009) recognized two subprovinces and six sectors (treated herein as districts). The Central subprovince includes the Central district, whereas the Caribbean–Atlantic subprovince includes the Bahoruco-Hottense, Neiba–Matheux–Northwest, Azua–San Juan–Hoya Enriquillo–Port-au-Prince–Artiobonite–Gonaivës, Caribeo–Cibense, and North districts. A more recent analysis by Cano-Ortiz et al. (2016) detected 19 areas of endemism, which could be evaluated as possible districts.

PUERTO RICO PROVINCE

Greater Antilles province (in part)—Udvardy, 1975: 42 (regional.).
Puerto Rico province—Samek et al., 1988: 32 (regional.); Muñiz, 1991: 284 (regional.);
    Morrone, 2001a: 54 (regional.), 2001e: 56 (regional.); Huber and Riina, 2003: 260
    (glossary); Morrone, 2006: 479 (regional.), 2014b: 44 (regional.).
Puerto Rico sector—Cano et al., 2009: 543 (cluster anal. and regional.).
Puerto Rican district—Huber and Riina, 2003: 265 (glossary).
Puerto Rico–Virgin Islands ecoregion (in part)—Abell et al., 2008: 408 (ecoreg.).

Definition

The Puerto Rico province comprises the island of Puerto Rico (Morrone, 2001e, 2006, 2014b).

Endemic and Characteristic Taxa

I (Morrone, 2014b) provided a list of endemic and characteristic taxa. Some examples include *Pictetia punicea* and *Poitea florida* (Fabaceae), *Antilliscaris danforthi*, *Barylaus estriatus*, and *Oxydrepanus coamensis* (Carabidae), *Mayagueza argentifera* (Drosophilidae), *Eleutherodactylus portoricensis* (Eleutherodactylidae), *Nesospingus speculiferus* (Phaenicophilidae; Sánchez-González, pers. comm.), and *Amazona vittata* (Psittacidae).

Vegetation

There are tropical humid forests and seasonally dry tropical forests (Dinerstein et al., 1995).

Biotic Relationships

The genus *Barylaus* (Coleoptera: Carabidae), with one species from Puerto Rico, is cladistically more closely related to Old World taxa rather than to other Neotropical taxa, emphasizing the isolation of this province, especially when highland taxa are considered (Liebherr, 1986). According to a cladistic biogeographic analysis based on vertebrate, crustacean, and insect taxa (Crother and Guyer, 1996), this province is closely related to the Hispaniola province.

Regionalization

Dinerstein et al. (1995) recognized two ecoregions, which were treated by Morrone (2014b) as districts. The Puerto Rican Dry Forests district corresponds to southern Puerto Rico, and the Puerto Rican Moist Forests district corresponds to northern and central Puerto Rico.

LESSER ANTILLES PROVINCE

Lesser Antilles province—Samek et al., 1988: 32 (regional.); Muñiz, 1991: 284 (regional.); Morrone, 2001a: 54 (regional.), 2001e: 56 (regional.), 2006: 479 (regional.), 2014b: 44 (regional.).
Leeward Islands Dry Forests ecoregion—Dinerstein et al., 1995: 94 (ecoreg.).
Leeward Islands Moist Forests ecoregion—Dinerstein et al., 1995: 87 (ecoreg.).
Leeward Islands Xeric Scrub ecoregion—Dinerstein et al., 1995: 103 (ecoreg.).
Windward Islands Dry Forests ecoregion—Dinerstein et al., 1995: 93 (ecoreg.).
Windward Islands Moist Forests ecoregion—Dinerstein et al., 1995: 87 (ecoreg.).
Windward Islands Xeric Scrub ecoregion—Dinerstein et al., 1995: 103 (ecoreg.).
Lesser Antilles area—Porzecanski and Cracraft, 2005: 266 (PAE).
Puerto Rico–Virgin Islands ecoregion (in part)—Abell et al., 2008: 408 (ecoreg.).
Windward and Leeward Islands ecoregion—Abell et al., 2008: 408 (ecoreg.).
Lesser Antilles sector—Cano et al., 2009: 543 (cluster anal. and regional.).
Lesser Antillean province—Rivas-Martínez et al., 2011: 26 (regional.).

Definition

The Lesser Antilles province comprises the archipelagos of the Lesser Antilles and the Virgin Islands (Morrone, 2001e, 2006, 2014b; Trejo-Torres and Ackerman, 2001). The Lesser Antilles belong to a small volcanic arc with about 21 main islands and numerous islets and keys dating from mid-Eocene, with an approximate total area of 8,320 square kilometers. They are divided into the Windward and Leeward Islands. The Winward islands include Grenada, The Grenadines,

St. Vincent, Barbados, St. Lucia, and Martinique; and the Leeward Islands include Dominica, Marie Galante, Guadeloupe, La Desirade, Montserrat, Antigua, Nevis, St. Kitts, Barbuda, St. Eustatius, Saba, St. Barthélemy, St. Martin, and Anguilla (Acevedo-Rodríguez and Strong, 2008). The Virgin Islands include Anegada, Culebra, St. Croix, St. John, St. Thomas, Tortola, Vieques, and Virgin Gorda. These islands are divided into the Windward and Leeward islands, according to their position relative to the prevailing winds, with the former situated in the south and directly exposed to the south to north trade winds, and the latter situated in the north and sheltered from these currents (Acevedo-Rodríguez and Strong, 2008).

Endemic and Characteristic Taxa

I (Morrone, 2014b) provided a list of endemic and characteristic taxa. Some examples include *Cuphea crudyana* (Lythraceae), *Erithalis odorifera* (Rubiaceae), *Decuanellus brevicrus, D. buclavatus,* and *Sicoderus contiguus* (Curculionidae), *Ityphilus mauriesi* and *Taeniolinum setosum guadeloupensis* (Geophilidae), *Dythemis sterilis* (Libellulidae), *Liophis cursor* species group (Colubridae), *Cyclura pinguis* and *Iguana delicatissima* (Iguanidae), *Cinclocerthia gutturalis* and *Ramphocinclus brachyurus* (Mimidae), and *Myiarchus nugator* (Tyrannidae).

Vegetation

There are tropical humid forests, seasonally dry tropical forests, and xerophytic scrublands (Dinerstein et al., 1995).

Biotic Relationships

The Virgin Islands have been included in the Puerto Rico province by Samek et al. (1988). A parsimony analysis of endemicity based on species of Orchidaceae (Trejo-Torres and Ackerman, 2001) indicates that the Lesser Antilles constitute a natural unit, which is subordinate to the Greater Antillean group, whereas in a parsimony analysis of endemicity based on bird taxa (Vazquez-Miranda et al., 2007) they appeared as paraphyletic at the base of the cladogram.

CHAPTER 6

The Brazilian Subregion

The Brazilian subregion comprises central and southern Mexico, Central America, and northwestern South America (Morrone, 2014b,c). The northern portion of the Brazilian subregion (Mexico and Central America) was assigned previously to the Caribbean subregion (e.g., Cabrera and Willink, 1973; Morrone, 2001a, 2006), but it has been shown recently that the Antilles should be separated as a distinct subregion (Nihei and de Carvalho, 2007; Sigrist and Carvalho, 2009; Morrone, 2014b,c).

The South American portion of the Brazilian subregion corresponds approximately to the Amazonian subregion of previous authors. The naturalness of the traditional Amazonian subregion has been questioned by several authors. Amorim and Pires (1996) considered that Amazonia was comprised of two nonrelated areas: northwestern Amazonia and southeastern Amazonia. Amorim (2001) also considered that the Amazon forest did not correspond to a natural biogeographic area, being geographically delimited, based on the Amazon River basin. Nihei and de Carvalho (2007) undertook a cladistic biogeographic analysis and concluded that Amazonia should be regarded as a composite area, because northwestern Amazonia was closely related to the Caribbean subregion, whereas southeastern Amazonia was related to the Chacoan and Paraná subregions. Sigrist and de Carvalho (2009) undertook another cladistic biogeographic analysis to examine whether the inclusion of open area formations influence area relationships of the surrounding forests. They found a basal split between the Amazonian and Atlantic forests, suggesting that they have been isolated for a long period of time, and corroborated the hypothesis that Amazonia is a composite area; however, when they added two areas with open formations (Cerrado and Caatinga), internal relationships within Amazonia changed; so, they concluded that a biogeographic classification comprising both forests and open formations should be preferred given their complementary history. Pires and Marinoni (2010) found that when the Cerrado and Caatinga were included the Atlantic forest resulted to be monophyletic, whereas the Amazonian forest did not, and concluded that a single history of the current distribution of taxa in the area analyzed was unlikely.

BRAZILIAN SUBREGION

Brazilian subregion—Blyth, 1871: 428 (regional.); Wallace, 1876: 78 (regional.); Heilprin, 1887: 80 (regional.); Lydekker, 1896: 135 (regional.); Bartholomew et al., 1911: 9 (regional.); Mello-Leitão, 1937: 244 (regional.); Hershkovitz, 1969: 3 (regional.); Kuschel, 1969: 710 (regional.); Bănărescu and Boşcaiu, 1978: 258 (regional.); Almirón et al., 1997: 23 (regional.); Ojeda et al., 2002: 23 (biotic evol.); Morrone, 2014b: 45 (regional.); Klassa and Santos, 2015: 520 (endem. anal.).

Tropical American region (in part)—Engler, 1882: 345 (regional.).

Amazonian subregion (in part)—Sclater and Sclater, 1899: 65 (regional.); Morrone, 1999: 6 (regional.), 2000b: 102 (regional.), 2001a: 67 (regional.), 2005: 238; Viloria, 2005: 449 (regional.); López Ruf et al., 2006: 116 (track anal.); Quijano-Abril et al., 2006: 1268 (track anal.); Morrone, 2006: 480 (regional.); Asiain et al., 2010: 178 (track anal.); Morrone, 2010a: 37 (regional.); Mercado-Salas et al., 2012: 459 (track anal.); Morrone, 2014c: 206 (clad. biogeogr.); Pires and Marinoni, 2011: 8 (track anal.); Coulleri and Ferrucci, 2012: 105 (track anal.).

Amazonian district (in part)—Cabrera and Yepes, 1940: 14 (regional.).

Amazonian dominion—Orfila, 1941: 86 (regional.); Morrone and Coscarón, 1996: 1 (PAE); Huber and Riina, 1997: 150 (glossary); Ojeda et al., 2002: 24 (biotic evol.); Huber and Riina, 2003: 124 (glossary).

Guianan–Brazilian subregion (in part)—Mello-Leitão, 1943: 128 (regional.); Ringuelet, 1961: 156 (biotic evol. and regional.); Rapoport, 1968: 72 (biotic evol. and regional.); Ringuelet, 1978: 255 (biotic evol.); Paggi, 1990: 303 (regional.); Sánchez Osés and Pérez-Hernández, 2005: 168 (regional.).

Amazonian province (in part)—Schmidt, 1954: 328 (regional.); Huber and Riina, 1997: 23 (glossary).

Guianan–Brazilian region—Fittkau, 1969: 636 (regional.).

Amazonian Basin area—Sick, 1969: 451 (regional.).

Non-Andean East area (in part)—Sick, 1969: 451 (regional.).

Caribbean dominion (in part)—Cabrera and Willink, 1973: 32 (regional.); Huber and Riina, 1997: 151 (glossary); Zuloaga et al., 1999: 18 (regional.); Huber and Riina, 2003: 124 (glossary).

Caribbean Amazonian subkingdom (in part)—Rivas-Martínez and Tovar, 1983: 521 (regional.).

Caribbean region (in part)—Takhtajan, 1986: 251 (regional.); Samek et al., 1988: 26 (regional.); Huber and Riina, 1997: 119 (glossary); Huber and Riina, 2003: 97 (glossary).

Amazonian region—Rivas-Martínez and Navarro, 1994: map (regional.); Rangel et al., 1995b: 82 (veget.).

Amazonian area—Coscarón and Coscarón-Arias, 1995: 726 (areas of endem.).

Caribbean subregion (in part)—Morrone, 1999: 2 (regional.), 2001a: 30 (regional.), 2001e: 46 (regional.); Corona and Morrone, 2005: 38 (track anal.); Morrone, 2005: 238 (regional.); Viloria, 2005: 449 (regional.); Morrone, 2006: 478 (regional.); Quijano-Abril et al., 2006: 1268 (track anal.); Asiain et al., 2010: 178 (track anal.); Morrone, 2010a: 34 (regional.); Coulleri and Ferrucci, 2012: 105 (track anal.); Lamas et al., 2014: 955 (clad. biogeogr.).

Northwestern component—Nihei and de Carvalho, 2004: 271 (clad. biogeogr.).

Amazonic dominion—Donato, 2006: 422 (clad. biogeogr.).

Amazon dominion—Fiaschi and Pirani, 2009: 480 (regional.).

Amazonian component—Sigrist and de Carvalho, 2009: 81 (clad. biogeogr.).
South American subregion (in part)—Echeverry and Morrone, 2013: 1628 (track anal.).
Amazon subregion (in part)—Lamas et al., 2014: 955 (clad. biogeogr.).

Endemic and Characteristic Taxa

Taxa endemic to the Brazilian subregion include many plant and animal genera and species. Morrone (2000b, 2001a) provided a list of Amazonian taxa, which are mostly restricted to the South American portion of the Brazilian subregion. Examples of endemic taxa are *Calyptrion* (Violaceae; Figure 6.1; Paula-Souza and Pirani, 2014), *Herpailurus yagouaroundi*, *Leopardus pardalis*, *L. wiedii*, and *Panthera onca* (Felidae), *Potos flavus* (Procyonidae), and *Tayassu pecari* (Tayassuidae; Ceballos and Oliva, 2005).

Biotic Relationships

A parsimony analysis of endemicity based on insect taxa (Morrone and Coscarón, 1996) showed that the diagonal of open areas of the Chacoan dominion

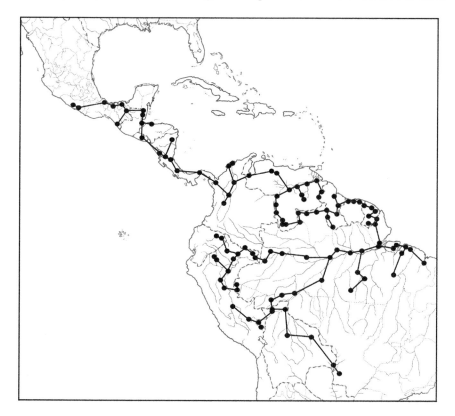

Figure 6.1 Map with the individual track of *Calyptrion* (Violaceae) in the Brazilian subregion.

developed gradually, separating the former forest into the Brazilian subregion and the Paraná dominion. According to the cladistic biogeographic analysis of Morrone (2014c), the Brazilian subregion is the sister area of the Chacoan subregion (including both the Chacoan and Paraná dominions). Their separation might have begun with the formation of a lake along the Amazon, Madeira, and Mamoré rivers, in the Late Cretaceous, and finished with the formation of a wide epicontinental sea by water invasion through the northern, eastern, and southern portal seaways, in the Miocene (Amorim, 2001; Frailey, 2002; Nihei and de Carvalho, 2004, 2007). This scenario implies an ancient history for the diversification of the Neotropics that contrasts with the Pleistocene refugia hypothesis (e.g., Haffer, 1969, 1974, 1981). Evidence of ancient (Neogene) speciation events in the Neotropical dry forests was found by Pennington et al. (2004), although these authors did not discount more recent (Pleistocene) events, for example, for plant taxa that dispersed to Mesoamerica after the Panama Isthmus closed. The phylogeographic analysis of the *Crotalus durissus* complex (Viperidae) by Wüster et al. (2005) also showed a similar northern–southern sequence (Figure 6.2) that was interpreted as indicating a gradual series of dispersal events from Central America to southern Brazil. It is interesting to note that Löwenberg-Neto et al. (2008) found that the

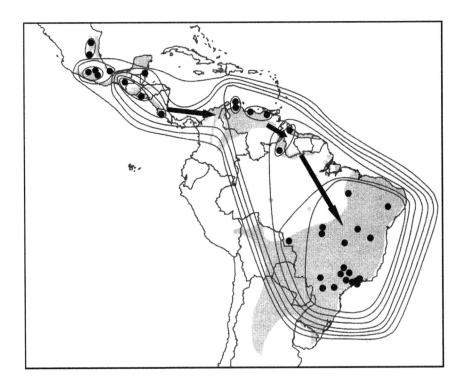

Figure 6.2 Dispersal events of the *Crotalus durissus* complex (Viperidae), estimated by the phylogeographic analysis of Wüster et al. (2005).

richness of basal and derived taxa of Muscidae (Diptera) was strongly correlated with the basic separation (Antillean [Brazilian, Chacoan]). The basal richness was higher in the Brazilian subregion, whereas the derived richness was higher in the Chacoan subregion. They concluded that species richness can be linked to deep history, and that geologic events may be important in the history of biotic diversification. Evidence of a former (Paleogene) forest, covering most of the Neotropical region, fragmented during Eocene–Miocene, and with recurrent connections during Quaternary was provided by Sobral-Souza et al. (2015) through ecological niche modeling.

According to the cladistic biogeographic analysis of Morrone (2014c), the dominions of the Brazilian subregion split following the sequence (Boreal Brazilian [South Brazilian (Pacific, Mesoamerican)]). The vicariance between the Boreal Brazilian dominion and the remaining areas can be associated with the Romeral Fault Zone and/or the final uplift of the northern Andes. The Romeral Fault of Cretaceous age is an active and continuous fault system of almost 700 kilometers long that comprises three or four parallel regional faults, which form the boundary between autochthonous continental rocks to the east and accreted oceanic arc rocks related to Caribbean terranes in the west (Kennan and Pindell, 2009; Heads, 2012; Echeverry and Morrone, 2013). The uplift of the Andes began in the Cretaceous, but has been more conspicuous since the Miocene, 23–27 mya and finished in the Pliocene (Lundberg et al., 1998; Kennan, 2000; Cortés-Ortiz et al., 2003; Garzione et al., 2008; Hoorn et al., 2010). The vicariance between the South Brazilian dominion and the Pacific–Mesoamerican dominions can be associated with the formation of an epicontinental sea by water invasion through the Maracaibo and Amazon basins, in the Pliocene (Rodríguez-Olarte et al., 2011).

Regionalization

The Brazilian subregion (Figure 6.3) comprises the Boreal Brazilian, South Brazilian, Pacific, and Mesoamerican dominions (Morrone, 2014b).

BOREAL BRAZILIAN DOMINION

Boreal Brazilian subarea—Clarke, 1892: 381 (regional.).
Hyléa province (in part)—Mello-Leitão, 1937: 246 (regional.); Fittkau, 1969: 642 (regional.).
Cariba, Guianan, or Amazonian center (in part)—Lane, 1943: 414 (regional.).
Amazonian province (in part)—Mello-Leitão, 1943: 129 (regional.); Rizzini, 1963: 49 (regional.); Cabrera and Willink, 1973: 48 (regional.); Ávila-Pires, 1974a: 133 (regional.), 1974b: 159 (regional.); Ringuelet, 1975: 107 (regional.); Udvardy, 1975: 41 (regional.); Fernandes and Bezerra, 1990: 77 (regional.); Huber and Riina, 1997: 23 (glossary); Rizzini, 1997: 623 (regional.); Fernandes, 2006: 46 (regional.).
Amazon region—Good, 1947: 235 (regional.); Huber and Riina, 1997: 23 (glossary).
Amazonian dominion (in part)—Cabrera, 1971: 6 (regional.), 1976: 3 (regional.); Huber and Riina, 1997: 150 (glossary); Huber and Riina, 2003: 124 (glossary).

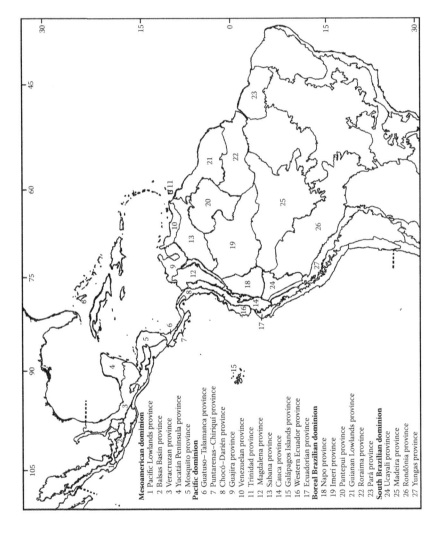

Figure 6.3 Map of the dominions and provinces of the Brazilian subregion.

Amazonian Equatorial dominion (in part)—Ab'Sáber, 1977: map (climate).
Amazonian region (in part)—Takhtajan, 1986: 251 (regional.); Huber and Riina, 1997: 23 (glossary).
Amazonia bioregion (in part)—Dinerstein et al., 1995: map 1 (ecoreg.); Huber and Riina, 1997: 37 (glossary).
Amazonia ecoregion—Salazar Bravo et al., 2002: 78 (ecoreg.).
Northwest Amazonia subregion (in part)—Nihei and de Carvalho, 2007: 497 (clad. biogeogr.).
Amazonian–Guyanan superegion (in part)—Rivas-Martínez et al., 2011: 26 (regional.).
Guyanan–Orinoquian region—Rivas-Martínez et al., 2011: 26 (regional.).
Northern Amazonian dominion—Morrone, 2014c: 206 (clad. biogeogr.).
Boreal Brazilian dominion—Morrone, 2014b: 63 (regional.); Daniel et al., 2016: 1167 (track anal. and regional.).

Definition

The Boreal Brazilian dominion comprises the Amazonian forest, basically north of the Amazon River (Morrone, 2014b,c).

Endemic and Characteristic Taxa

Taxa endemic to this dominion are *Pionites melanocephalus melanocephalus* (Psittacidae; Restall et al., 2006) and *Cheracebus* (Pitheciidae; Byrne et al., 2016).

Biotic Relationships

Several models have been proposed to explain the biotic diversification in the Amazon area (Haffer, 1997; Leite and Rogers, 2013). Some of them are:

1. Riverine barriers: This model was originally proposed by Wallace (1852). It assumes that widespread ancestral populations were divided and isolated when the network of major rivers developed from Late Miocene onwards.
2. Refugia: This model postulates past cyclic changes in the distribution of forest and non-forest areas, breaking up into isolated patches and coalescing due to dry to humid climatic changes in the Neogene and Quaternary (Haffer, 1969, 1974, 1977; Vanzolini and Williams, 1970; Prance, 1982; Lourenço, 1986). Geological studies have established that astronomical Milankovitch cycles causing global climatic-vegetational fluctuations have influenced the distribution of forest and non-forest vegetation on earth during the Quaternary, Paleogene, Neogene, and even before (Haffer, 1997).
3. River–refugia: This combines aspects of the riverine barrier and the refuge hypotheses. Ancient populations have been isolated in semi-refugia, by a combination of the broad lower courses of the Amazonian rivers and by extensive unsuitable areas in the headwaters regions that remained more or less unforested during the dry glacial climatic periods, when the forests were contracted on broad latitudinal forests (Haffer, 1993).

4. Gradients: Diversification was promoted by environmental gradients. It predicts parapatric speciation across steep environmental gradients without separation of the populations and the existence of centers of diversity in environmentally uniform zones (Endler, 1982).
5. Disturbance–vicariance: The main environmental force during glacial times was cooling rather than aridity (Bush and Colinvaux, 1990; Colinvaux, 1996, 1997, 1998; Colinvaux et al., 1996, 2000; Colinvaux and Oliveira, 1999). Barriers correspond to unsuitable cold-related rainforest conditions, temperature oscillations affecting vertical distribution ranges, and fragmenting populations into patches.
6. Paleogeography: Some authors (Croizat, 1958, 1976; Cracraft and Prum, 1988; Bush, 1994; Patton et al., 2000) considered that ancient, pre-Quaternary changes repeatedly separated populations due to paleogeographic changes in forest areas and continental seas during Neogene and Quaternary. Paleogeographic hypotheses include marine incursions (Miocene), structural geological arches (Pliocene and older), and the *Amazonian Lake* (Frailey et al., 1988) that covered most of Amazonia (Late Pleistocene–Early Holocene).

A strict evaluation of these models has not been done yet. Some partial results, however, indicate that it is possible that a combination of them has caused the Amazonian biotic diversification. Leite and Rogers (2013) explored the possibilities of such tests, using previously published phylogeographic analyses.

Regionalization

The Boreal Brazilian dominion comprises the Napo, Imerí, Pantepui, Guianan Lowlands, Roraima, and Pará provinces.

NAPO PROVINCE

Napo subcenter—Müller, 1973: 83 (regional.).
El Cóndor unit—Lamas, 1982: 349 (regional.).
Napo unit—Lamas, 1982: 343 (regional.).
Napo center—Beven et al., 1984: 386 (refugia).
North Amazon (Napo) center—Cracraft, 1985: 69 (areas of endem.).
Napo area—Haffer, 1985: 121 (refugia); Silva and Oren, 1996: 430 (PAE); Bates et al., 1998: 785 (PAE); Ron, 2000: 387 (PAE); Marks et al., 2002: 155 (biotic evol.); Racheli and Racheli, 2003: 36 (PAE), 2004: 347 (PAE); Silva et al., 2005: 692 (areas of endem.); Borges, 2007: 924 (PAE).
Northwestern Amazonian area—Cracraft, 1988: 223 (clad. biogeogr.).
Western or Andean sector—Fernandes and Bezerra, 1990: 94 (regional.); Fernandes, 2006: 61 (regional.).
Amazonian province—Hernández et al., 1992a: 138 (regional.).
Loreto province (in part)—Rivas-Martínez and Navarro, 1994: map (regional.).
Eastern Cordillera Real Montane Forests ecoregion—Dinerstein et al., 1995: 92 (ecoreg.); Huber and Riina, 1997: 75 (glossary).

Napo Moist Forests ecoregion—Dinerstein et al., 1995: 89 (ecoreg.); Huber and Riina, 1997: 65 (glossary).
Varzea Forests ecoregion (in part)—Dinerstein et al., 1995: 90 (ecoreg.).
Amazonian region (in part)—Rangel et al., 1995b: 82 (veget.).
Napo province—Morrone, 1999: 7 (regional.), 2000b: 109 (regional.), 2001a: 70 (regional.), 2006: 480 (regional.); Quijano-Abril et al., 2006: 1270 (track anal.); Mercado-Salas et al., 2012: 459 (track anal.); Morrone, 2014b: 64 (regional.).
Varzea province (in part)—Morrone, 1999: 7 (regional.), 2000b: 112 (regional.), 2001a: 76 (regional.), 2006: 480 (regional.).
Western Amazon Piedmont ecoregion—Abell et al., 2008: 408 (ecoreg.).
West Amazonian province (in part)—Rivas-Martínez et al., 2011: 26 (regional.).

Definition

The Napo province comprises southwestern Colombia, eastern Ecuador, and northern Peru (Morrone, 2000b, 2006, 2014b). It is defined by the Marañón and Amazon rivers in the south, the Andes in the west, and the limit of the lowland rainforest in the north (Cracraft, 1985). The Napo province harbors one of the richest biotas in the world (Dinerstein et al., 1995).

Endemic and Characteristic Taxa

Morrone (2014b) provided a list of endemic and characteristic taxa. Some examples include *Gymnosiphon capitatus* (Burmanniaceae), *Piper corpuientispicum*, *P. calan-yanum*, and *P. purulentum* (Piperaceae), *Tuberodesmus* (Chelodesmidae), *Saguinus nigricollis* and *S. tripartitus* (Callithrichidae), *Akodon latebricola* (Cricetidae), *Cacicus sclateri* (Icteridae), and *Selenidera reinwardtii reinwardtii* (Ramphastidae).

Vegetation

There are moist forests, with an extensive system of meandering rivers that create habitat mosaics (Dinerstein et al., 1995; Palacios et al., 1999).

Biotic Relationships

Müller (1973) considered this province to be related to the Ucayali province. According to a parsimony analysis of endemicity based on primate taxa (Silva and Oren, 1996), it is closely related to the Madeira and Imerí provinces; to two parsimony analyses of endemicity based on bird taxa (Bates et al., 1998; Borges, 2007), it is closely related to the Madeira province; and to a third analysis based on butterfly taxa (Racheli and Racheli, 2003), it is closely related to the Imerí province. A cladistic biogeographic analysis based on bird taxa (Cracraft and Prum, 1988) found a close relationship of this province with the Madeira province.

Regionalization

Hernández et al. (1992a) identified six districts in the Colombian portion of this province: Alto Putumayo, Caguan, Florencia, Huitoto, Kofán, and Ticuna (Morrone, 2014b).

IMERÍ PROVINCE

Rio Negro subprovince—Rizzini, 1963: 50 (regional.), 1997: 624 (regional.).
Imerí center—Beven et al., 1984: 386 (refugia); Cracraft, 1985: 69 (areas of endem.).
Imerí area Haffer, 1985: 121 (refugia); Cracraft, 1988: 223 (clad. biogeogr.); Silva
    and Oren, 1996: 430 (PAE); Bates et al., 1998: 785 (PAE); Ron, 2000: 387 (PAE);
    Marks et al., 2002: 155 (biotic evol.); Racheli and Racheli, 2003: 36 (PAE), 2004:
    347 (PAE); Silva et al., 2005: 692 (areas of endem.); Borges, 2007: 924 (PAE).
Imerí refuge—Lourenço, 1986: 580 (refugia).
Rio Negro province—Huber and Alarcón, 1988: map (regional.); Rivas-Martínez and
    Navarro, 1994: map (regional.).
Guianan province—Hernández et al., 1992a: 131 (regional.).
Amazonian region (in part)—Rangel et al., 1995b: 82 (veget.).
Loreto province (in part)—Rivas-Martínez and Navarro, 1994: map (regional.).
Amazonian Savannas ecoregion—Dinerstein et al., 1995: 99 (ecoreg.); Huber and
    Riina, 1997: 298 (glossary).
Japura/Negro Moist Forests ecoregion—Dinerstein et al., 1995: 89 (ecoreg.); Huber
    and Riina, 1997: 64 (glossary).
Macarena Montane Forests ecoregion—Dinerstein et al., 1995: 89 (ecoreg.).
Varzea Forests ecoregion—(in part) Dinerstein et al., 1995: 90 (ecoreg.).
Imerí province—Morrone, 1999: 7 (regional.), 2000b: 110 (regional.), 2001a: 71
    (regional.); Viloria, 2005: 450 (regional.); Morrone, 2006: 480 (regional.); Quijano-
    Abril et al., 2006: 1270 (track anal.); Mercado-Salas et al., 2012: 459 (track anal.);
    Morrone, 2014b: 65 (regional.); Daniel et al., 2016: 1169 (track anal. and regional.).
Varzea province (in part)—Morrone, 1999: 7 (regional.), 2000b: 112 (regional.), 2001a:
    76 (regional.), 2006: 480 (regional.).
Río Negro ecoregion—Abell et al., 2008: 408 (ecoreg.).
Guyanese Brazilian province—Rivas-Martínez et al., 2011: 26 (regional.).

Definition

The Imerí province comprises the lowlands of southern Venezuela, southwestern Colombia, northeastern Peru, and northern Brazil (Cracraft, 1985; Morrone, 2000b, 2006, 2014b). Some of the world's largest blackwater river ecosystems occur in the Imerí province (Dinerstein et al., 1995).

Endemic and Characteristic Taxa

Morrone (2014b) provided a list of endemic and characteristic taxa. Some examples include *Pseudoconnarus rhynchosioides*, *Rourea cuspidata*, and *R. neglecta*

(Connaraceae), *Chactopsis anduzei* (Chactidae), *Charis iquitos* (Riodinidae), *Gypopsitta barrabandi* (Psittacidae), and *Crypturellus casiquiare* (Tinamidae).

Vegetation

It has a great diversity of forest types, namely *terra firme*, *igapó* forests, *varzea* forests, and swamp forests, as well as savannas (Dinerstein et al., 1995).

Biotic Relationships

According to a parsimony analysis of endemicity based on primate taxa (Silva and Oren, 1996), it is closely related to the Napo and Madeira provinces; and to a parsimony analysis of endemicity based on butterfly taxa (Racheli and Racheli, 2003), it is closely related to the Napo province. A cladistic biogeographic analysis based on bird taxa (Cracraft and Prum, 1988) found a close relationship of this province with the Napo and Madeira provinces.

Regionalization

Hernández et al. (1992a) identified five districts in the Colombian portion of this province: Ariari-Guayabero, Macarena, Northern Guaviare Forests, Vaupes Complex, and Yarí-Mirití districts (Morrone, 2014b).

PANTEPUI PROVINCE

Region of Venezuela and Guiana (in part)—Good, 1947: 235 (regional.).
Pantepui area—Mayr and Phelps, 1967: 276 (biotic evol.); Borges, 2007: 924 (PAE).
Highlands of Guiana and Brazil area (in part)—Sick, 1969: 454 (regional.).
Guianan dominion—Cabrera and Willink, 1973: 67 (regional.); Huber and Riina, 1997: 151 (glossary).
Guianan province—Cabrera and Willink, 1973: 67 (regional.); Ringuelet, 1975: 107 (regional.); Huber and Riina, 1997: 273 (glossary); Morrone, 1999: 6 (regional.), 2001a: 72 (regional.); Viloria, 2005: 450 (regional.); Morrone, 2006: 480 (regional.).
Pantepui center—Müller, 1973: 64 (regional.); Cracraft, 1985: 59 (areas of endem.).
Guianan district (in part)—Ávila-Pires, 1974b: 176 (regional.).
Campos Limpos province—Udvardy, 1975: 42 (regional.).
Roraima–Guianan dominion (in part)—Ab'Sáber, 1977: map (climate).
Roraima Sandstone formation—Maguire, 1979: 223 (veget.).
Duida subcenter—Cracraft, 1985: 60 (areas of endem.).
Great Savanna subcenter—Cracraft, 1985: 60 (areas of endem.).
Guiana area (in part)—Haffer, 1985: 121 (refugia); Silva et al., 2005: 692 (areas of endem.); Borges, 2007: 924 (PAE).
Guianan Highlands province—Huber and Alarcón, 1988: map (regional.).
Septentrional or Guianan sector (in part)—Fernandes and Bezerra, 1990: 92 (regional.); Fernandes, 2006: 59 (regional.).

Pantepui province—Huber, 1994: 53 (regional.); Huber and Riina, 1997: 246 (glossary); Rull, 2005: 923 (biotic evol.); Costa et al., 2013: 200 (areas of endem.); Morrone, 2014b: 66 (regional.).

Tepuis province—Rivas-Martínez and Navarro, 1994: map (regional.).

Guianan area (in part)—Coscarón and Coscarón-Arias, 1995: 726 (areas of endem.); Bates et al., 1998: 785 (PAE); Naka, 2011: 682 (ordination anal.).

Guianan Highlands Moist Forests ecoregion—Dinerstein et al., 1995: 88 (ecoreg.); Huber and Riina, 1997: 64 (glossary).

Tepuis ecoregion—Dinerstein et al., 1995: 88 (ecoreg.).

Uatama Moist Forests ecoregion—Dinerstein et al., 1995: 89 (ecoreg.); Huber and Riina, 1997: 65 (glossary).

Guianan Highlands region—Huber and Riina, 1997: 284 (glossary).

Guianan province—Morrone, 2000b: 106 (regional.).

Tepuis area—Porzecanski and Cracraft, 2005: 266 (PAE).

Orinoco Guiana Shield ecoregion—Abell et al., 2008: 408 (ecoreg.).

Pantepui region—De Marmels, 2007: 117 (track anal.); Désamoré et al., 2010: 255 (biotic evol.); Leite et al., 2015: 706 (biotic evol.).

Tepuyan province—Rivas-Martínez et al., 2011: 26 (regional.).

Definition

The Pantepui province is situated in northwestern South America, in the Guianan Shield, between Venezuela, Colombia, Guyana, Suriname, and northern Brazil (Cracraft, 1985; Morrone, 2000b, 2006, 2014b). It consists of sandy plateaus or *tepuis*, higher than 2,000 meters elevation, derived from Precambrian rocks that have been eroded since the Jurassic–Cretacic (Bonaccorso and Guayasamin, 2013; Leite et al., 2015). Tepuis are tabular mountains reaching 1,200–2,800 meters of elevation from a surrounding lowland/highland tropical rainforest (Figure 6.4), which belong to the Precambrian Guiana Shield that separated from the African Shield with the opening of the South Atlantic Ocean (80–100 mya). They represent the remains of a gigantic erosion surface that has been dissected by the Amazon and Orinoco basins; being Tepuis' summits it is mostly flat and their surface area ranges between 0.2 and 1,096.3 square kilometers (Steyermark, 1986; Désamoré et al., 2010). The height of the tepuis increases toward the east, with many tepuis reaching several thousand meters above the surrounding lowlands (Dinerstein et al., 1995).

Endemic and Characteristic Taxa

Morrone (2014b) provided a list of endemic and characteristic taxa. Some examples include *Achnopogon*, *Cardonaea*, *Chimantaea*, *Duidaea*, *Stenopadus*, and *Tyleropappus* (Asteraceae), *Partamona mourei* (Apidae), *Naupactus viloriai* (Curculionidae; Figure 6.5), *Gigantofalca duida* and *Parides phosphorus laurae* (Lycaenidae), *Parides klagesi* and *phosphorus laurae* (Papilionidae), *Podoxymys roraimae* and *Rhipidomys macconelli* (Cricetidae), and *Aulacorhynchus duidae*, *A. osgoodi*, and *A. whitelianus* (Ramphastidae).

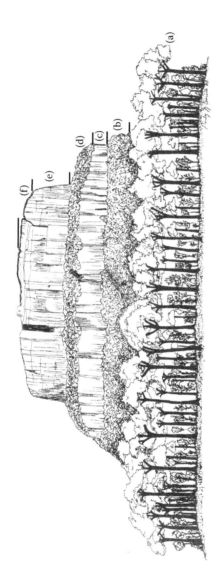

Figure 6.4 Aspect of a typical tepui. (a) Lowland forest; (b) talus slope; (c) escarpment; (d) talus slope; (e) escarpment; and (f) summit. (Modified from Steyermark, J. A., "Speciation and endemism in the flora of the Venezuelan tepuis." In: Vuilleumier, F. and M. Monasterio (eds.), *High altitude tropical biogeography*. Oxford University Press and American Museum of Natural History, New York and Oxford, pp. 317–373, 1986.)

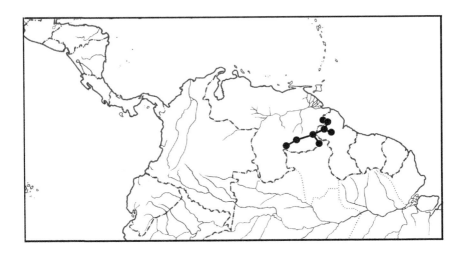

Figure 6.5 Map with the individual track of *Naupactus viloriai* (Curculionidae) in the Pantepui province.

Vegetation

There are savannas alternating with moist forests (Cabrera and Willink, 1973; Dinerstein et al., 1995). Vegetation at the base of the tepuis is constituted by *Caryocar microcarpum*, *Macrolobium acaciaefolium*, and *Abuta grandifolia*; on the talus slopes by species of *Korchubaea*, *Clusia*, *Moronobea*, *Miconia*, *Graffenrieda*, *Magnolia*, *Myrcia*, *Schefflera*, *Ocotea*, and *Nectandra*; at the base of the escarpments *Ichnanthus duidensis*, *I. longifolius*, *Axonopus steyermarkii*, *Xyris ptariana*, *Paepalanthus scopolorum*, *Stegolepis gleasoniana*, *Nautilocalyx resioides*, *Utricularia auremaculata*, and *Alomia ballotaefolia*; and on the summits by species of *Hymenophyllopsis*, *Salpinctes*, *Crepinella*, *Tyleropappus*, *Guaicaia*, *Roraimanthus*, *Wurdackia*, *Myriocladus*, *Adenarake*, *Saccifolium*, *Cephalodendron*, *Pagameopsis*, *Neotatea*, *Neogleasomia*, and *Achlyphila* (Steyermark, 1986).

Biotic Relationships

This province has been separated from the remaining Amazonian provinces in a distinct dominion (Cabrera and Willink, 1973), but their relationships with them are strong (Müller, 1973; Cortés and Franco, 1997). There are several hypotheses proposed to explain the biotic relationships of the Pantepui (Mayr and Phelps, 1967; Pérez-Hernández and Lew, 2001; Rull, 2004a,b, 2005; Désamoré et al., 2010; Bonaccorso and Guayasamin, 2013; Leite et al., 2015):

1. Dispersal: The oldest one postulated the existence of past mountainous connections between the currently separated "subtropical islands" and other areas, as the Andes, from where long-distance dispersal events occurred in Quaternary and previous times.

2. Lost World: It postulated that the Pantepui biota is the remnant of an ancient biota that was widespread on a plateau and was dissected by erosion into the separate tepuis. This theory was initially favored by the high levels of plant and animal endemism. For example, Maguire (1970) considered that half of the 8,000 known vascular plant species were endemic to the province. As exploration intensified, however, this apparent endemicity decreased to 30 percent (Rull, 2004a).

3. Vertical displacement: It assumed that during some times of the glacial–interglacial cycles of the Pleistocene, the lowlands between the tepuis and the Guyanan Shield had a colder weather that allowed the dispersal of species in both directions. Rull (2004a) found that the biota of the Pantepui responded to climate shifts with vertical displacements, supporting the hypothesis. However, he noted that physiographic data indicated that around half of the tepuis would have never been connected by lowlands, so the Lost World hypothesis should not be abandoned, at least for some taxa.

Rull (2005) proposed a model to explain the biotic diversification of the Pantepui, based on palynological findings, which shows a downward biotic migration of ca. 1,100-meter altitude during glacials and the subsequent interglacial upward shift, in response to colder and warmer climates, respectively. This model (Figure 6.6) predicts that during glacials, biotic mixing promoting sympatric speciation, hybridization, and polyploidy is expected in the lowlands. At the mountaintops, cold-adapted (possibly Páramo-like) communities might have been affected by vicariance. During the interglacials, many taxa had the opportunity for ascending to the mountains again, allowing genetic interchange between their slopes and summits while others would have adapted to lowlands. The interglacial highland communities, where vicariance still predominated, experienced some extinction owing to habitat loss by upland displacement. According to Rull's (2005) model, the succession of glacials and interglacials resulted in a net increase of diversity and endemism, favored by the complex topography and habitat heterogeneity, which allowed high niche diversification.

Cenocrons

Bonaccorso and Guayassamin (2013) analyzed *Aulacorhynchus* toucanets (Ramphastidae) from the Pantepui and other areas, including the Andes and northern Venezuela. They compared alternative phylogenetic hypotheses based on mitochondrial and nuclear DNA sequences and estimated a time framework for the diversification of the species. They found that a sister relationship between the Pantepui and the Andes–Cordillera de la Costa was significantly more likely than alternative topologies, estimating that the group has diversified since the late Miocene. They concluded that the patterns found, where the Andes were the biotic source of other areas, are consistent with those found on other bird taxa, not supporting the hypothesis of the geologically old Pantepui as a source of Neotropical mountain diversity.

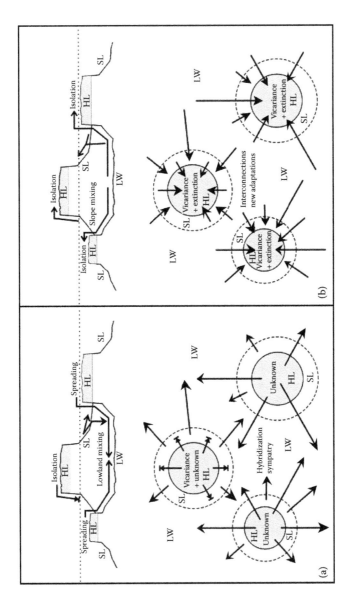

Figure 6.6   Schematic representation of Rull's (2005) model explaining the diversification in the Pantepui. (a) Glacial period; (b) interglacial period. HL: highlands; LW: lowlands; SL: slopes. Events represented in cross sections (top) and their corresponding aerial views (bottom). Straight dotted lines indicate the 1,100 meters altitudinal limit, above which the tepui summits have been isolated from lowlands, even during glacial phases. Solid arrows indicate the sense of biotic migration in each case.

GUIANAN LOWLANDS PROVINCE

Upper Rio Branco subprovince—Rizzini, 1963: 49 (regional.), 1997: 623 (regional.).
Guyanan center—Müller, 1973: 69 (regional.).
Guianan district (in part)—Ávila-Pires, 1974b: 176 (regional.).
Guianan province (in part)—Udvardy, 1975: 41 (regional.).
Roraima–Guianan dominion (in part)—Ab'Sáber, 1977: map (climate).
Guiana center—Beven et al., 1984: 386 (refugia).
Guianan center (in part)—Cracraft, 1985: 68 (areas of endem.).
Eastern Guianan refuge—Lourenço, 1986: 580 (refugia).
Imataca refuge—Lourenço, 1986: 580 (refugia).
Western Guianan refuge—Lourenço, 1986: 580 (refugia).
Northwestern Amazonian area (in part)—Cracraft, 1988: 223 (clad. biogeogr.).
Guianan Lowlands province—Huber and Alarcón, 1988: map (regional.); Morrone,
    2014b: 66 (regional.).
Septentrional or Guianan sector (in part)—Fernandes and Bezerra, 1990: 92 (regional.);
    Fernandes, 2006: 59 (regional.).
Guyanas province—Rivas-Martínez and Navarro, 1994: map (regional.).
Guianan area (in part)—Coscarón and Coscarón-Arias, 1995: 726 (areas of endem.);
    Bates et al., 1998: 785 (PAE); Silva et al., 2005: 692 (areas of endem.); Naka, 2011:
    682 (ordination anal.).
Guianan Moist Forests ecoregion—Dinerstein et al., 1995: 89 (ecoreg.); Huber and
    Riina, 1997: 64 (glossary).
Orinoco Delta Swamp Forests ecoregion—Dinerstein et al., 1995: 88 (ecoreg.); Huber
    and Riina, 1997: 71 (glossary).
Paramaribo Swamp Forests ecoregion—Dinerstein et al., 1995: 89 (ecoreg.).
Humid Guyana province—Morrone, 1999: 7 (regional.), 2000b: 108 (regional.), 2001a:
    74 (regional.); Viloria, 2005: 450 (regional.); Morrone, 2006: 480 (regional.);
    Quijano-Abril et al., 2006: 1270 (track anal.).
Guyana area (in part)—Marks et al., 2002: 155 (biotic evol.).
Amazonia North area (in part)—Porzecanski and Cracraft, 2005: 266 (PAE).
Guianan ecoregion—Abell et al., 2008: 408 (ecoreg.).
Deltaic Orinoquian province—Rivas-Martínez et al., 2011: 26 (regional.).
Guyanan province—Rivas-Martínez et al., 2011: 26 (regional.).
Guianian Lowlands province—Daniel et al., 2016: 1169 (track anal. and regional.).

Definition

The Guianan Lowlands province comprises southwestern Venezuela, northern
Brazil, Suriname, and Guyana (Morrone, 2000b, 2006, 2014b).

Endemic and Characteristic Taxa

Morrone (2014b) provided a list of endemic and characteristic taxa. Some
examples include *Piper nematanthera*, *P. saramaccanum*, and *P. pulleanum*
(Piperaceae), *Naupactus vilmae*, *Pileophorus procerus*, *Prosicoderus gyllen-*
hali, and *Sicoderus nodieri* (Curculionidae; Figure 6.7), *Rhipidita primogenita*

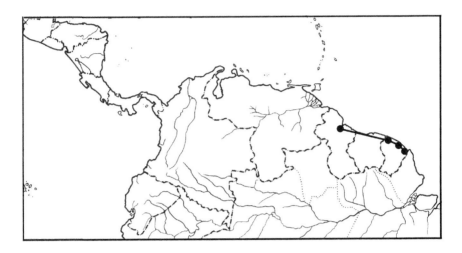

Figure 6.7   Map with the individual track of *Sicoderus nodieri* (Curculionidae) in the Guianan Lowlands province.

(Ditomyiidae), *Charis cleonus* and *Theope brevignoni* (Riodinidae), *Simulium pintoi* (Simuliidae), *Alouatta macconelli* (Cebidae), and *Phrynops nasutus* (Chelidae).

Vegetation

There are moist and swamp forests; there are flooded grasslands and mangroves in mosaic habitats (Dinerstein et al., 1995).

Biotic Relationships

Müller (1973) considered this province to be related to the Napo, Ucayali, and Pará provinces. According to a parsimony analysis of endemicity based on primate species (Silva and Oren, 1996), it is closely related to the Guianan Lowlands province.

RORAIMA PROVINCE

Jari–Trombetas subprovince—Rizzini, 1963: 50 (regional.), 1997: 623 (regional.).
Oceanic sector—Rizzini, 1963: 50 (regional.).
Roraima center—Müller, 1973: 62 (regional.).
Guianan district (in part)—Ávila-Pires, 1974b: 176 (regional.).
Guianan province (in part)—Udvardy, 1975: 41 (regional.).
Roraima–Guianan dominion (in part)—Ab'Sáber, 1977: map (climate).
Guianan center (in part)—Cracraft, 1985: 68 (areas of endem.).
Guiana area (in part)—Haffer, 1985: 121 (refugia); Silva and Oren, 1996: 430 (PAE); Ron, 2000: 387 (PAE); Silva et al., 2005: 692 (areas of endem.); Borges, 2007: 924 (PAE).
Northwestern Amazonian area (in part)—Cracraft, 1988: 223 (clad. biogeogr.).
Amazonas Delta province (in part)—Rivas-Martínez and Navarro, 1994: map (regional.).

Roraima–Trombetas province—Rivas-Martínez and Navarro, 1994: map (regional.).
Amapá Moist Forests ecoregion—Dinerstein et al., 1995: 89 (ecoreg.).
Eastern Amazonian Flooded Grasslands ecoregion—Dinerstein et al., 1995: 101 (ecoreg.); Huber and Riina, 1997: 71 (glossary).
Varzea Forests ecoregion (in part)—Dinerstein et al., 1995: 90 (ecoreg.).
Guianan area (in part)—Bates et al., 1998: 785 (PAE); Racheli and Racheli, 2003: 36 (PAE), 2004: 347 (PAE); Naka, 2011: 682 (ordination anal.).
Amapá province—Morrone, 1999: 7 (regional.), 2000b: 111 (regional.), 2001a: 75 (regional.), 2006: 480 (regional.); Quijano-Abril et al., 2006: 1270 (track anal.).
Roraima province—Morrone, 1999: 6 (regional.), 2000b: 110 (regional.), 2001a: 75 (regional.); Viloria, 2005: 450 (regional.); Morrone, 2006: 480 (regional.); Quijano-Abril et al., 2006: 1270 (track anal.); Morrone, 2014b: 67 (regional.).
Varzea province (in part)—Morrone, 1999: 7 (regional.), 2000b: 112 (regional.), 2001a: 76 (regional.), 2006: 480 (regional.).
Guyana area (in part)—Marks et al., 2002: 155 (biotic evol.).
Amazonia North area (in part)—Porzecanski and Cracraft, 2005: 266 (PAE).
Amazonas Guiana Shield ecoregion—Abell et al., 2008: 408 (ecoreg.).
North Amazonian province—Rivas-Martínez et al., 2011: 26 (regional.).

Definition

The Roraima province comprises northern Brazil, southeastern Venezuela, Suriname, and Guyana (Morrone, 2000b, 2006, 2014b).

Endemic and Characteristic Taxa

Morrone (2014b) provided a list of endemic and characteristic taxa. Some examples include *Cecropia silvae* (Cecropiaceae), *Broteochactas sissomi* (Chactidae), *Caprimulgus longirostris roraimae* (Caprimulgidae; Figure 6.8), *Proechimys guyanensis arabupu* (Echimyidae), and *Myiophobus roraimae roraimae* (Tyrannidae).

Vegetation

Basically consisting of savannas, there are also gallery forests bordering the rivers (Dinerstein et al., 1995).

Biotic Relationships

Müller (1973) considered this province to be related to the Sabana province.

PARÁ PROVINCE

Southeast sector—Rizzini, 1963: 51 (regional.).
Pará center—Müller, 1973: 75 (regional.); Cracraft, 1985: 71 (areas of endem.).

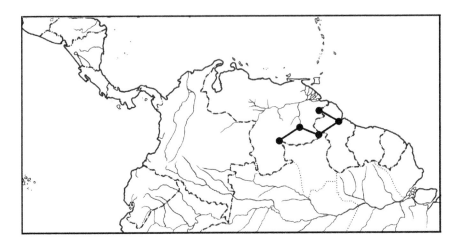

Figure 6.8   Map with the individual track of *Caprimulgus longirostris roraimae* (Caprimulgidae) in the Roraima province.

Babacu province—Udvardy, 1975: 42 (regional.).
Belém center—Beven et al., 1984: 386 (refugia).
Belém (Maranhão) center—Cracraft, 1985: 72 (areas of endem.).
Belém area—Haffer, 1985: 121 (refugia); Silva and Oren, 1996: 430 (PAE); Bates et al., 1998: 785 (PAE); Ron, 2000: 387 (PAE); Racheli and Racheli, 2003: 36 (PAE), 2004: 347 (PAE); Silva et al., 2005: 692 (areas of endem.); Borges, 2007: 924 (PAE).
Southwestern Amazonian area (in part)—Cracraft, 1988: 223 (clad. biogeogr.).
Eastern or Northern sector—Fernandes and Bezerra, 1990: 96 (regional.); Fernandes, 2006: 64 (regional.).
Amazonas Delta province (in part)—Rivas-Martínez and Navarro, 1994: map (regional.).
Pará province—Morrone, 1999: 8 (regional.), 2000b: 115 (regional.), 2001a: 80 (regional.), 2006: 481 (regional.), 2014b: 68 (regional.).
Amazonas Estuary and Coastal Drainages ecoregion—Abell et al., 2008: 408 (ecoreg.).
Deltaic Amazonian province—Rivas-Martínez et al., 2011: 27 (regional.).
Tocantins province (in part)—Rivas-Martínez et al., 2011: 27 (regional.).

Definition

The Pará province comprises northwestern Brazil, limiting to the north and west with the Tocantins and Araguaia rivers, to the south with the Serra do Gurupí and the Grajau River, and to the east with the Guana River (Müller, 1973; Morrone, 2000b, 2006, 2014b).

Endemic and Characteristic Taxa

Morrone (2014b) provided a list of endemic and characteristic taxa. Some examples include *Geotrigona aequinoctialis* (Figure 6.9), *Partamona chapadicola*

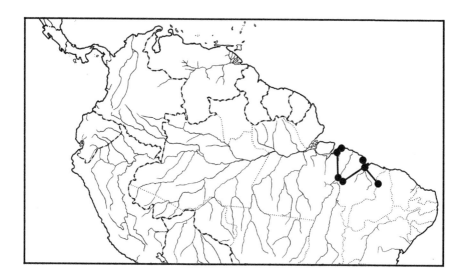

Figure 6.9 Map with the individual track of *Geotrigona aequinoctialis* (Apidae) in the Pará province.

(Apidae), *Xipholena lamellipennis* (Cotingidae), *Proechimys leioprimma, P. nesiotes,* and *P. oris* (Echimyidae), and *Pteroglossus bitorquatus* (Ramphastidae).

Vegetation

There are moist forests and flooded grasslands (Dinerstein et al., 1995).

Biotic Relationships

Müller (1973) considered this province to be related to the Madeira province. According to a parsimony analysis of endemicity based on primate species (Silva and Oren, 1996), it is closely related to the Guianan Lowlands province; to another parsimony analysis of endemicity based on bird taxa (Bates et al., 1998), it is closely related to the Rondônia province; and to a third analysis based on bird species (Borges, 2007), it is related to the Xingu–Tapajós province. A cladistic biogeographic analysis based on bird taxa (Cracraft and Prum, 1988) found a close relationship of this province with the Atlantic province.

SOUTH BRAZILIAN DOMINION

South Brazilian province—Engler, 1882: 345 (regional.).
Austral Brazilian subarea (in part)—Clarke, 1892: 381 (regional.).
Cariba, Guianan, or Amazonian center (in part)—Lane, 1943: 414 (regional.).

South Brazilian region—Good, 1947: 235 (regional.).

Amazonian dominion (in part)—Cabrera, 1971: 6 (regional.), 1976: 3 (regional.); Huber and Riina, 1997: 150 (glossary), 2003: 124 (glossary).

Amazonian province (in part)—Cabrera and Willink, 1973: 48 (regional.); Ringuelet, 1975: 107 (regional.); Udvardy, 1975: 41 (regional.); Fernandes and Bezerra, 1990: 77 (regional.); Huber and Riina, 1997: 23 (glossary); Rizzini, 1997: 623 (regional.); Fernandes, 2006: 46 (regional.).

Amazon center—Müller, 1973: 80 (regional.).

Western district—Ávila-Pires, 1974b: 177 (regional.).

Amazonian Equatorial dominion (in part)—Ab'Sáber, 1977: map (climate).

Meridional or Brazilian sector (in part)—Fernandes and Bezerra, 1990: 93 (regional.); Fernandes, 2006: 61 (regional.).

Amazonia bioregion (in part)—Dinerstein et al., 1995: map 1 (ecoreg.); Huber and Riina, 1997: 37 (glossary).

Northwest Amazonia subregion (in part)—Nihei and de Carvalho, 2007: 497 (clad. biogeogr.).

Amazonas Lowlands ecoregion—Abell et al., 2008: 408 (ecoreg.).

Amazonian–Guyanan superegion (in part)—Rivas-Martínez et al., 2011: 26 (regional.).

Amazonian region (in part)—Rivas-Martínez et al., 2011: 26 (regional.).

Southwestern Amazonian dominion—Morrone, 2014c: 206 (clad. biogeogr.).

South Brazilian dominion—Morrone, 2014b: 68 (regional.); Daniel et al., 2016: 1167 (track anal. and regional.).

Definition

The South Brazilian dominion comprises the Amazonian forest, southwest of the Amazon River (Morrone, 2014b,c).

Endemic and Characteristic Taxa

Taxa endemic to this dominion are *Cyphocharax spiluropsis* (Curimatidae; Figure 6.10; Vari, 1992) and *Plecturocebus* (Pitheciidae; Byrne et al., 2016).

Regionalization

The South Brazilian dominion comprises the Ucayali, Madeira, Rondônia, and Yungas provinces (Morrone, 2014b).

UCAYALI PROVINCE

West sector—Rizzini, 1963: 51 (regional.).

Ucayali subcenter—Müller, 1973: 83 (regional.).

Loreto province (in part)—Rivas-Martínez and Navarro, 1994: map (regional.).

Tocantins Moist Forests ecoregion—Dinerstein et al., 1995: 91 (ecoreg.).

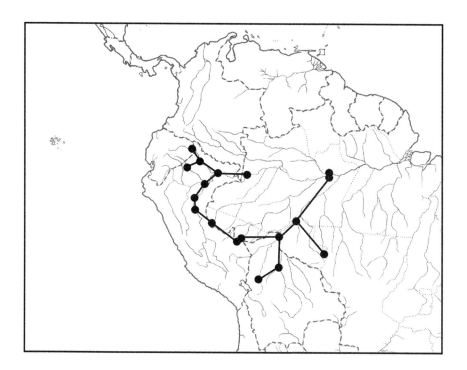

Figure 6.10 Map with the individual track of *Cyphocharax spiluropsis* (Curimatidae) in the South Brazilian dominion.

Ucayali Moist Forests ecoregion—Dinerstein et al., 1995: 90 (ecoreg.); Huber and Riina, 1997: 65 (glossary).

Varzea Forests ecoregion (in part)—Dinerstein et al., 1995: 90 (ecoreg.).

Western Amazonian Flooded Grasslands ecoregion—Dinerstein et al., 1995: 100 (ecoreg.); Huber and Riina, 1997: 71 (glossary).

Western Amazonian Swamp Forests ecoregion—Dinerstein et al., 1995: 90 (ecoreg.).

Ucayali province—Morrone, 1999: 7 (regional.), 2000b: 112 (regional.), 2001a: 77 (regional.), 2006: 480 (regional.); Quijano-Abril et al., 2006: 1270 (track anal.); Morrone, 2014b: 68 (regional.).

Varzea province (in part)—Morrone, 1999: 7 (regional.), 2000b: 112 (regional.), 2001a: 76 (regional.), 2006: 480 (regional.).

West Amazonian province (in part)—Rivas-Martínez et al., 2011: 26 (regional.).

Definition

The Ucayali province comprises eastern Peru, northern Bolivia, and western Brazil (Morrone, 2000b, 2006, 2014b). The Ucayali province harbors one of the richest biotas in the world (Dinerstein et al., 1995).

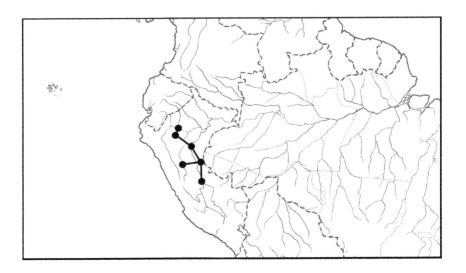

Figure 6.11   Map with the individual track of *Proechimys brevicauda* (Echimyidae) in the Ucayali province.

Endemic and Characteristic Taxa

Morrone (2014b) provided a list of endemic and characteristic taxa. Some examples include *Chactopsis insignis* (Chactidae), *Innoxius magnus* and *Stygnus klugi* (Stygnidae), *Proechimys brevicauda* (Echimyidae; Figure 6.11), *Crypturellus bartletti* and *C. strigulosus* (Tinamidae), and *Muscisaxicola fluviatilis* (Tyrannidae).

Vegetation

It consists of moist forests and flooded grasslands (Dinerstein et al., 1995).

Biotic Relationships

Müller (1973) considered that the Ucayali province is related to the Napo province.

Regionalization

Lamas (1982) identified eight units within the Peruvian portion of this province that may be treated as districts: Carpish, Chachapoyas, Huallaga, Huánuco, La Peca, Marañón, Molinopampa, and Tocache.

MADEIRA PROVINCE

Tertiary Plain subprovince—Rizzini, 1963: 50 (regional.), 1997: 623 (regional.).
Madeira center—Müller, 1973: 80 (regional.).
Madeiran province (in part)—Udvardy, 1975: 41 (regional.); Huber and Riina, 1997: 206 (glossary).
Loreto unit (in part)—Lamas, 1982: 352 (regional.).
Yurimaguas unit—Lamas, 1982: 343 (regional.).
Inambari center—Beven et al., 1984: 386 (refugia).
South Amazon (Inambari) center—Cracraft, 1985: 69 (areas of endem.).
Inambari area—Haffer, 1985: 121 (refugia); Silva and Oren, 1996: 430 (PAE); Bates et al., 1998: 785 (PAE); Ron, 2000: 387 (PAE); Marks et al., 2002: 155 (biotic evol.); Racheli and Racheli, 2003: 36 (PAE), 2004: 347 (PAE); Silva et al., 2005: 692 (areas of endem.).
Southwestern Amazonian area (in part)—Cracraft, 1988: 223 (clad. biogeogr.).
Madeira province—Rivas-Martínez and Navarro, 1994: map (regional.); Morrone, 1999: 8 (regional.), 2000b: 114 (regional.), 2001a: 77 (regional.), 2006: 481 (regional.); Mercado-Salas et al., 2012: 459 (track anal.); Morrone, 2014b: 69 (regional.); Daniel et al., 2016: 1169 (track anal. and regional.).
Juruá Moist Forests ecoregion—Dinerstein et al., 1995: 90 (ecoreg.).
Purus/Madeira Moist Forests ecoregion—Dinerstein et al., 1995: 90 (ecoreg.).
Rondônia/Mato Grosso Moist Forests ecoregion—Dinerstein et al., 1995: 90 (ecoreg.); Huber and Riina, 1997: 65 (glossary).
São Luis Flooded Grasslands ecoregion—Dinerstein et al., 1995: 101 (ecoreg.).
Varzea Forests ecoregion (in part)—Dinerstein et al., 1995: 90 (ecoreg.).
Varzea province (in part)—Morrone, 1999: 7 (regional.), 2000b: 112 (regional.), 2001a: 76 (regional.), 2006: 480 (regional.).
Amazonia South area—Porzecanski and Cracraft, 2005: 266 (PAE).

Definition

The Madeira province comprises southwestern Brazil, limiting to the north with the Amazon River, to the west with the Madeira and Beni rivers, to the east with the Xingu River, and to the west with the eastern cordillera of Bolivia (Müller, 1973; Morrone, 2000b, 2006, 2014b).

Endemic and Characteristic Taxa

Morrone (2014b) provided a list of endemic and characteristic taxa. Some examples include *Geotrigona subgrisea subfulva* and *Partamona batesi* (Apidae), *Callithrix humeralifera* and *Saguinus labiatus* (Callithrichidae), *Conopophaga melanogaster* (Conopophagidae; Figure 6.12), and *Gypopsitta aurantiocephala* and *Pyrrhura rhodogaster* (Psittacidae).

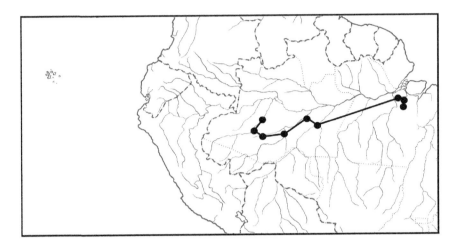

Figure 6.12   Map with the individual track of *Conopophaga melanogaster* (Conopophagidae) in the Madeira province.

Vegetation

There is a wide range of forest types, with many transitional formations located southwards to the Cerrado and Rondônia provinces (Dinerstein et al., 1995).

Biotic Relationships

According to a parsimony analysis of endemicity based on primate species (Silva and Oren, 1996), it is closely related to the Rondônia and Napo provinces; and to two parsimony analyses of endemicity based on bird taxa (Bates et al., 1998; Borges, 2007), it is closely related to the Napo province. A cladistic biogeographic analysis based on bird taxa (Cracraft and Prum, 1988) found a close relationship of this province with the Napo province.

RONDÔNIA PROVINCE

Southwest sector—Rizzini, 1963: 51 (regional.).
Rondônia center—Beven et al., 1984: 386 (refugia); Cracraft, 1985: 71 (areas of endem.).
Rondônia area—Haffer, 1985: 121 (refugia); Silva and Oren, 1996: 430 (PAE); Bates et al., 1998: 785 (PAE); Ron, 2000: 387 (PAE); Marks et al., 2002: 155 (biotic evol.); Racheli and Racheli, 2003: 36 (PAE), 2004: 347 (PAE); Silva et al., 2005: 692 (areas of endem.); Borges, 2007: 924 (PAE).
Guapore refuge—Lourenço, 1986: 580 (refugia).
Southwestern Amazonian area (in part)—Cracraft, 1988: 223 (clad. biogeogr.).
Acre–Madre de Dios province—Rivas-Martínez and Navarro, 1994: map (regional.).
Beni province—Rivas-Martínez and Navarro, 1994: map (regional.).

Pantanal province Rivas-Martínez and Navarro, 1994: map (regional.); Morrone, 1999: 9 (regional.), 2000b: 116 (regional.), 2001a: 80 (regional.), 2006: 481 (regional.); Mercado-Salas et al., 2012: 459 (track anal.).

Beni Savannas ecoregion—Dinerstein et al., 1995: 99 (ecoreg.); Huber and Riina, 1997: 300 (glossary).

Beni Swamp and Gallery Forests ecoregion—Dinerstein et al., 1995: 91 (ecoreg.).

Bolivian Lowland Dry Forests ecoregion—Dinerstein et al., 1995: 95 (ecoreg.); Huber and Riina, 1997: 94 (glossary).

Pantanal ecoregion—Dinerstein et al., 1995: 101 (ecoreg.); Huber and Riina, 1997: 245 (glossary).

Patía Valley Dry Forests ecoregion—Dinerstein et al., 1995: 95 (ecoreg.); Huber and Riina, 1997: 95 (glossary).

Southwestern Amazonian Moist Forests ecoregion—Dinerstein et al., 1995: 90 (ecoreg.); Huber and Riina, 1997: 64 (glossary).

Rondônia province—Morrone, 1999: 8 (regional.), 2014b: 70 (regional.); Daniel et al., 2016: 1160 (track anal. and regional.).

Chiquitania ecoregion—Salazar Bravo et al., 2002: 78 (ecoreg.).

Inambari area (in part)—Silva et al., 2005: 692 (areas of endem.); Borges, 2007: 924 (PAE).

Benian province—Rivas-Martínez et al., 2011: 27 (regional.).

Pantanalian province—Rivas-Martínez et al., 2011: 27 (regional.).

Southwest Amazonian province—Rivas-Martínez et al., 2011: 27 (regional.).

West Amazonian province (in part)—Rivas-Martínez et al., 2011: 26 (regional.).

Definition

The Rondônia province comprises southern and central Brazil, southern Peru, northwestern Bolivia, and northern Paraguay (Morrone, 2000b, 2006, 2014b). It is delimited to the north by the Amazon River, to the east by the Tapajós River, and to the south by the limit of the tropical rainforest (Cracraft, 1985).

Endemic and Characteristic Taxa

Morrone (2014b) provided a list of endemic and characteristic taxa. Some examples include *Piper moense* and *P. udisilvestre* (Piperaceae), *Prosicoderus xingu* (Curculionidae), *Charis tefe* (Curculionidae), *Protimesius albilineatus*, *Stygnus marthae*, and *S. weyrauchi* (Stygnidae), *Alouatta sara* (Cebidae), *Steindachnerina fasciata* (Curimatidae; Figure 6.13), and *Sciurus sanborni* (Sciuridae).

Vegetation

Typical vegetation, usually known as *pantanal*, consists of a mosaic of flooded grasslands, gallery and dry forests, and several transitional types (Rizzini, 1997). During the rainy season, more than 80 percent of the province is flooded (Dinerstein et al., 1995). Dry forests of the Rondônia province may be among the richest dry forest ecosystems of the world (Dinerstein et al., 1995). Common plant species include

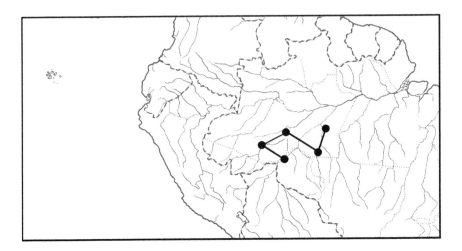

Figure 6.13 Map with the individual track of *Steindachnerina fasciata* (Curimatidae) in the Rondônia province.

Tabebuia caraiba, T. ipe, Copernicia alba, Triplaris formicosa, Ipomoea fistulosa, Paratheria prostrata, Genipa americana, Astronium urundeuva, Magonia glabrata, M. pubescens, and *Myroxylon balsanum* (Rizzini, 1997).

Biotic Relationships

According to a parsimony analysis of endemicity based on primate species (Silva and Oren, 1996), the Rondônia province is closely related to the Pará, Xingu–Tapajós, and Madeira provinces; and to another parsimony analysis of endemicity based on bird species (Borges, 2007), it is related to the Xingu–Tapajós and Pará provinces.

Regionalization

Lamas (1982) identified 11 units within the Peruvian portion of this province, which were treated by Morrone (2014b) as districts. These are Apurímac, Atalaya, Chachapoyas, Chanchamayo, Loreto, Madre de Dios, Marcapata, Oxapampa, Quincemil, Titicaca, and Unini districts. Daniel et al. (2016) recently suggested excluding the Pantanal area of this province and treating it as a district of the Chaco province.

YUNGAS PROVINCE

Subtropical formation—Holmberg, 1898: 440 (regional.).
Hygrophyllous Subtropical Forests area—Hauman, 1920: 46 (regional.).
Tucumán–Bolivian Forest province—Hauman, 1931: 60 (regional.); Castellanos and Pérez-Moreau, 1944: 98 (regional.).

Tucumán–Bolivian Forest area—Parodi, 1934: 171 (regional.); Castellanos and Pérez-Moreau, 1941: 379 (regional.); Parodi, 1945: 127 (regional.).
Western Subtropical province—Cabrera, 1951: 24 (regional.), 1953: 109 (regional.).
Montane district—Cabrera, 1953: 113 (regional.).
Oranense district—Cabrera, 1953: 110 (regional.).
Tucumán district—Cabrera, 1953: 112 (regional.).
Tucumán–Bolivian Forests subregion—Hueck, 1957: 40 (regional.).
Tucumán–Oranense Forest area—Ragonese, 1967: 121 (regional.).
Yunga province—Cabrera, 1971: 8 (regional.); Huber and Riina, 1997: 271 (glossary).
Yungas province—Cabrera and Willink, 1973: 54 (regional.); Udvardy, 1975: 42 (regional.); Cabrera, 1976: 3 (regional.); Rivas-Martínez and Navarro, 1994: map (regional.); Morales et al., 1995: 163 (veget.); Huber and Riina, 1997: 272 (glossary); Morrone, 1999: 8 (regional.), 2001a: 81 (regional.), 2006: 481 (regional.); Zuloaga et al., 1999: 37 (regional.); Morrone, 2000b: 105 (regional.); Ojeda et al., 2002: 24 (biotic evol.); Donato, 2006: 422 (clad. biogeogr.); López Ruf et al., 2006: 116 (track anal.); Aagesen et al., 2009: 310 (endem. anal.); Navarro et al., 2009: 517 (endem. anal.); Ferretti et al., 2012: 1 (track anal.); Ferro, 2013: 324 (biotic evol.); Ferretti et al., 2014: 1089; Morrone, 2014b: 71 (regional.); del Río et al., 2015: 1294 (track anal. and clad. biogeogr.).
Marañón center—Müller, 1973: 97 (regional.).
Yungas center—Müller, 1973: 89 (regional.).
Equatorial Andes dominion—Ab'Sáber, 1977: map (climate).
Yunga province—Rivas-Martínez and Navarro, 1994: map (regional.).
Yungas area—Coscarón and Coscarón-Arias, 1995: 726 (areas of endem.).
Andean Yungas ecoregion—Dinerstein et al., 1995: 93 (ecoreg.).
Bolivian Montane Dry forests ecoregion—Dinerstein et al., 1995: 96 (ecoreg.); Huber and Riina, 1997: 96 (glossary).
Bolivian Yungas ecoregion—Dinerstein et al., 1995: 92 (ecoreg.).
Peruvian Yungas ecoregion—Dinerstein et al., 1995: 92 (ecoreg.); Huber and Riina, 1997: 364 (glossary).
Transition Forests district—Huber and Riina, 1997: 146 (glossary).
Yungas region—Huber and Riina, 1997: 285 (glossary).
Yungas ecoregion—Burkart et al., 1999: 15 (ecoreg.); Salazar Bravo et al., 2002: 78 (ecoreg.).
Amazonas High Andes ecoregion (in part)—Abell et al., 2008: 408 (ecoreg.).
Yungenian province—Rivas-Martínez et al., 2011: 27 (regional.).

Definition

The Yungas province comprises the eastern slopes of the Andes, between 500 and 3,500 meters, from northern Peru to northwestern Argentina (Cabrera and Willink, 1973; Cabrera, 1976; Morrone, 2000b, 2006, 2014b). The climate of the Yungas is mild and humid, with mountain slopes often covered by clouds (Aagesen et al., 2009).

Endemic and Characteristic Taxa

Morrone (2014b) provided a list of endemic and characteristic taxa. Some examples include *Podocarpus parlatorei* (Podocarpaceae), *Holocheilus fabrisii*, *Jungia pauciflora, Perezia carduncelloides, Trixis grisebachii,* and *T. ragonesei* (Asteraceae), *Chlorus bolivianus* (Acrididae), *Americhernes andinus* and *Maxchernes birabeni* (Chernetidae), *Brachystylodes pilosus, Ericydeus argentinensis, Naupactus hirsutus, Pantomorus minutus, Parapantomorus quatuordecimpunctatus,* and *Sicoderus tringa* (Curculionidae), *Leucothyreus* (Scarabaeidae), *Oligosarcus bolivianus* (Characidae), and *Calomys fecundus* (Cricetidae).

Vegetation

There are dry and cloud forests, especially rich in Lauraceae and Myrtaceae, alternating with forests of *Alnus acuminata* and *Podocarpus* spp. and grasslands (Cabrera, 1971; Cabrera and Willink, 1973; Aagesen et al., 2009). The composition varies along the Yungas province: in Peru *Cabralea weberbaueris, Cedrela angustifolia, C. lilloi, Weinmannia microphylla, W. nebularum,* and *Persea crassifolia* are common; and in Argentina *Blepharocalyx gigantea, Pseudocaryophyllus guilii, Eugenia mato, Phoebe porphyria, Ocotea pubescens, Parapiptadenia excelsa, Enterolobium contortisiliquum,* and *Tabebuia avellanedae* (Cabrera and Willink, 1973). Three basic vegetation types have been recognized (Cabrera, 1971; Morales et al., 1995; Aagesen et al., 2009; Navarro et al., 2009): premontane, Subandean piedmont forest, or transition forest (300–600 m); montane forest (600–1,600 m) that comprises the lower montane forest or *selva basal* from 900 to 1,200 meters, the upper montane forest or *selva de mirtáceas* from 1,200 to 1,600 meters, and montane cloud forest (1,200–2,500 m).

Biotic Relationships

Müller (1973) considered this province to be related to the Cauca and Guianan Lowlands provinces.

Regionalization

Cabrera and Willink (1973) considered that it was not possible to identify districts in the whole province. Three districts for the Argentinean portion of the Yungas province have been identified: Transition Forests, Montane Jungles, and Montane Forests (Cabrera, 1971, 1976; Huber and Riina, 1997; Ferro, 2013; Morrone, 2014b). The Transition Forests district is a strip on the foothills, from 350 meters to 500 meters; the 700-millimeter isohyet represents the eastern limit of this formation (Cabrera, 1976; Ferro, 2013). The Montane Jungles district corresponds to the lower eastern slopes from 500 meters to 1,500 meters (Cabrera, 1976; Ferro, 2013). The Montane Forests district constitutes a ribbon on the upper slopes of the mountains from 1,500 meters to 3,500 meters (Cabrera, 1976; Ferro, 2013). Lamas (1982) identified nine units within the Peruvian portion of this province that may be treated as districts: Ancash, Chuquibamba, Cutervo, Huamachuco, Lima, Mantaro, Pampas, Parinacochas, and Santa Ana.

PACIFIC DOMINION

North Brazilian–Guianan province (in part)—Engler, 1882: 345 (regional.).
Tropical Andean subarea (in part)—Clarke, 1892: 381 (regional.).
Colombian subregion (in part)—Sclater and Sclater, 1899: 65 (regional.).
Caribbean province (in part)—Mello-Leitão, 1937: 232 (regional.), 1943: 129 (regional.).
Incasic district (in part)—Cabrera and Yepes, 1940: 16 (regional.).
Sabana district (in part)—Cabrera and Yepes, 1940: 14 (regional.).
Incasic dominion (in part)—Orfila, 1941: 86 (regional.).
Incasic center (in part)—Lane, 1943: 414 (regional.).
Incasic province (in part)—Fittkau, 1969: 642 (regional.).
Pacific province—Cabrera and Willink, 1973: 52 (regional.); Huber and Riina, 1997: 273 (glossary), 2003: 263 (glossary).
Orinoco–Venezuelan dominion—Ringuelet, 1975: 107 (regional.).
Pacific area—Coscarón and Coscarón-Arias, 1995: 726 (areas of endem.).
Northern Andes bioregion—Dinerstein et al., 1995: map 1 (ecoreg.); Huber and Riina, 1997: 37 (glossary).
Orinoco bioregion—Dinerstein et al., 1995: map 1 (ecoreg.); Huber and Riina, 1997: 37 (glossary).
Northwestern South American dominion—Morrone, 2004a: 157 (track anal.), 2006: 479 (regional.); Asiain et al., 2010: 178 (track anal.); Morrone, 2014c: 207 (clad. biogeogr.).
Neogranadian region—Rivas-Martínez et al., 2011: 26 (regional.).
Pacific dominion—Morrone, 2014b: 53 (regional.).

Definition

The Pacific dominion comprises southern Central America (southeastern Nicaragua and Panama), northwestern South America (Colombia, Ecuador, Peru, Venezuela, and Trinidad and Tobago) and the Galapagos Islands (Morrone, 2014b).

Endemic and Characteristic Taxa

Plant species belonging to Fabaceae (genera *Inga*, *Dussia*, *Macrolobium*, *Pentaclethra*, *Swarsia*, and *Andira*), Moraceae (*Brossium*, *Castilla*, *Cecropia*, *Coussapoa*, *Helicostylis*, *Ficus*, and *Pourouma*), Annonaceae (*Anaxagorea*, *Crematosperma*, *Pseudoxandra*, and *Xylopia*), Bombaceae, Burseraceae, Hypericaceae, Poaceae (*Arberella*), and Myristicaceae have been cited as characteristic taxa (Cabrera and Willink, 1973; Soderstrom et al., 1988). Vertebrate species include representatives of Soricormorpha (genus *Cryptotis*), Primates (*Alouatta*, *Cebus*, *Saimiri*, *Aotus*, *Ateles*, and *Saguinus*), Xenarthra (*Tamandua*, *Cyclops*, *Myrmecophaga*, *Bradypus*, *Choloepus*, and *Dasypus*), Rodentia (*Microsciurus*, *Sciurus*, *Glaucoys*, *Neocomys*, *Rhypidomys*, *Aporodon*, *Dasyprocta*, and *Coendou*), Carnivora (*Speothos*, *Urocyon*, *Galictis*, *Eira*, and *Conepatus*), Perissodactyla (*Tapirus*), Tinamiformes (*Crypturellus* and *Nothocercus*), Galliformes (*Crax* and *Ortalis*), Psittaciformes (*Amazona, Ara,* and *Aratinga*), Trogoniformes (*Pharomachrus*), Piciformes (*Ramphastos, Pteroglossus,* and *Aulacorhynchus*), Crocodilia (*Crocodylus* and *Caiman*), and Squamata *(Crotalus, Lachesis, Micrurus, Boa, Anolis, Iguana,* and *Sceloporus*; Cabrera and Willink, 1973). Some species endemic to this dominion include *Philodendron ligulatum* (Araceae; Echeverry and Morrone, 2010), *Cecropia angustifolia* (Cecropiaceae; Figure 6.14a; Franco-Rosselli and Berg, 1997), *Saltator albicollis* (Fringillidae; Müller, 1973), *Pionus chalcopterus* and *P. tumultuosus* species groups (Psittacidae; Ribas et al., 2007), and *Transandinomys talamancae* (Cricetidae; Figure 6.14b; Musser and Williams, 1985).

Biotic Relationships

The Pacific dominion is biotically related to the Mesoamerican dominion (Morrone, 2014c).

Regionalization

The Pacific dominion comprises the Guatuso–Talamanca, Puntarenas–Chiriquí, Chocó–Darién, Guajira, Venezuelan, Trinidad, Magdalena, Sabana, Cauca, Galapagos Islands, Western Ecuador, and Ecuadorian provinces (Morrone, 2014b).

GUATUSO–TALAMANCA PROVINCE

Guatuso–Talamanca province—Ryan, 1963: 31 (regional.); Morrone, 2014b: 54 (regional.).
Costa Rica–Panama Highlands province—Stuart, 1964: 358 (regional.).
San Juan province—Miller, 1966: 782 (regional.); Smith and Bermingham, 2005: 1843 (cluster anal. and regional.); Hulsey and López-Fernández, 2011: 280 (regional.).
Isthmian province (in part)—Miller, 1966: 782 (regional.); Savage, 1966: 736 (biotic evol.).
Talamancan province—Savage, 1966: 736 (biotic evol.).
Costa Rican center—Müller, 1973: 23 (regional.).
Talamanca Páramo center—Müller, 1973: 26 (regional.).
Talamanca Montane Forest subcenter—Müller, 1973: 14 (regional.).
Guatemala–Panama province (in part)—Samek et al., 1988: 29 (regional.); Huber and Riina, 2003: 151 (glossary).
Central American Atlantic Moist Forests ecoregion (in part)—Dinerstein et al., 1995: 87 (ecoreg.).
Talamancan Montane Forests ecoregion—Dinerstein et al., 1995: 88 (ecoreg.).
Costa Rican Páramo ecoregion—Dinerstein et al., 1995: 101 (ecoreg.).
Central American province (in part)—Brown et al., 1998: 31 (veget.); Huber and Riina, 2003: 99 (glossary).
Eastern Panama Highlands province—Campbell, 1999: 116 (regional.).
Talamancan Cordillera area—Marshall and Liebherr, 2000: 206 (clad. biogeogr.).
Talamanca Ridge area—Flores-Villela and Goyenechea, 2001: 174 (clad. biogeogr.); Flores-Villela and Martínez-Salazar, 2009: 820 (clad. biogeogr.).

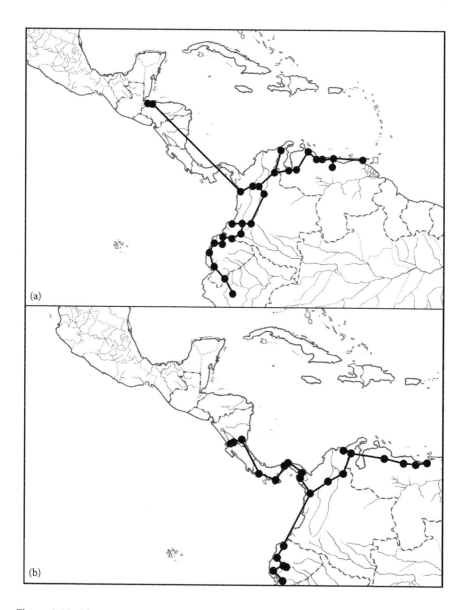

Figure 6.14 Maps with individual tracks in the Pacific dominion. (a) *Cecropia angustifolia* (Cecropiaceae); (b) *Transandinomys talamancae* (Cricetidae).

Eastern Central America province (in part)—Morrone, 2001a: 46 (regional.), 2001e: 53 (regional.), 2006: 479 (regional.); Quijano-Abril et al., 2006: 1269 (track anal.); Escalante et al., 2011: 32 (track anal.); Mercado-Salas et al., 2012: 459 (track anal.).

Highlands of Costa Rica and Western Panama region—Huber and Riina, 2003: 156 (glossary).

Chiriquí–Darién Highlands area—Porzecanski and Cracraft, 2005: 266 (PAE).

Bocas province—Smith and Bermingham, 2005: 1843 (cluster anal. and regional.).
Chagres province—Smith and Bermingham, 2005: 1843 (cluster anal. and regional.).
Chagres ecoregion—Abell et al., 2008: 408 (ecoreg.).
Cocos Island ecoregion—Abell et al., 2008: 408 (ecoreg.).
Panamanian–Costa Rican province (in part)—Rivas-Martínez et al., 2011: 26
    (regional.).

Definition

The Guatuso–Talamanca province comprises eastern Central America, between
southeastern Nicaragua and eastern Panama (Morrone, 2014b). The montane forests
of the Guatuso–Talamanca province are notable for their rich biotas and high number of endemic species (Dinerstein et al., 1995).

Endemic and Characteristic Taxa

Morrone (2014b) provided a list of endemic and characteristic taxa. Some
examples include *Notorhopalotria panamensis* and *N. taylori* (Belidae; O'Brien
and Tang, 2015), *Craugastor bransfordii*, *Pristimantis altae*, and *P. talamancae*
(Craugastoridae), *Micrurus multifasciatus* (Elapidae), *Dendropsophus phlebodes*
(Hylidae; Figure 6.15), and *Zeledonia coronata* (Parulidae).

Vegetation

It consists of different types of forests and grasslands with pines (*Pinus caribaea*;
Dinerstein et al., 1995).

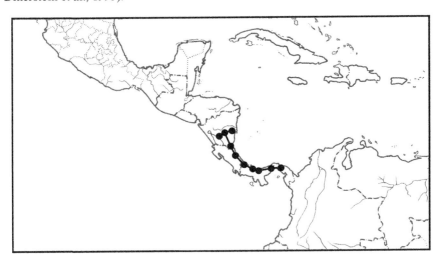

Figure 6.15 Map with the individual track of *Dendropsophus phlebodes* (Hylidae) in the
Guatuso–Talamanca province.

Biotic Relationships

Müller (1973) considered this province to be related to the Chiapas Highlands, Chocó–Darién, Cuban, Jamaica, and Hispaniola provinces. A parsimony analysis of endemicity based on bird taxa (Porzecanski and Cracraft, 2005) postulated a close relationship of this province with the Pantepui, Páramo, and Puna provinces.

PUNTARENAS–CHIRIQUÍ PROVINCE

Puntarenas–Chiriquí province—Ryan, 1963: 30 (regional.); Morrone, 2014b: 54 (regional.).
Pacific Costa Rica–Panama province—Stuart, 1964: 349 (regional.).
Azuero Rainforest area—West, 1964: 368 (regional.).
Rainforest of Southwestern Costa Rica area—West, 1964: 368 (regional.).
Savanna of Central Panama area—West, 1964: 368 (regional.).
Chiapas–Nicaraguan province—Miller, 1966: 781 (regional.); Smith and Bermingham, 2005: 1843 (cluster anal. and regional.).
Isthmian province (in part)—Miller, 1966: 782 (regional.).
Golfo Dulcean province—Savage, 1966: 736 (biotic evol.).
Chiriquí subcenter—Müller, 1973: 24 (regional.).
Pacific Southern Middle America area—Haffer, 1985: 120 (refugia).
Costa Rican Seasonal Moist Forests ecoregion—Dinerstein et al., 1995: 88 (ecoreg.).
Isthmian–Pacific Moist Forests ecoregion—Dinerstein et al., 1995: 88 (ecoreg.).
Panamanian Dry Forests ecoregion—Dinerstein et al., 1995: 94 (ecoreg.).
Northern subregion—Rangel et al., 1995d: 123 (veget.).
Central American province (in part)—Brown et al., 1998: 31 (veget.); Huber and Riina, 2003: 99 (glossary).
Costa Rica and Western Panama province—Campbell, 1999: 116 (regional.).
Western Central American Pacific Lowlands area—Flores-Villela and Goyenechea, 2001: 174 (clad. biogeogr.).
Western Panamanian Isthmus province—Morrone, 2001a: 48 (regional.), 2001e: 53 (regional.); Morrone and Márquez, 2001: 637 (track anal.); Corona and Morrone, 2005: 38 (track anal.); Morrone, 2006: 479 (regional.); Quijano-Abril et al., 2006: 1269 (track anal.); Escalante et al., 2011: 32 (track anal.); Mercado-Salas et al., 2012: 459 (track anal.).
Chiriquí province—Smith and Bermingham, 2005: 1843 (cluster anal. and regional.).
Santa Marta province—Smith and Bermingham, 2005: 1843 (cluster anal. and regional.).
Santa María ecoregion—Abell et al., 2008: 408 (ecoreg.).
Western Lowlands area (in part)—Flores-Villela and Martínez-Salazar, 2009: 820 (clad. biogeogr.).
Panamanian–Costa Rican province (in part)—Rivas-Martínez et al., 2011: 26 (regional.).

Definition

The Puntarenas–Chiriquí province comprises western Central America, from Costa Rica to western Panama (Morrone, 2001e, 2006, 2014b). Kohlmann and

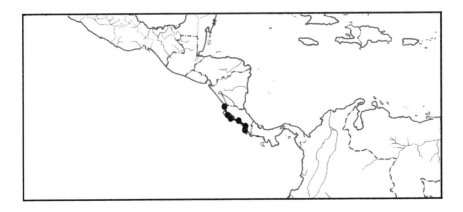

Figure 6.16 Map with the individual track of *Tikuna atramentum* (Leptophlebiidae) in the Puntarenas–Chiriquí province.

Wilkinson (2007) found a clear boundary between the northern portion of this province and the Pacific Lowlands province. It is situated at the Grande de Tárcoles River and is coincident with the Costa Rica–Panama microplate boundary.

Endemic and Characteristic Taxa

Morrone (2014b) provided a list of endemic and characteristic taxa. Some examples include *Tikuna atramentum* (Leptophlebiidae; Figure 6.16), *Hemiphileurus jamesoni, H. youngi,* and *Onthophagus orphnoides* (Scarabaeidae), *Gansia obscura, Homalolinus gracilis, H. sharpi,* and *Misantlius gebieni* (Staphylinidae), and *Poeciliopsis paucimaculata* and *P. retropinna* (Poeciliidae).

Vegetation

There are tropical humid forests and seasonally dry tropical forests (Dinerstein et al., 1995).

Biotic Relationships

Müller (1973) considered that this province was related to the Veracruzan and Chocó–Darién provinces. According to a track analysis based on species of Coleoptera (Morrone and Márquez, 2001), this province is related to the Pacific Lowlands, Veracruzan, and Chiapas Highlands provinces.

CHOCÓ–DARIÉN PROVINCE

Chocó–Darién province—Ryan, 1963: 31 (regional.); Morrone, 2014b: 55 (regional.).
Chocoan province—Savage, 1966: 736 (biotic evol.).

Chocó subcenter—Müller, 1973: 38 (regional.).

Colombian Pacific center—Müller, 1973: 38 (regional.).

Northern Pacific province—Ringuelet, 1975: 107 (regional.).

Colombian Coastal province—Udvardy, 1975: 41 (regional.); Huber and Riina, 1997: 130 (glossary).

Panamanian province—Udvardy, 1975: 41 (regional.).

Chocó Rainforest center (in part)—Cracraft, 1985: 53 (areas of endem.).

Pacific Colombia (Chocó) area—Haffer, 1985: 121 (refugia).

Chocó area—Cracraft, 1988: 223 (clad. biogeogr.); Bates et al., 1998: 785 (PAE); Ippi and Flores, 2001: 54 (PAE); Marks et al., 2002: 155 (biotic evol.).

Chocó–Magdalena province (in part)—Hernández et al., 1992a: 118 (regional.).

Chocó sector—Hernández et al., 1992a: 118 (regional.).

Chocó/Darién Moist Forests ecoregion—Dinerstein et al., 1995: 91 (ecoreg.); Huber and Riina, 1997: 65 (glossary).

Eastern Panamanian Montane Forests ecoregion—Dinerstein et al., 1995: 91 (ecoreg.); Huber and Riina, 1997: 75 (glossary).

Pacific Coast region—Rangel et al., 1995a: 21 (veget.), 1995d: 121 (veget.).

Colombian province (in part)—Rivas-Martínez and Navarro, 1994: map (regional.).

Chocó province—Morrone, 1999: 5 (regional.), 2001a: 55 (regional.), 2001e: 58 (regional.), 2006: 479 (regional.); Quijano-Abril et al., 2006: 1269 (track anal.); Mercado-Salas et al., 2012: 459 (track anal.).

Chocó Lowlands area—Porzecanski and Cracraft, 2005: 266 (PAE).

Tiura province—Smith and Bermingham, 2005: 1843 (cluster anal. and regional.).

Río Tiura ecoregion—Abell et al., 2008: 408 (ecoreg.).

Colombian–Pacific province—Rivas-Martínez et al., 2011: 26 (regional.).

Definition

The Chocó–Darién province comprises the Pacific coast of northern Ecuador, western Colombia, and eastern lowlands of Panama (Cracraft, 1985; Rangel et al., 1995a, 1995d; Morrone, 2001e, 2006, 2014b; Rangel, 2004a). The Chocó–Darién province is considered to harbor one of the world's richest biotas, with exceptional richness and endemism in several taxa (Dinerstein et al., 1995).

Endemic and Characteristic Taxa

Morrone (2014b) provided a list of endemic and characteristic taxa. Some examples include *Wettinia panamensis* (Arecaceae), *Aristolochia colossifolia* and *A. trianae* (Aristolochiaceae; Figure 6.17a), *Dicranopygium arusisense*, *D. odoratum*, and *D. trilobulata* (Cyclanthaceae), *Ananteris gorgonae* (Buthidae), *Geotrygon goldmani*, *Patagioenas goodsoni*, and *P. subvinacea berlepschi* (Columbidae), *Micrurus ancoralis* (Elapidae), *Galbula ruficauda melanogenia* (Galbulidae), *Platyrrhinus chocoensis* (Phyllostomatidae), *Androdon aequatorialis* (Figure 6.17b), and *Coeligena wilsoni* and *Thalurania fannyi* (Trochilidae).

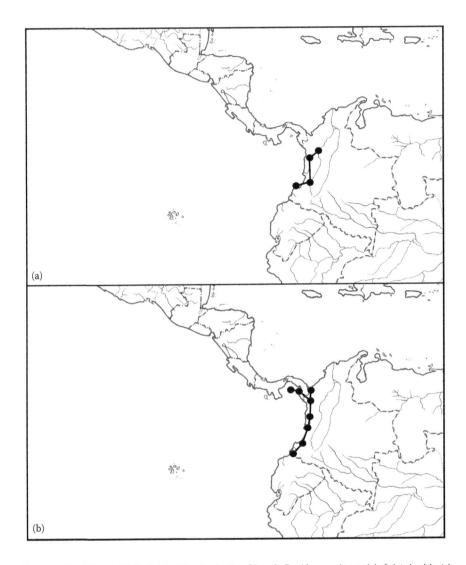

Figure 6.17 Maps with individual tracks in the Chocó–Darién province. (a) *Aristolochia tri-anae* (Aristolochiaceae); (b) *Androdon aequatorialis* (Trochilidae).

Vegetation

There are moist forests; there are cloud forests at the top of the Serranía de San Blas, Darién, Maje, and Pirre in central and eastern Panama (Dinerstein et al., 1995). Dominant plant species include *Anacardium excelsum, Avicennia* spp., *Brunnellia* spp., *Castilla elastica, Cavanillesia platanifolia, Chrysophyllum* spp., *Elaegia uti-lis, Erythrina fusca, Inga* spp., *Mora megistosperma, Panicum* spp., *Polygonum acuminatum*, *Prioria copaifera*, *Raphia taedigera*, *Rhyzophora harrisoni*, *R. mangle*, and *Sorocea* spp. (Rangel et al., 1995a; Rangel, 2004b).

Biotic Relationships

According to a parsimony analysis of endemicity based on anurans (Ron, 2000), the Chocó–Darién province is closely related to the biogeographic provinces of Central America.

Regionalization

Hernández et al. (1992a) identified districts in the Colombian portion of this province and Rangel et al. (1995d) identified subregions, which were treated by Morrone (2014b) as three subprovinces and 12 districts. The Pacific Basin subprovince corresponds to the littoral areas of the province to 500 meters (Rangel et al., 1995d). Within this province, Hernández et al. (1992a) described the Aspavé–El Limón–Pirre, Baudó, Gorgona, Juradó, Micay, Tumaco, and Utria districts. The Central subprovince corresponds to the central part of the province (Rangel et al., 1995d); within it, Hernández et al. (1992a) described the Murrí and Sucio River districts. The Highlands subprovince corresponds to the eastern highlands of the province (Rangel et al., 1995); within it, Hernández et al. (1992a) described the Acandí–San Blas, Alto Atrato–San Juan, and Tacarcuna districts.

GUAJIRA PROVINCE

Guajira province—Cabrera and Willink, 1973: 46 (regional.); Huber and Riina, 1997: 273 (glossary); Morrone, 2014b: 56 (regional.); del Río et al., 2015: 1294 (track anal. and clad. biogeogr.).
Caribbean center—Müller, 1973: 57 (regional.).
Caribe–Guajira Subequatorial dominion—Ab'Sáber, 1977: map (climate).
Guajiran center—Cracraft, 1985: 57 (areas of endem.).
Meridan Montane center—Cracraft, 1985: 58 (areas of endem.).
Colombia and Northern Venezuela subregion—Samek et al., 1988: 32 (regional.); Huber and Riina, 2003: 316 (glossary).
Guajira/Barranquilla Xeric Scrub ecoregion—Dinerstein et al., 1995: 105 (ecoreg.).
Paraguaná Xeric Scrub ecoregion—Dinerstein et al., 1995: 105 (ecoreg.).
Guajira subregion—Rangel et al., 1995c: 21 (veget.).
Maracaibo ecoregion—Abell et al., 2008: 408 (ecoreg.).
Guajiran–Caribbean province (in part)—Rivas-Martínez et al., 2011: 26 (regional.).

Definition

The Guajira province comprises the arid coastal areas of northern Colombia and northwestern Venezuela (Cabrera and Willink, 1973; Cracraft, 1985; Rangel et al., 1995b, 1995c; Morrone, 2001e, 2006, 2014b).

Endemic and Characteristic Taxa

Morrone (2014b) provided a list of endemic and characteristic taxa. Some examples include *Crotalaria vitellina* and *Prosopis juliflora* (Fabaceae), *Passiflora bracteosa* and *P. schlimiana* (Passifloraceae), *Condalia henriquezii* (Rhamnaceae), *Schendylops colombianus* (Schendylidae), *Patagioenas corensis* (Columbidae), *Micrurus dissoleucus nigrirostris* (Elapidae; Figure 6.18), and *Anisognathus melanogenys* and *Conirostrum rufum* (Thraupidae).

Vegetation

There are tropical humid forests; xerophytic scrublands with small trees, shrubs, and columnar cacti; and swamps caused by periodic floods (Cabrera and Willink, 1973; Dinerstein et al., 1995). Most frequent plant species include *Acacia tortuosa, Astronium graveolens, Azorella julianii, Batis maritima, Bursera glabra, Caesalpinia coriacea, Calamagrostis effusa, Castela erecta, Cercidium praecox, Cereus margaritensis, Erythrina velutina, Haematoxylon brasiletto, Libidibia coraria, Lippia* sp., *Lonchocarpus punctatus, Melochia* sp., *Oritrophium peruvianum, Phyloxerus vermicularis, Pithecelobium ligustrinum,* and *Prosopis juliflora* (Cabrera and Willink, 1973; Rangel et al., 1995c). There are mangroves with *Rhyzophora mangle, Avicennia germinas, Laguncularia racemosa,* and *Conocarpus erecta* on the coast of the Caribbean Sea (Reyes and Campos, 1992).

Biotic Relationships

Müller (1973) considered that the Guajira province is related to the Magdalena province.

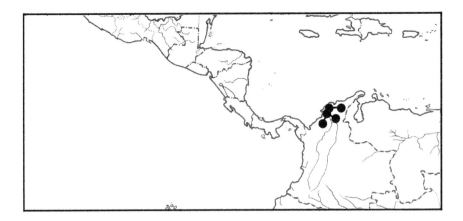

Figure 6.18   Map with the individual track of *Micrurus dissoleucus nigrirostris* (Elapidae) in the Guajira province.

Regionalization

Müller (1973) and Hernández et al. (1992a) identified 15 nested units, which were treated by Morrone (2014b) as districts. These are the Alta Guajira, Aracataca, Ariguaní–César, Baja Guajira and Alto César, Barranquilla, Caracolicito, Cartagena, Chundúa, Guachaca, Macuira, Maracaibo, María and Piojó Hills, Marocaso, Santa Marta, and Sierra Nevada districts.

VENEZUELAN PROVINCE

Venezuelan province—Cabrera and Willink, 1973: 56 (regional.); Huber and Riina, 1997: 275 (glossary); Morrone, 2014b: 58 (regional.); del Río et al., 2015: 1294 (track anal. and clad. biogeogr.).
Venezuelan Coastal Forest center—Müller, 1973: 54 (regional.).
Venezuelan Montane Forest center—Müller, 1973: 56 (regional.).
Caribbean Coast province—Ringuelet, 1975: 107 (regional.).
Venezuelan Dry Forest province (in part)—Udvardy, 1975: 41 (regional.).
Parian Montane center—Cracraft, 1985: 57 (areas of endem.).
Venezuelan Montane center—Cracraft, 1985: 58 (areas of endem.).
Septentrional Venezuelan province (in part)—Rivas-Martínez and Navarro, 1994: map (regional.).
Araya and Paría Xeric Scrub ecoregion—Dinerstein et al., 1995: 105 (ecoreg.).
Aruba/Curaçao/Bonaire Cactus Scrub ecoregion—Dinerstein et al., 1995: 105 (ecoreg.).
Cordillera La Costa Montane Forests ecoregion—Dinerstein et al., 1995: 88 (ecoreg.).
La Costa Xeric Shrublands ecoregion—Dinerstein et al., 1995: 105 (ecoreg.).
Lara/Falcón Dry Forests ecoregion—Dinerstein et al., 1995: 95 (ecoreg.); Huber and Riina, 1997: 94 (glossary).
Paraguaná Restingas ecoregion—Dinerstein et al., 1995: 105 (ecoreg.).
Venezuelan Coast province—Morrone, 1999: 4 (regional.), 2001a: 58 (regional.), 2001e: 58 (regional.); Viloria, 2005: 449 (regional.); Morrone, 2006: 479 (regional.); Quijano-Abril et al., 2006: 1270 (track anal.); Escalante et al., 2011: 32 (track anal.).
South America Caribbean Drainages–Trinidad ecoregion (in part)—Abell et al., 2008: 408 (ecoreg.).
Guajiran–Caribbean province (in part)—Rivas-Martínez et al., 2011: 26 (regional.).

Definition

The Venezuelan province comprises northern Venezuela and Colombia, including the islands of Aruba, Curaçao, and Bonaire (Morrone, 2001e, 2006, 2014b). Montane forests, found on several isolated mountains near the coast harbor biotas, have been long isolated from one another by drier habitats in the surrounding lowlands (Dinerstein et al., 1995). The climate is tropical, with a mean annual precipitation of 1,200–1,800 millimeters and a temperature of 24–28°C (Cabrera and Willink, 1973).

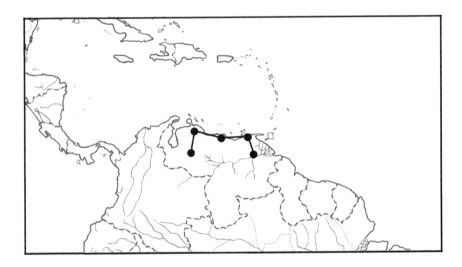

Figure 6.19 Map with the individual track of *Clivina oblita* (Carabidae) in the Venezuelan province.

Endemic and Characteristic Taxa

Morrone (2014b) provided a list of endemic and characteristic taxa. Some examples include *Montanoa fragans* (Asteraceae), *Piper schlimii* and *P. sierra-aroense* (Piperaceae), *Clivina oblita* (Carabidae; Figure 6.19), *Rhinacloa puertoricensis* (Miridae), *Rhipidita brevicornis* and *R. vespertilio* (Sciaridae), *Pipreola formosa rubidior* (Cotingidae), *Proechimys guairae* and *P. urichi* (Echimyidae), and *Aulacorhynchus sulcatus erythrognathus* (Ramphastidae).

Vegetation

There are tropical humid forests and seasonally dry tropical forests, mixed with xerophytic scrublands and savannas. Characteristic plant species include *Cedrela mexicana*, *Swietenia macrophylla*, *Astronium graveolens*, *Hymenaea courbaril*, *Tabebuia* spp., *Erythrina* spp., and *Platamiscium pinnatum* (Cabrera and Willink, 1973). On the coast, there are dunes known as *restingas* (Dinerstein et al., 1995).

Biotic Relationships

Müller (1973) considered that the Venezuelan province was related to the Cauca and Pantepui provinces.

TRINIDAD PROVINCE

Trinidad province—Ringuelet, 1975: 107 (regional.); Morrone, 2014b: 59 (regional.).
Trinidad and Tobago Dry Forests ecoregion—Dinerstein et al., 1995: 95 (ecoreg.).

Trinidad and Tobago Moist Forests ecoregion—Dinerstein et al., 1995: 88 (ecoreg.).
Trinidad and Tobago province—Morrone, 1999: 4 (regional.), 2001a: 59 (regional.),
   2001e: 58 (regional.), 2006: 479 (regional.).
South America Caribbean Drainages–Trinidad ecoregion (in part)—Abell et al., 2008:
   408 (ecoreg.).

Definition

The Trinidad province comprises the islands of Trinidad and Tobago (Morrone, 2001e, 2006, 2014b).

Endemic and Characteristic Taxa

Morrone (2014b) provided a list of endemic and characteristic taxa. Some examples include *Nexophallus popei* (Coccinellidae), *Sicoderus propinquus* (Curculionidae), *Protoptila ignera* (Glossosomatidae), and *Proechimys trinitatus* (Echimyidae).

Vegetation

It consists of tropical humid forests and seasonally dry tropical forests (Dinerstein et al., 1995).

Biotic Relationships

Müller (1973) considered this province to be related to the Guianan Lowlands provinces. Samek et al. (1988) and Borhidi (1996) considered Trinidad and Tobago as part of the Venezuelan–Colombia Caribbean subregion. A parsimony analysis of endemicity based on species of Orchidaceae (Trejo-Torres and Ackerman, 2001) found that Trinidad was related to the Greater Antilles and Tobago with the Lesser Antilles–Virgin Islands.

MAGDALENA PROVINCE

Magdalena center—Müller, 1973: 32 (regional.); Cracraft, 1985: 56 (areas of endem.).
Magdalena dominion—Ringuelet, 1975: 107 (regional.).
Perijan Montane center—Cracraft, 1985: 59 (areas of endem.).
Cauca–Magdalena (Nechí) area (in part)—Haffer, 1985: 121 (refugia).
Chocó–Magdalena province (in part)—Hernández et al., 1992a: 118 (regional.).
Magdalena sector—Hernández et al., 1992a: 125 (regional.).
Colombian province (in part)—Rivas-Martínez and Navarro, 1994: map (regional.).
Catatumbo Moist Forests ecoregion—Dinerstein et al., 1995: 92 (ecoreg.); Huber and
   Riina, 1997: 65 (glossary).
Cordillera Oriental Montane Forests ecoregion—Dinerstein et al., 1995: 92 (ecoreg.).
Magdalena/Urabá Moist Forests ecoregion—Dinerstein et al., 1995: 92 (ecoreg.);
   Huber and Riina, 1997: 65 (glossary).

Magdalena Valley Dry Forests ecoregion—Dinerstein et al., 1995: 95 (ecoreg.); Huber
   and Riina, 1997: 95 (glossary).
Magdalena Valley Montane Forests ecoregion—Dinerstein et al., 1995: 92 (ecoreg.);
   Huber and Riina, 1997: 75 (glossary).
Venezuelan Andes Montane Forests ecoregion—Dinerstein et al., 1995: 92 (ecoreg.);
   Huber and Riina, 1997: 75 (glossary).
Magdalena province—Morrone, 1999: 4 (regional.), 2001a: 60 (regional.), 2001e: 60
   (regional.), 2006: 479 (regional.); Quijano-Abril et al., 2006: 1269 (track anal.);
   Escalante et al., 2011: 32 (track anal.); Mercado-Salas et al., 2012: 459 (track anal.);
   Morrone, 2014b: 59 (regional.).
Magdalena Valley province—Morrone, 1999: 5 (regional.).
Magdalena–Sinu ecoregion—Abell et al., 2008: 408 (ecoreg.).
Colombian–Andean province—Rivas-Martínez et al., 2011: 26 (regional.).

Definition

   The Magdalena province comprises western Venezuela and northwestern
Colombia (Dinerstein et al., 1995; Morrone, 2001e, 2006, 2014b).

Endemic and Characteristic Taxa

   Morrone (2014b) provided a list of endemic and characteristic taxa. Some
examples include *Plagiogyria semicordata* (Plagiogyriaceae), *Gunnera saint-johnii*
(Gunneraceae), *Bulnesia carrapo* (Zygophyllaceae), *Eutimesius ornatus* and *Phareus
raptator* (Stygnidae), *Columbina passerina parvula* (Columbidae), *Proechimys magdalenae* (Echimyidae; Figure 6.20), and *Sciurus granatensis norosiensis* (Sciuridae).

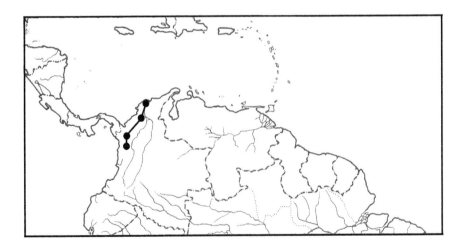

Figure 6.20 Map with the individual track of *Proechimys magdalenae* (Echymidae) in the
Magdalena province.

Vegetation

There are tropical humid forests and seasonally dry tropical forests (Dinerstein et al., 1995).

Biotic Relationships

Müller (1973) considered this province to be related to the Guajira province.

Regionalization

Hernández et al. (1992a) identified nine districts in the Colombian portion of this province. These are Barbacoas, Carare, Catatumbo, La Gloria, Lebrija, Magdalena Delta, Nechí, Sinú–San Jorge, and Turbo districts (Morrone, 2014b).

SABANA PROVINCE

Sabana district—Orfila, 1941: 86 (regional.).
Caquetio province (in part)—Fittkau, 1969: 642 (regional.).
Sabana province—Cabrera and Willink, 1973: 63 (regional.); Huber and Riina, 1997: 271 (glossary); Morrone, 2014b: 60 (regional.).
Orinoquia province—Ringuelet, 1975: 107 (regional.); Hernández et al., 1992a: 129 (regional.).
Llanos province—Udvardy, 1975: 42 (regional.); Huber and Alarcón, 1988: map (regional.); Rivas-Martínez and Navarro, 1994: map (regional.); Huber and Riina, 1997: 203 (glossary); Morrone, 1999: 5 (regional.).
Llanos de Orinoco dominion (in part)—Ab'Sáber, 1977: map (climate).
Septentrional Venezuelan province (in part)—Rivas-Martínez and Navarro, 1994: map (regional.).
Llanos ecoregion—Dinerstein et al., 1995: 98 (ecoreg.).
Llanos Dry Forests ecoregion—Dinerstein et al., 1995: 95 (ecoreg.); Huber and Riina, 1997: 94 (glossary).
Orinoco Wetlands ecoregion—Dinerstein et al., 1995: 100 (ecoreg.); Huber and Riina, 1997: 193 (glossary).
Venezuelan Llanos province—Morrone, 2001a: 60 (regional.), 2001e: 61 (regional.); Viloria, 2005: 449 (regional.); Morrone, 2006: 479 (regional.); Escalante et al., 2011: 32 (track anal.); Mercado-Salas et al., 2012: 459 (track anal.).
Northern South America area—Porzecanski and Cracraft, 2005: 266 (PAE).
Orinoco Llanos ecoregion—Abell et al., 2008: 408 (ecoreg.).
Guaviarean–Orinoquian province—Rivas-Martínez et al., 2011: 26 (regional.).
Llaneran province—Rivas-Martínez et al., 2011: 26 (regional.).

Definition

The Sabana province comprises the plains of a great part of Venezuela and north-western Colombia (Cabrera and Willink, 1973; Morrone, 2001e, 2006, 2014b).

Endemic and Characteristic Taxa

Morrone (2014b) provided a list of endemic and characteristic taxa. Some examples include *Centrosema tetragonolobum* (Fabaceae), *Opsiphanes quiteria phylas* (Brassolidae), *Parides eurymedes agathokles* and *Protrographium dioxippus marae* (Papilionidae), *Forsteria venezuelensis* (Trichodactylidae; Figure 6.21), and *Phelpsia inornata* (Tyrannidae).

Vegetation

Mainly savannas, there are also tropical humid forests, seasonally dry tropical forests, gallery forests, grasslands, and wetlands. The Sabana or Llanos represents the largest savanna ecosystem in northern South America (Dinerstein et al., 1995). Dominant plant species include *Acrocomia sclerocarpa*, *Byrsonina coriacea*, *Copernicia tectorum*, *Curatella americana*, *Hemicrepidospermum rhoifolium*, *Ocotea* sp., and *Pourouma guianensis*, as well as several species of Poaceae of the genera *Andropogon*, *Gymnopogon*, *Imperata*, *Leptocoryphium*, *Panichum*, *Paspalum*, and *Trachypogon* (Cabrera and Willink, 1973).

Biotic Relationships

Müller (1973) considered this province to be related to the Roraima, Cerrado, and Caatinga provinces.

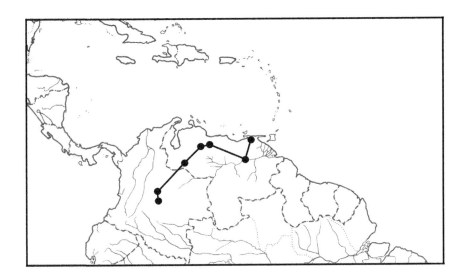

Figure 6.21   Map with the individual track of *Forsteria venezuelensis* (Trichodactylidae) in the Sabana province.

Regionalization

Hernández et al. (1992a) identified six districts in the Colombian portion of this province: Arauca–Apure, Casanare, Maipures, Piedemonte Casanare–Arauca, Piedemonte Meta, and Sabanas Altas (Morrone, 2014b).

CAUCA PROVINCE

Cauca center—Müller, 1973: 33 (regional.).
Colombian Montane Forest center—Müller, 1973: 34 (regional.).
East Andean subcenter—Müller, 1973: 35 (regional.).
West Andean subcenter—Müller, 1973: 35 (regional.).
Northern Andes province—Udvardy, 1975: 42 (regional.).
Amotape unit—Lamas, 1982: 345 (regional.).
Ayabaca unit—Lamas, 1982: 354 (regional.).
El Caucho unit—Lamas, 1982: 343 (regional.).
Huancabamba unit—Lamas, 1982: 349 (regional.).
Marañón unit (in part)—Lamas, 1982: 345 (regional.).
Marañón center—Cracraft, 1985: 68 (areas of endem.).
Cauca–Magdalena (Nechí) area (in part)—Haffer, 1985: 121 (refugia).
Cauca Valley Dry Forests ecoregion—Dinerstein et al., 1995: 95 (ecoreg.); Huber and Riina, 1997: 95 (glossary).
Cauca Valley Montane Forests ecoregion—Dinerstein et al., 1995: 92 (ecoreg.); Huber and Riina, 1997: 75 (glossary).
Guayaquil Flooded Grasslands ecoregion—Dinerstein et al., 1995: 101 (ecoreg.).
Marañón Dry Forests ecoregion—Dinerstein et al., 1995: 95 (ecoreg.); Huber and Riina, 1997: 94 (glossary).
Northwestern Andean Montane Forests ecoregion—Dinerstein et al., 1995: 91 (ecoreg.); Huber and Riina, 1997: 75 (glossary).
Cauca province—Morrone, 1999: 5 (regional.), 2001a: 62 (regional.), 2001e: 59 (regional.), 2006: 479 (regional.); Quijano-Abril et al., 2006: 1270 (track anal.); Mercado-Salas et al., 2012: 459 (track anal.); Morrone, 2014b: 61 (regional.); del Río et al., 2015: 1294 (track anal. and clad. biogeogr.).
Amazonas High Andes ecoregion (in part)—Abell et al., 2008: 408 (ecoreg.).
Guayaquilean–Ecuadorian province (in part)—Rivas-Martínez et al., 2011: 26 (regional.).

Definition

The Cauca province comprises western Colombia (between the West and Central Cordilleras), Ecuador, and northern Peru (Morrone, 2001e, 2006, 2014b). Submontane and montane forests of this province are exceptionally rich in species and have a high level of endemism, which have been favored by the complex topography, climate, geology, and biogeographic history of the northern Andes (Dinerstein et al., 1995).

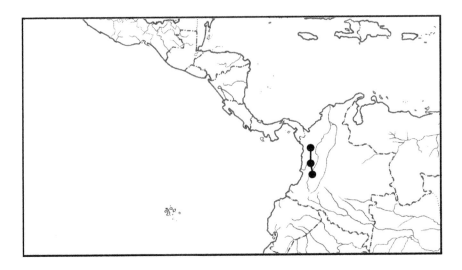

Figure 6.22   Map with the individual track of *Cecropia megastachya* (Cecropiaceae) in the Cauca province.

Endemic and Characteristic Taxa

Morrone (2014b) provided a list of endemic and characteristic taxa. Some examples include *Matisia longiflora* (Bombacaceae), *Cecropia megastachya* (Cecropiaceae; Figure 6.22), *Fuchsia dependens* and *F. loxenis* species groups (Onagraceae), *Peucetia cauca* (Oxyopidae), *Pompilocalus ruminahui* (Pompilidae), *Caenolestes condorensis* (Caenolestidae), and *Aulacorhynchus griseigularis* (Ramphastidae).

Vegetation

It consists of tropical humid forests, seasonally dry tropical forests, and montane scrubland (Dinerstein et al., 1995; Valencia et al., 1999).

Biotic Relationships

Müller (1973) suggested a close relationship between the Cauca province and the Yungas and Pantepui provinces.

GALAPAGOS ISLANDS PROVINCE

Galapagos Islands subregion—Mello-Leitão, 1937: 240 (regional.).
Galapagos Islands province—Cabrera and Willink, 1973: 46 (regional.); Udvardy, 1975: 42 (regional.); Rivas-Martínez and Navarro, 1994: map (regional.); Morrone, 1999: 5 (regional.), 2001a: 63 (regional.), 2001e: 62 (regional.), 2006: 479 (regional.), 2014b: 62 (regional.); del Río et al., 2015: 1294 (track anal. and clad. biogeogr.).

Galapagos center—Müller, 1973: 106 (regional.).

Galapageian province—Takhtajan, 1986: 251 (regional.); Huber and Riina, 1997: 182
    (glossary).

Galapagos Islands Xeric Scrub ecoregion—Dinerstein et al., 1995: 105 (ecoreg.).

Galapagos province—Huber and Riina, 1997: 271 (glossary).

Galapagos Islands ecoregion—Abell et al., 2008: 408 (ecoreg.).

Insular Galapagos province—Rivas-Martínez et al., 2011: 26 (regional.).

Definition

The Galapagos Islands province corresponds to the archipelago of Colón
(Figure 6.23), which is situated in the Pacific Ocean, 950 kilometers west of the coast
of Ecuador. It is comprised of 16 major islands (Pinta, Marchena, Tower, Fernandina,
Isabela, San Salvador, Daphne Major, Seymour, Baltra, Rabida, Pinzon, Santa Cruz,
Santa Fe, San Cristóbal, Floreana, and Española) and several smaller islands (Kuschel,
1961; Johnson and Raven, 1973; Peck and Kukalová-Peck, 1990; Bisconti et al., 2001;
Parent et al., 2008). The archipelago has a total land area of 7,880 square kilometers;

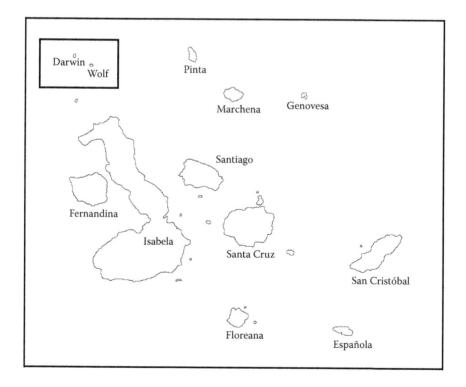

Figure 6.23   Map of the Galapagos Islands.

the largest island is Isabela, one of the two youngest, making up more than half the total land area at 4,855 square kilometers (Neall and Trewick, 2008). With the exception of Isabela that comprises six shield volcanoes, all the other islands represent single shield volcanoes in various stages of erosion. The westernmost islands, Fernandina and Isabela, represent the current location of the Galapagos hot spot. The oldest exposed rocks date from only 2 mya, but seamounts on the Cocos Ridge (on the Cocos plate) and the Carnegie Ridge (on the Nazca plate) extend back to 14 mya, and may have been subaerial islands (Neall and Trewick, 2008).

Ali and Aitchison (2014) provided a series of paleogeographic maps for the archipelago for the last 700,000 years, accommodating thermal subsidence of the islands, eustatic sea-level changes, and seafloor loading. According to these maps, in the recent geological past, a sizeable region within the archipelago must have experienced significant changes in its geography due to major shifts in sea level, periodically connecting and then isolating the islands. Several large central and western islands regularly connected for periods of 5,000–10,000 years before becoming isolated for ca. 90,000 years. Ali and Aitchison (2014) concluded that their "oscillating geography forcing mechanism" is amenable to testing using lineages dated with molecular clocks.

Endemic and Characteristic Taxa

Morrone (2014b) provided a list of endemic and characteristic taxa. Some examples include *Darwiniothamnus alternifolius*, *D. lancifolius*, *Scalesia affinis*, *S. aspera*, *S. atractyloides*, *S. peduncolata*, and *S. villosa* (Asteraceae), *Tiquilia darwini*, *T. fusca*, and *T. galapagoa* (Boraginaceae), *Nolana galapagensis* (Solanaceae), *Ormiscus variegatus* (Anthribidae), *Bembidion galapagoensis*, *Calosoma galapageium*, and *Selenophorus obscuricornis* (Carabidae), *Anchonus galapagoensis*, *Galapaganus ashlocki*, *G. darwini*, *G. galapagoensis*, *Neopentarthrum mutchleri*, and *Pagiocerus frontalis* (Curculionidae), *Camillina galapagoensis*, *Neozimiris pinzon*, *Poecilochroa bifasciata*, *Trachyzelotes kulckzynskii*, and *Zelotes reformans* (Gnaphosidae), *Buteo galapagoensis* (Accipitridae), *Pseudalsophis biserialis*, *P. dorsalis*, and *P. slevini* (Dipsadidae), *Nesomimus macdonaldi* and *N. trifasciatus* (Mimidae), *Spheniscus mendiculus* (Spheniscidae), *Chelonoidis nigra* (Testudinidae), and *Conolophus pallidus*, *Microlophus albermarlensis*, *M. grayii*, and *Tropidurus albermarlensis* (Tropiduridae).

Vegetation

There are five basic vegetation zones (Figure 6.24; Cabrera and Willink, 1973; Johnson and Raven, 1973; Peck, 1991; Peck and Kukalová-Peck, 1995; Huber and Riina, 1997): littoral zone with mangroves of *Avicennia germinans*, *Conocarpus erecta*, *Laguncularia racemosa*, and *Rhizophora mangle;* xeric scrubland, arid zone or arid coastal zone, from the littoral to 120–180 meters altitude, with *Acacia macracantha*, *A. rorudiana*, *Brachycereus nesioticus*, *Erythrina velutina*, *Jasminocereus thouarsii*, *Opuntia echios*, *O. galapageia*, *Parkinsonia aculeata*, *Prosopis juliflora*, and *Scutia pauciflora;* Scalesia zone, between 180–550 meters, with *Pisonia floribunda*, *Psidium galapageium*, *Scalesia cordata*, *S. microcephala*,

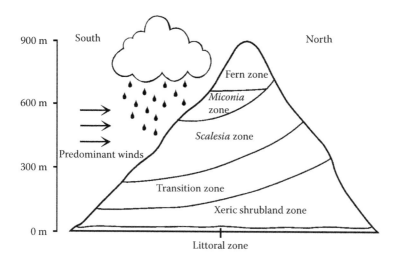

Figure 6.24   Vegetation zones in the Galapagos Islands. (Modified from Peck and Kukalová-Peck, 1990.)

and *S. pedunculata; Miconia* or moist zone, the highest and most humid, with *Darwiniothamnus tenuifolius*, *Miconia robinsoniana*, and *Scalesia microcephala*; and fern, fern sedge or grassy zone: on the highest tops, with ferns and Cyperaceae.

Biotic Relationships

It has been commonly assumed that the Galapagos biota is basically Neotropical, with Antillean, Mexican, Central American, and South American biotic relationships (Cabrera and Willink, 1973; Peck and Kukalová-Peck, 1990; Bisconti et al., 2001; Parent et al., 2008). Kuschel (1961) considered that the connections with Mexico, Central America, and the Antilles were due more than anything to the lack of data between Panama and northern Peru, and that the relationships of the Galapagos Islands were with the Ecuadorian coast. The track of the genus *Galapaganus* (Figure 6.25) shows the relationship of the Galapagos with the coasts of Ecuador and northern Peru (Lanteri, 1992). The cladogram of *Galapaganus* allowed Lanteri (1992, 2001) to infer two or three dispersal events from South America to the archipelago.

Track analyses (Craw et al., 1999; Grehan, 2001) have postulated that the Galapagos Islands constitute a node, where three generalized tracks intersect, connecting them with the Pacific coast of North America, the Antilles, and the Pacific coast of Ecuador. This would be correlated with the tectonic placement of the islands near the union of the Nazca, Cocos, and Pacific plates, and the Panama fracture zone, being possible that the original Galapagos biota inhabited an island arc in the eastern Pacific. Other arguments that may corroborate this hypothesis are the analyses that show that iguanas (Wiles and Sarich, 1983) and weevils (Sequeira et al., 2000) from the islands are much older than the oldest island, which is four million years old.

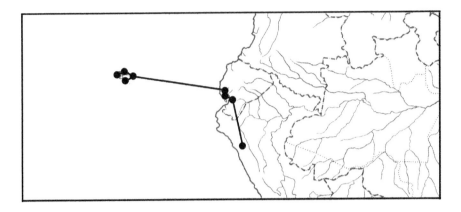

Figure 6.25 Map with the individual track of *Galapaganus* (Curculionidae) in the Galapagos Islands, Ecuador, and Peru.

WESTERN ECUADOR PROVINCE

Guayas province (in part)—Ringuelet, 1975: 107 (regional.).
Ecuadorian Dry Forest province (in part)—Udvardy, 1975: 41 (regional.); Huber and Riina, 1997: 155 (glossary).
Pacific Equatorial dominion (in part)—Ab'Sáber, 1977: map (climate).
Chocó Rainforest center (in part)—Cracraft, 1985: 53 (areas of endem.).
Tumbesan center (in part)—Cracraft, 1985: 66 (areas of endem.).
Ecuadorian province (in part)—Rivas-Martínez and Navarro, 1994: map (regional.).
Ecuadorian Dry Forests ecoregion—Dinerstein et al., 1995: 95 (ecoreg.); Huber and Riina, 1997: 95 (glossary).
Western Ecuador Moist Forests ecoregion—Dinerstein et al., 1995: 91 (ecoreg.).
Arid Ecuador province—Morrone, 1999: 6 (regional.), 2001a: 65 (regional.), 2006: 479 (regional.); Quijano-Abril et al., 2006: 1270 (track anal.).
Western Ecuador province—Morrone, 1999: 6 (regional.), 2001a: 64 (regional.), 2001e: 61 (regional.), 2006: 479 (regional.); Quijano-Abril et al., 2006: 1270 (track anal.); Morrone, 2014b: 63 (regional.); del Río et al., 2015: 1294 (track anal. and clad. biogeogr.); Daniel et al., 2016: 1169 (track anal. and regional.).
Dry Ecuador province—Morrone, 2001e: 62 (regional.).
North Andean Pacific Slopes–Río Atrato ecoregion (in part)—Abell et al., 2008: 408 (ecoreg.).
Guayaquilean–Ecuadorian province (in part)—Rivas-Martínez et al., 2011: 26 (regional.).

Definition

The Western Ecuador province comprises western Ecuador and southwestern Colombia (Morrone, 2001e, 2006, 2014b). Dry forests of the Western Ecuador province are known for their high levels of both regional and local endemism (Dinerstein et al., 1995).

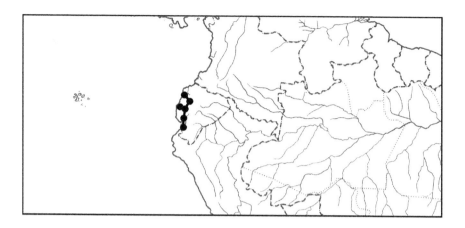

Figure 6.26 Map with the individual track of *Proechimys decumanus* (Echimyidae) in the Western Ecuador province.

Endemic and Characteristic Taxa

Morrone (2014b) provided a list of endemic and characteristic taxa. Some examples include *Cleistocactus sepium var. morleyanus*, *Echinopsis pachanoi*, *Opuntia cylindrica*, *O. soederstromiana*, and *O. pubescens* (Cactaceae), *Lithachne pauciflora* (Poaceae), *Galapaganus femoratus* species group (Curculionidae), *Proechimys decumanus* (Echimyidae; Figure 6.26), and *Gypopsitta pulchra* (Psittacidae).

Vegetation

There are tropical humid forests, seasonally dry tropical forests, xerophytic scrublands, and mangroves (Dinerstein et al., 1995; Cerón et al., 1999).

ECUADORIAN PROVINCE

Ecuadorian subcenter—Müller, 1973: 101 (regional.).
Guayas province (in part)—Ringuelet, 1975: 107 (regional.).
Ecuadorian Dry Forest province (in part)—Udvardy, 1975: 41 (regional.).
Illescas unit—Lamas, 1982: 352 (regional.).
Piura unit—Lamas, 1982: 352 (regional.).
Tumbesan center (in part)—Cracraft, 1985: 66 (areas of endem.).
Ecuadorian province—Rivas-Martínez and Navarro, 1994: map (regional.); Morrone, 2014b: 63 (regional.).
Tumbes/Piura Dry Forests ecoregion—Dinerstein et al., 1995: 95 (ecoreg.); Huber and Riina, 1997: 94 (glossary).
Tumbes–Piura province—Morrone, 1999: 6 (regional.), 2001a: 66 (regional.), 2001e: 63 (regional.), 2006: 479 (regional.).

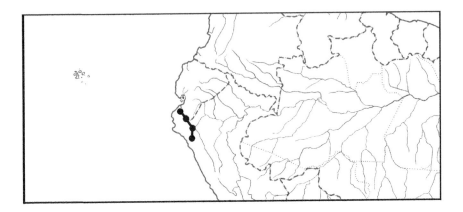

Figure 6.27 Map with the individual track of *Fuchsia ayavacensis* (Onagraceae) in the Ecuadorian province.

Equatorial Pacific area—Porzecanski and Cracraft, 2005: 266 (PAE).
North Andean Pacific Slopes–Río Atrato ecoregion (in part)—Abell et al., 2008: 408
   (ecoreg.).

Definition

The Ecuadorian province comprises southern Ecuador and northern Peru (Morrone, 2001e, 2006, 2014b). Dry forests of the Ecuadorian province are known for their high levels of both regional and local endemism (Dinerstein et al., 1995).

Endemic and Characteristic Taxa

Morrone (2014b) provided a list of endemic and characteristic taxa. Some examples include *Fuchsia ayavacensis* (Figure 6.27) and *F. scherffiana* (Onagraceae), *Stygnus mediocris* (Stygnidae), *Proechimys decumanus* species group (Echimyidae), and *Piezorhina cinerea* (Thraupidae).

Vegetation

There are seasonally dry tropical forests, xerophytic scrublands, and mangroves (Dinerstein et al., 1995; Cerón et al., 1999).

MESOAMERICAN DOMINION

Neomeridional subregion (in part)—Blyth, 1871: 427 (regional.).
Mexican subregion (in part)—Wallace, 1876: 78 (regional.); Heilprin, 1887: 80
   (regional.); Lydekker, 1896: 135 (regional.); Bartholomew et al., 1911: 9 (regional.);
   Mello-Leitão, 1937: 222 (regional.).

Central American zone (in part)—Engler, 1882: 345 (regional.).
Central American subarea (in part)—Clarke, 1892: 381 (regional.).
Central American subregion (in part)—Sclater and Sclater, 1899: 65 (regional.).
Central American center (in part)—Lane, 1943: 413 (regional.).
Caribbean region (in part)—Good, 1947: 232 (regional.); Takhtajan, 1986: 251
    (regional.); Samek et al., 1988: 26 (regional.); Rangel et al., 1995d: 21 (veget.);
    Huber and Riina, 1997: 119 (glossary), 2003: 97 (glossary); Procheş and Ramdhani,
    2012: 263 (cluster anal. and regional.).
Tropical Lowlands kingdom (in part)—West, 1964: 365 (regional.).
Mesoamerican region—Savage, 1966: 736 (biotic evol.); Ferrusquía-Villafranca, 1990:
    map (regional.).
Central American province (in part)—Fittkau, 1969: 642 (regional.); Udvardy, 1975:
    41 (regional.); Huber and Riina, 2003: 99 (glossary).
Mesoamerican subregion—Bănărescu and Boşcaiu, 1978: 259 (regional.); Sánchez
    Osés and Pérez-Hernández, 2005: 168 (regional.); Echeverry and Morrone, 2013:
    1628 (track anal.).
Caribbean Northern Middle America area—Haffer, 1985: 120 (refugia).
Central American subregion (in part)—Samek et al., 1988: 27 (regional.); Rapoport,
    1968: 71 (biotic evol. and regional.); Huber and Riina, 2003: 315 (glossary); Procheş
    and Ramdhani, 2012: 263 (cluster anal. and regional.).
Colombian Mesoamerican region (in part)—Rivas-Martínez and Navarro, 1994: map
    (regional.).
Mesoamerican province—Rivas-Martínez and Navarro, 1994: map (regional.).
Central America bioregion (in part)—Dinerstein et al., 1995: map 1 (ecoreg.).
Mesoamerican dominion—Morrone, 2004a: 157 (track anal.), 2006: 478 (regional.);
    Asiain et al., 2010: 178 (track anal.); Morrone, 2014b: 45 (regional.), 2014c: 207
    (clad. biogeogr.).
Chiapan–Honduran province—Rivas-Martínez et al., 2011: 26 (regional.).
Panamanian region (in part)—Holt et al., 2013: 77 (cluster anal. and regional.).

Definition

The Mesoamerican dominion comprises the lowlands of southern and central
Mexico and most of Central America, in Guatemala, Belize, Honduras, El Salvador,
and northern Nicaragua (Morrone, 2014b,c). The name *Mesoamerica* was coined
originally in anthropology by Kirchhoff (1943) to refer to the area of influence of the
Mayan people, and applied in biogeography by Vivó (1943). There is no consensus,
however, on its boundaries, because the area named Mesoamerica may include all
Mexico and Central America (e.g., Flores and Gerez, 1994; Arroyo-Cabrales et al.,
2007), span from southeastern Mexico to Panama (e.g., Navarro et al., 2001; Ford,
2005) or from the Yucatán peninsula to Panama (e.g., Cavers et al., 2003). Winker
(2011) proposed that the name *Middle America* is more appropriate in a biogeographic context. The name Middle America, however, relies on geopolitical boundaries ("… the lands between the United States of America and South America";
Winker, 2011: p. 5), and the use of Mesoamerica is widespread in biogeography—so
should be preferred (Sánchez-González et al., 2013). As here used for a biogeographic

dominion, it corresponds to southern and central Mexico and most of Central America, reaching northern Nicaragua (Morrone, 2014b,c).

Mesoamerica is a tectonically complex area. After the separation of the North American plate from the South American–African plates in the Jurassic (180 mya), four blocks (Figure 6.28) assembled connecting them (Donnelly, 1988, 1992; Gutiérrez-García and Vázquez-Domínguez, 2013). The Maya block is the northern-most block and corresponds to southern Mexico (Yucatán Peninsula and Chiapas), Belize, and Guatemala north of the Motagua River; its northern boundary corresponds to the Isthmus of Tehuantepec. The Chortis block corresponds to Guatemala south of the Motagua River, Honduras, northern El Salvador, and a portion of Nicaragua; it is separated from the Maya block by the Motagua–Polochic–Jocotlán fault system (eastern Guatemala) and from the Chorotega–Chocó blocks by the Hess escarpment. The Chorotega block comprises the rest of Central America, from Costa Rica to the northern Panama highlands; it is separated from the Chortis block by the Hess escarpment and from the Chocó block by the North Panama fracture zone. The Chocó block was the latest region to emerge, transforming Central America into a continuous landmass. The tectonic history of Mesoamerica may be summarized as (Donnelly, 1988, 1992):

1. Early Cretaceous (125 mya): The Chortis block has detached from western Mexico and the North American and South American plates are diverging.
2. Middle Cretaceous (119 mya): The Caribbean basal plateau begins to move into the growing gap of the American plates and the proto-Antillean arc (largely subma-rine) spans the Americas.
3. Late Cretaceous (84 mya): Compressive island arcs develop around the margin of the Caribbean plate and subduction on the Pacific extends south of the Chortis block to form a new island arc.
4. End of the Cretaceous (65 mya): The Chortis and Maya blocks collide along the Motagua suture zone.
5. Eocene (38 mya): The Yucatán basin opens, Cuba is formed of a collage of tectonic fragments and the gap between the Chortis block and South America narrows.
6. Recent: The Mesoamerican isthmian gap closed 15 mya and the eastern margin of the Chortis block subsided beneath the water.

This complex history implies that there were possible terrestrial connections between the Americas at various stages (Savage, 1982; Crawford and Smith, 2005; Gutiérrez-García and Vázquez-Domínguez, 2013). Mesoamerica was shaped by tectonic activity, including changes in sea level, island emersion, volcanism, mas-sive earthquakes, faults along the continental edge, and tsunamis, as a result of movements of the Cocos, North American, Caribbean, and Nazca tectonic plates. The movement of the tectonic plates, together with processes like glacial and inter-glacial periods during the Miocene, caused sea-level changes that exposed groups of islands, which have been called GAARlandia, proto-Antilles or proto-Greater Antilles (Iturralde-Vinent and MacPhee, 1999; Crawford and Smith, 2005). This island chain might have emerged during the Late Cretaceous (80–70 mya), when there was a decrease in sea level of ca. 60 meters below the present level. Later, the

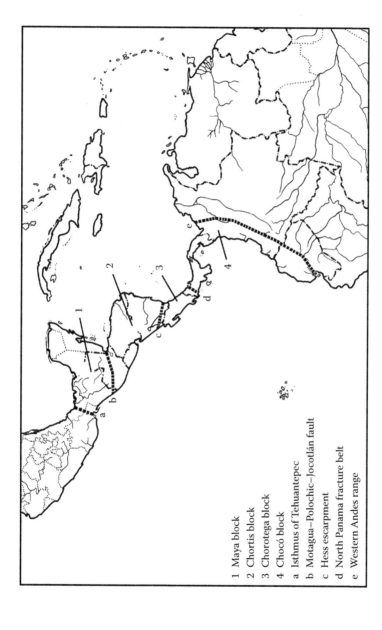

1 Maya block
2 Chortis block
3 Chorotega block
4 Chocó block
a Isthmus of Tehuantepec
b Motagua–Polochic–Jocotlán fault
c Hess escarpment
d North Panama fracture belt
e Western Andes range

Figure 6.28 Map with the tectonic blocks recognized in Mesoamerica.

emergent Greater Antilles connected to northern South America during the Late
Eocene–Early Oligocene transition (35–33 mya). After a period of land exposure
from the Miocene to the Early Pliocene (24–5 mya), most of the islands emerged
and during the Late Miocene (9 mya), "stepping stone" connections between North
and South America became available (Gutiérrez-García and Vázquez-Domínguez,
2013).

Endemic and Characteristic Taxa

Taxa endemic to this dominion include *Ptychoderes bivittatus* (Anthribidae;
Figure 6.29), *Ototylomys phyllotis*, *Rheomys thomasi*, and *Tlacuatzin canescens* (Cricetidae), *Balantiopteryx plicata* (Emballonuridae), *Dermanura wat-
soni* (Phyllostomidae), *Lontra longicaudis* (Mustelidae), *Conepatus leuconotus*
(Mephitidae), and *Sciurus deppei* (Sciuridae; Ceballos and Oliva, 2005; Mermudes
and Napp, 2006; Gutiérrez-García and Vázquez-Domínguez, 2012).

Biotic Relationships

A track analysis of the species of *Bombus* (Apidae; Abrahamovich et al., 2004)
has shown two generalized tracks: Northern Mesoamerican generalized track in the
Mexican coast of the Pacific Ocean, and Southern Mesoamerican generalized track
in Central America, which overlap in a node in the Isthmus of Tehuantepec. Another

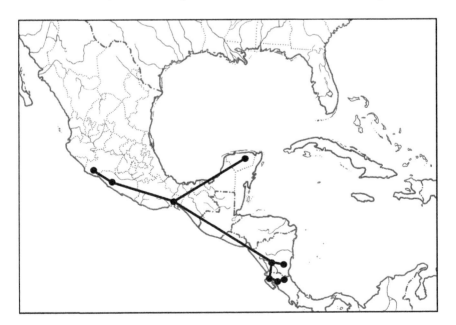

Figure 6.29 Map with the individual track of *Ptychoderes bivittatus* (Anthribidae) in the
Mesoamerican dominion.

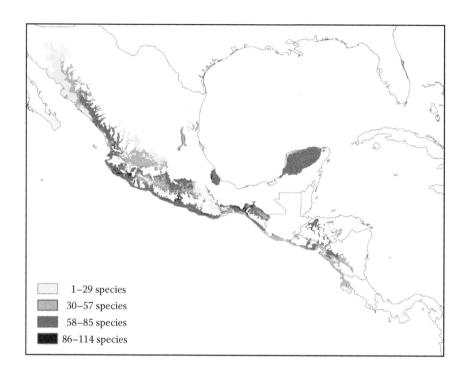

Figure 6.30 Map indicating the number of bird species associated to seasonally dry forests of Mesoamerica. (Modified from Ríos-Muñoz, C. A. and A. G. Navarro-Sigüenza, *Studies on Neotropical Fauna and Environment*, 47: 171–182, 2012.)

track analysis of Charaxinae (Nymphalidae; Maya-Martínez et al., 2011) showed the classical *Y* in the tropical lowlands of Central America and the Pacific and Mexican Gulf coasts. Some authors have analyzed the patterns of richness and endemicity of different taxa of this area. For example, Ríos-Muñoz and Navarro-Sigüenza (2012) examined 650 resident species of birds associated with seasonally dry tropical forests from Mexico to Panama (Figure 6.30), finding two general groups east and west of the Isthmus of Tehuantepec.

A cladistic biogeographic analysis based on plant and animal taxa (Morrone, 2014c) has shown that the Mesoamerican dominion is the sister area to the Pacific dominion. This vicariance event might have occurred more recently than detected in some previous analyses (Amorim and Pires, 1996; Amorim, 2001; Camargo and Pedro, 2003), where it was considered as an older vicariant event. In the present analysis, this event has a younger age, as suggested previously by Camargo (1996) and Camargo and Moure (1996).

Regionalization

The Mesoamerican dominion comprises the Pacific Lowlands, Balsas Basin, Veracruzan, Yucatán Peninsula, and Mosquito provinces (Morrone, 2014b).

Cenocrons

Savage (1982) postulated that the Mesoamerican herpetofauna was comprised basically of endemic genera, with a small number of genera of Nearctic origin and a somewhat larger number of Neotropical genera. He identified four elements, which are assimilated to cenocrons:

1. Middle American cenocron: Genera that are basically tropical Mesoamerican in distribution, being mostly endemic to Central America and Mexico and have their closest relatives in South America. According to fossil evidence, some members of this cenocron have a more extensive distribution in North America, in the Paleogene, when humid warm climates occurred as far north as southern United States.
2. Old Northern cenocron: Genera primarily extratropical in Eurasia and/or North America, but are represented by several tropical species in the Neotropics. Their ancestors were distributed continuously and circumpolarly in the Paleogene, but were forced southward and fragmented by the increased cooling during the Cenozoic.
3. South American cenocron: Genera of South American origin, where most of its evolution occurred during the Cenozoic and recently dispersed to Mesoamerica.
4. Young Northern cenocron: Represented by a few genera that are primarily extratropical in distribution, and associated with arid and semiarid areas of southwestern United States and northern Mexico.

Savage (1982) proposed a dispersal–vicariance model to explain the complex biotic evolution of the Mesoamerican herpetofauna. He assumed a generalized tropical herpetofauna that ranged over tropical North, Central, and most of South America in the Cretaceous–Paleocene (South and Middle American cenocrons); to the north ranged subtropical–temperate Laurasian taxa (Old Northern cenocron). By the Eocene, northern and southern fragments of both tropical faunas had become isolated in Central and South America, respectively, and differentiation in situ until Pliocene produced the distinctive herpetofaunas that became intermixed with the establishment of the Panama Isthmian. As the former continuity between Central America and eastern United States was affected by mountain building and subsequent climatic change during the Oligocene–Pliocene, these biotas became disjunct, originating the Central American component of the Old Northern cenocron that evolved in association with the Middle American cenocron for the rest of the Cenozoic. Thus, the initial biotic evolution of the Mesoamerican herpetofauna involved two vicariance events: a complete geographic isolation from South America, and fragmentation and isolation of the Central American biota from its northern congeners by a combination of physiographic and climatic factors. By Oligocene, most of the genera or their ancestors, which now form the Old Northern and Middle American cenocrons, were present in the area. The uplift of the Sierras Madre created two important additional vicariance events: the Sierra Madres of Mexico were present as upland areas in the Oligocene, and the highlands of Nuclear Central America developed in the Miocene. The final sequence of uplift was in lower Central America, leading to the closure of the Panamanian Portal in the Pliocene. As the mountains were uplifted, the distributions of other taxa, perhaps originally associated with the lowlands of earlier times, became fragmented in the three major highland areas today comprising Mesoamerica, originating endemic biotas in the southern Sierras of Mexico, Nuclear Central America, and the Talamanca area, which developed in situ from ancestors that "rode" the uplifted areas and evolved with them (Savage, 1982). The final major event was the complete emergence of the Panamanian Isthmus in the Pliocene to connect North and South America, which allowed the dispersal of many South American genera northward and of some Old Northern and many Middle American taxa to South America. These dispersal events are also well documented for other major groups and fully confirmed by the mammal fossil record.

There are several comparative phylogeographic analyses for taxa distributed in Mesoamerica. Daza et al. (2010) evaluated the distributions of divergence times of different lineages of snakes across the Isthmus of Tehuantepec, the Motagua–Polochic–Jocotlán fault, the Nicaraguan Depression, the Talamanca cordillera, and the Central American–South American transition. This multilineage comparison across multiple spatial and temporal scales provided evidence for a surprisingly high number of lineages showing coordinated divergence, which often fit previous expectations based on geological and tectonic data. In the case of the Nicaraguan Depression, Daza et al. (2010) detected multiple periods of vicariance (and probable dispersal), rejecting a model centered on a single discrete barrier. They concluded that this fits a model that has suggested the existence of a dynamic landmass that existed across the Nicaraguan Depression during the second half of the Miocene. Ornelas et al. (2013) analyzed plant, bird, and rodent lineages with fossil-calibrated genealogies and coalescent-based divergence time inferences, finding shared breaks that correspond to the Isthmus of Tehuantepec, Los Tuxtlas, and the Chiapas Central Depression. The Isthmus of Tehuantepec resulted in the most frequently shared break among different taxa, although the dating analyses suggested that the breaks occurred at different times in different taxa, the authors concluding that current divergence patterns are consistent with broad vicariance across the Isthmus of Tehuantepec derived from different mechanisms operating at different times. Bagley and Johnson (2014) reviewed previous phylogeographic analyses, finding consistent multitaxon breaks across the Nicaraguan depression, the Chorotega volcanic front, western and central Panama, and the Darién isthmus, which suggested a potentially shared history of responses to regional-scale geological processes.

Becerra (2005) analyzed temporal and spatial changes of diversity of the genus *Bursera*, using a time-calibrated phylogenetic hypothesis. This genus is old and highly adapted to warm and dry conditions, being a dominant member of the Mexican tropical dry forest, so it can help elucidate the origin and expansion of this forest. Becerra (2005) estimated diversification rates at different times over the last 60 mya, finding that between 30–20 mya *Bursera* began a relatively rapid diversification. She considered that this suggests conditions favorable for its radiation and very probably for the establishment of the dry forest, as well. As the oldest lineages diverged mostly in western Mexico and the more recent ones diverged in south central Mexico, Becerra (2005) considered that the tropical dry forest probably first established in the west, and then expanded south and east. The timing of this radiation corresponds to the formation of the mountainous systems in western and central Mexico, which have been recognized as critical for the persistence of the dry forest. De-Nova et al. (2012) provided another analysis, based on a more exhaustive sample of species of *Bursera*, and estimated divergence times, temporal diversification heterogeneity and geographical structure, and habitat shifts. They concluded that *Bursera* became differentiated in the Early Eocene, diversifying during independent Early Miocene consecutive radiations that took place in the dry forest. They also suggested that the Late Miocene average age of the species and the strong geographical structure and conservatism confirmed the South American origin of the genus.

PACIFIC LOWLANDS PROVINCE

Pacific Lowlands region—West, 1964: 368 (regional.).
Western Lowlands subregion—Savage, 1966: 736 (biotic evol.).
Mexican Xerophyllous province (in part)—Cabrera and Willink, 1973: 34 (regional.); Huber and Riina, 2003: 264 (glossary).
Central American Pacific center—Müller, 1973: 19 (regional.).
Pacific Coast province—Rzedowski, 1978: 107 (regional.); Rzedowski and Reyna-Trujillo, 1990: map (regional.); Arriaga et al., 1997: 62 (regional.); Morrone et al., 1999: 510 (PAE); Contreras-Medina et al., 2007b: 408 (PAE); Espinosa et al., 2000: 64 (PAE and regional.); Huber and Riina, 2003: 258 (glossary); Puga-Jiménez et al., 2013: 1181 (PAE).
Central American Pacific Coast province—Samek et al., 1988: 28 (regional.); Huber and Riina, 2003: 259 (glossary).
Pacific province—Ferrusquía-Villafranca, 1990: map (regional.); Espinosa Organista et al., 2008: 60 (regional.).
Western Mexican province—Casas-Andreu and Reyna-Trujillo, 1990: map (regional.).
Central American Pacific Dry Forests ecoregion—Dinerstein et al., 1995: 94 (ecoreg.).
South Eastern Coast province—Escalante et al., 1998: 285 (cluster anal. and regional.).
Pacific Lowlands province—Campbell, 1999: 117 (regional.); Morrone, 2014b: 46 (regional.).
Pacific Lowlands of Mexico and Balsas Depression area (in part)—Flores-Villela and Goyenechea, 2001: 174 (clad. biogeogr.).
Mexican Pacific Coast province—Morrone, 2001a: 40 (regional.), 2001e: 51 (regional.); Morrone and Márquez, 2001: 637 (track anal.); Morrone et al., 2002: 97 (regional.); Corona and Morrone, 2005: 38 (track anal.); Andrés Hernández et al., 2006: 901 (track anal.); Morrone, 2006: 478 (regional.); Quijano-Abril et al., 2006: 1270 (track anal.); Mariño-Pérez et al., 2007: 80 (track anal.); Morrone and Márquez, 2008: 20 (track anal.); Mercado-Salas et al., 2012: 459 (track anal.).
Pacific Arid Slope area—Porzecanski and Cracraft, 2005: 266 (PAE).
Western Lowlands area (in part)—Flores-Villela and Martínez-Salazar, 2009: 820 (clad. biogeogr.).
Pacific Coast of Mexico and Balsas Depression area (in part)—Flores-Villela and Martínez-Salazar, 2009: 820 (clad. biogeogr.).

Definition

The Pacific Lowlands province comprises a narrow strip in the Pacific coast of Mexico (states of Chiapas, Colima, Guerrero, Jalisco, Michoacán, Nayarit, Oaxaca, and Sinaloa), El Salvador, Honduras, Nicaragua, Costa Rica, and Guatemala, including the Revillagigedo archipelago (Morrone, 2001e, 2006, 2014b). Kohlmann and Wilkinson (2007) found a clear boundary between the southern portion of this province and the Puntarenas–Chiriquí province, situated at the Grande de Tárcoles River and coincident with the Costa Rica–Panama microplate boundary.

Endemic and Characteristic Taxa

Morrone (2014b) provided a list of endemic and characteristic taxa. Some examples include *Agave kewensis* (Asparagaceae), *Bursera arborea*, *B. excelsa* (Figure 6.31a), and *B. longicuspis* (Burseraceae), *Gouninia isabelensis* (Poaceae), *Euscelus rufiventris* (Attelabidae), *Pantomorus horridus*, *Phacepholis albicans*, and *P. viridicans* (Curculionidae), *Cyclocephala capitata* (Figure 6.31b) and *Viridimicus cyanochlorus* (Scarabaeidae), *Agalychnis dacnicolor* and *Tlalocohyla smithii* (Hylidae), *Cynomops mexicanus* and *Eumops underwoodi* (Molossidae), *Callipepla douglasii* (Odontophoridae), *Urosaurus bicarianatus* (Phrynosomatidae), *Deltarhynchus flammulatus* (Tyrannidae; Sánchez-González, pers. comm.), and *Crotalus basiliscus* (Viperidae).

Vegetation

There are tropical humid forests (Figure 6.32) and seasonally dry tropical forests, savannas, and *palmares* (Dinerstein et al., 1995; Ceballos et al., 2010). Lott and Atkinson (2010) distinguished the following distributional patterns for the dry tropical forests of the Pacific Lowlands province: northwestern (*Cephalocereus purpusii* and *Dicliptera resupinata*), Jalisco (*Bourreria rubra*, *Bursera palaciosii*, and *Lonchocarpus minor*), central coast (*Clowesia dodosoniana*, *Erycina echinata*, *Mexacanthus mcvaughii*, *Recchia mexicana*, and *Tillandsia diguettii*), Mexico (*Pachycereus pecten-aboriginum* and *Tetramerium glandulosum*), Chiapas–Guatemala (*Bursera heteresthes*), Central America (*Cladocolea oligantha* and *Sapranthus violaceus*), and widespread (*Crateva palmeri* and *Croton pseudoniveus*).

Biotic Relationships

Müller (1973) considered that the Pacific Lowlands province is related to the Yucatán Peninsula and Guajira provinces. According to a parsimony analysis of endemicity based on plant, insect, and bird taxa (Morrone et al., 1999), it is closely related to the Sierra Madre del Sur, Trans–Mexican Volcanic Belt, and Balsas Basin provinces. A track analysis based on species of Coleoptera (Morrone and Márquez, 2001)

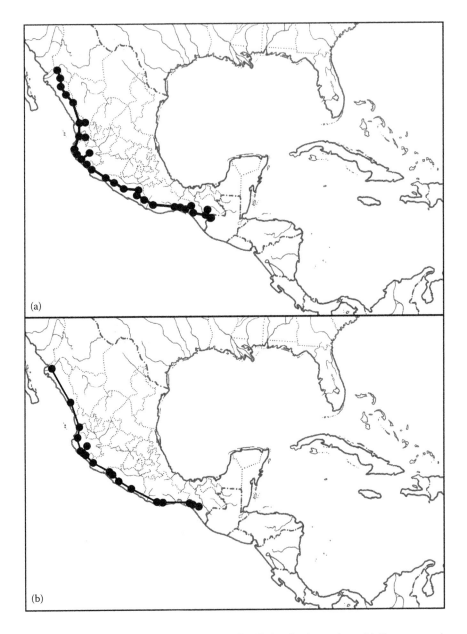

Figure 6.31   Maps with individual tracks in the Pacific Lowlands province. (a) *Bursera excelsa* (Burseraceae); (b) *Cyclocephala capitata* (Scarabaeidae).

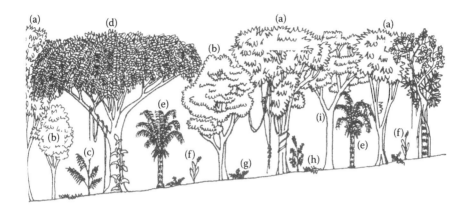

Figure 6.32 Vegetation of the Pacific Lowlands province. (a) *Brosimum alicastrum*; (b) *Celtis monoica*; (c) *Chamaedorea* sp.; (d) *Enterolobium cyclocarpum*; (e) *Orbignya cohune*; (f) *Canna indica*; (g) *Tectaria heracleifolia*; (h) *Tradescantia zebrina*; and (i) *Bursera simaruba*. (Modified from Challenger, A., *Utilización y conservación de los ecosistemas terrestres de México: Pasado, presente y futuro*. Conabio, Mexico City, 1998.)

showed that this province is related to the Puntarenas–Chiriquí, Veracruzan, and Chiapas Highlands provinces.

Regionalization

Six nested units that have been identified within this province (Goldman and Moore, 1945; Stuart, 1964; West, 1964; Rzedowski, 1978; Ferrusquía-Villafranca, 1990; Ramírez-Pulido and Castro-Campillo, 1990; Escalante et al., 1998; García-Trejo and Navarro, 2004) were treated by Morrone (2014b) as districts. For a preliminary delimitation of these districts, I (Morrone, 2014b) considered Stuart's (1964) provinces.

The Sinaloan district corresponds to the northernmost portion of the province, in the Mexican state of Sinaloa (Goldman and Moore, 1945). Endemic taxa include *Peromyscus simulus* (Cricetidae), *Chaetodipus artus* (Heteromyidae), *Poeciliopsis viriosa* (Poeciliidae), *Sciurus colliaei* (Sciuridae), and *Notiosorex evotis* (Soricidae; Mateos et al., 2002; Ceballos and Oliva, 2005).

The Revillagigedo Islands district corresponds to the Revillagigedo archipelago, situated in the Pacific Ocean, 350 kilometers from the Mexican coast (Rzedowski, 1978; Brattstrom, 1990; CONANP, 2004). It consists of four volcanic, oceanic islands, named Socorro, San Benedicto, Clarión, and Roca Partida, which rise independently from the ocean floor along the Clarion Fracture Zone. These islands are situated in the southern tip of the northwestward-moving Baja California Peninsula (Pacific plate), west of the junction of the Pacific, Rivera, and Cocos plates, the East Pacific Rise, and the Middle America Trench (Brattstrom, 1990). Endemic taxa include *Buteo socorroensis* (Accipitridae), *Columbina passerina socorroensis*,

Zenaida macroura clarionensis, and *Z. graysoni* (Columbidae), *Mimus graysoni* (Mimidae), *Setophaga pitiayumi graysoni* (Parulidae), *Urosaurus auriculatus* and *U. clarionensis* (Phrynosomatidae), *Aratinga brevipes* (Psittacidae), and *Troglodytes sissoni* (Troglodytidae; Brattstrom, 1990; CONANP, 2004; Arbogast et al., 2006; Feldman et al., 2011; Lovette et al., 2012; Ríos-Muñoz and Navarro-Sigüenza, 2012). Brattstrom (1990) concluded that the terrestrial vertebrate fauna (reptiles and birds) show major affinities to Mexico, north of the latitude of the islands.

The Nayarit–Guerrero district corresponds to the coastal area of Nayarit, Jalisco, Colima, Michoacán, and Guerrero (Goldman and Moore, 1945); endemic taxa include *Misantlius rufiventris* (Staphylinidae), *Salvadora mexicana* (Colubridae), *Neotoma palatina*, *Oryzomys melanotis*, and *Xenomys nelsoni* (Cricetidae), *Cratogeomys fumosus* and *Pappogeomys bulleri* (Geomyidae), *Sceloporus pyrocephalus* (Phrynosomatidae), *Spermophilus annulatus* (Sciuridae), and *Megasorex gigas* (Soricidae; Asiain and Márquez, 2003; Ceballos and Oliva, 2005; García, 2010).

The Tehuanan district corresponds to the Pacific coast of the Isthmus of Tehuantepec area (Goldman and Moore, 1945); endemic taxa include *Parallocorynus salasae* and *P. schiblii* (Belidae), *Passerina rositae* (Cardinalidae), *Reithrodontomys tenuirostris* (Cricetidae), *Dasyprocta punctata chiapensis* (Dasyproctidae), *Liomys salvini* (Heteromyidae), *Lepus flavigularis* (Leporidae), *Noctilio albiventris* (Noctilionidae), *Sciurus variegatoides* (Sciuridae), and *Rhogeessa genowaysi* (Vespertilionidae; García-Trejo and Navarro, 2004; Ceballos and Oliva, 2005; O'Brien and Tang, 2015).

The Tres Marías Islands district corresponds to the archipelago of Tres Marías, which includes the María Madre, María Magdalena, and María Cleofas islands, situated 112 kilometers of the Mexican Pacific coast (State of Nayarit); endemic taxa include *Oryzomys nelsoni* and *Peromyscus madrensis* (Cricetidae), *Icterus pustulatus graysonii* (Icteridae), *Sylvilagus graysoni* (Leporidae), *Procyon insularis* (Procyonidae), *Amazona tresmariae* (Psittacidae), and *Myotis findleyi* (Vespertilionidae; Ceballos et al., 2002; Eberhard and Bermingham, 2004; Ceballos and Oliva, 2005; Helgen and Wilson, 2005; Cortés-Rodríguez et al., 2008).

The Tapachultecan district corresponds to the southernmost portion of the province, in Central America (Smith, 1941); endemic taxa are *Cyclocephala melolonthida* (Scarabaeidae; Ratcliffe et al., 2013) and *Campylorhynchus chiapensis* (Troglodytidae; Sánchez-González, pers. comm.).

BALSAS BASIN PROVINCE

Balsas–Tepalcatepec Basin area—West, 1964: 368 (regional.).
Balsas Basin province—Rzedowski, 1978: 108 (regional.); Ramírez-Pulido and Castro-Campillo, 1990: map (regional.); Rzedowski and Reyna-Trujillo, 1990: map (regional.); Arriaga et al., 1997: 62 (regional.); Escalante et al., 1998: 285 (cluster anal. and regional.); Morrone et al., 1999: 510 (PAE); Espinosa et al., 2000: 64 (PAE and regional.); Morrone, 2001a: 38 (regional.), 2001e: 49 (regional.); Morrone and Márquez, 2001: 636 (track anal.); Morrone et al., 2002: 94 (regional.); Huber and Riina, 2003: 259 (glossary); Corona and Morrone, 2005: 38 (track anal.); Escalante

et al., 2005: 202 (PAE); Morrone, 2005: 237 (regional.); Andrés Hernández et al.,
    2006: 901 (track anal.); Morrone, 2006: 477 (regional.); Mariño-Pérez et al., 2007:
    80 (track anal.); Espinosa Organista et al., 2008: 61 (regional.); Morrone and
    Márquez, 2008: 20 (track anal.); Escalante et al., 2009: 473 (endem. anal.); Coulleri
    and Ferrucci, 2012: 105 (track anal.); Mercado-Salas et al., 2012: 459 (track anal.);
    Puga-Jiménez et al., 2013: 1181 (PAE); Morrone, 2014b: 48 (regional.).
Jaliscan–Guerreran province—Ferrusquía-Villafranca, 1990: map (regional.).
Balsas Dry Forests ecoregion—Dinerstein et al., 1995: 94 (ecoreg.).
Guerreran Cactus Scrub ecoregion—Dinerstein et al., 1995: 104 (ecoreg.).
Río Balsas Depression province—Anderson and O'Brien, 1996: 332 (biotic evol.).
Balsas province—Ayala et al., 1996: 429 (regional.).
Pacific Lowlands of Mexico and Balsas Depression area (in part)—Flores-Villela and
    Goyenechea, 2001: 174 (clad. biogeogr.).
Pacific Coast of Mexico and Balsas Depression area (in part)—Flores-Villela and
    Martínez-Salazar, 2009: 820 (clad. biogeogr.).

Definition

The Balsas Basin province comprises central Mexico (states of Guerrero, Jalisco,
Mexico, Michoacán, Morelos, Oaxaca, and Puebla), at elevations below 2,000 meters.
It corresponds to the basin of the Balsas River and is situated between the Trans–
Mexican Volcanic Belt and the Sierra Madre del Sur provinces (Rzedowski, 1978;
Morrone, 2001e, 2014b). The rainforests of the Balsas Basin province possess high
levels of regional and local endemism in a wide range of taxa (Dinerstein et al., 1995).

Endemic and Characteristic Taxa

Morrone (2014b) provided a list of endemic and characteristic taxa. Some exam-
ples include *Notholaena lemmonnii* var. *australis* (Pteridaceae), *Montanoa liebman-
nii* (Figure 6.33a) and *M. reveali* (Asteraceae), *Bursera bolivarii*, *B. chemapodicta*,
B. mirandae, *B. rzedowskii*, *B. sarukhanii*, and *B. xochipalensis* (Burseraceae),
Cotinis pueblensis (Scarabaeidae; Figure 6.33b), *Peucaea humeralis* (Emberizidae),
Spermophilus adocetus (Sciuridae), and *Megascops seductus* (Strigidae).

Vegetation

There are seasonally dry tropical forests (Figure 6.34) and grasslands (Dinerstein
et al., 1995). Dominant plant genera include *Backenbergia*, *Bursera*, *Castela*,
Haplocalymma, and *Pseudolopezia* (Rzedowski, 1978).

Biotic Relationships

A track analysis based on species of Coleoptera (Morrone and Márquez, 2001)
showed that this province is related to the Sierra Madre Occidental, Sierra Madre del
Sur, Sierra Madre Oriental, and Trans–Mexican Volcanic Belt provinces.

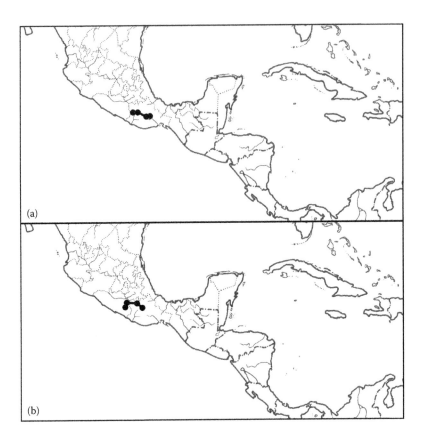

Figure 6.33 Maps with individual tracks in the Balsas Basin province. (a) *Montanoa liebmannii* (Asteraceae); (b) *Cotinis pueblensis* (Scarabaeidae).

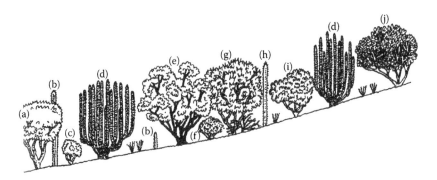

Figure 6.34 Vegetation of the Balsas Basin province. (a) *Bursera morelensis*; (b) *Neobuxbamia* sp.; (c) *Euphorbia* sp.; (d) *Stenocereus weberi*; (e) *Bursera longipes*; (f) *Lysiloma microphylla*; (g) *Ceiba parvifolia*; (h) *Pachycereus* sp.; (i) *Bursera fagaroides*; and (j) *Bursera copalifera*. (Modified from Challenger, A., *Utilización y conservación de los ecosistemas terrestres de México: Pasado, presente y futuro.* Conabio, Mexico City, 1998.)

Regionalization

Smith (1941) and Ferrusquía-Villafranca (1990) have identified nested units, which were treated by Morrone (2014b) as two districts; their preliminary delimitation is based on Smith's (1941) provinces. The Lower Balsas Basin district corresponds to the lower (western) part of the Balsas River basin and the Upper Balsas Basin district corresponds to the upper (eastern) part of the Balsas River basin.

VERACRUZAN PROVINCE

Veracruzan province—Smith, 1941: 110 (regional.); Dice, 1943: 63 (regional.); Goldman and Moore, 1945: 357 (regional.); Barrera, 1962: 101 (regional.); Stuart, 1964: 355 (regional.); Savage, 1966: 736 (biotic evol.); Casas-Andreu and Reyna-Trujillo, 1990: map (regional.); Ferrusquía-Villafranca, 1990: map (regional.); Huber and Riina, 2003: 264 (glossary); Morrone, 2014b: 49 (regional.).

Lempira–Tegucigalpa province—Ryan, 1963: 25 (regional.).

Caribbean–Gulf Lowlands region (in part)—West, 1964: 368 (regional.); Huber and Riina, 2003: 97 (glossary).

Usumacinta province—Miller, 1966: 778 (regional.); Hulsey and López-Fernández, 2011: 280 (regional.).

Eastern Lowland subregion—Savage, 1966: 736 (biotic evol.).

Mexican Xerophyllous province (in part)—Cabrera and Willink, 1973: 34 (regional.); Huber and Riina, 2003: 264 (glossary).

Central American Rainforest center—Müller, 1973: 10 (regional.).

Campechean province—Udvardy, 1975: 41 (regional.); Huber and Riina, 2003: 95 (glossary).

Mexican Gulf province Rzedowski, 1978: 109 (regional.); Samek et al., 1988: 28 (regional.); Rzedowski and Reyna-Trujillo, 1990: map (regional.); Arriaga et al., 1997: 63 (regional.); Morrone et al., 1999: 510 (PAE); Espinosa et al., 2000: 64 (PAE and regional.); Morrone, 2001a: 42 (regional.), 2001e: 49 (regional.); Morrone and Márquez, 2001: 637 (track anal.); Morrone et al., 2002: 96 (regional.); Huber and Riina, 2003: 259 (glossary); Corona and Morrone, 2005: 38 (track anal.); Morrone, 2006: 478 (regional.); Quijano-Abril et al., 2006: 1270 (track anal.); Mariño-Pérez et al., 2007: 80 (track anal.); Espinosa Organista et al., 2008: 61 (regional.); Morrone and Márquez, 2008: 20 (track anal.); Puga-Jiménez et al., 2013: 1181 (PAE).

Campechean–Petén province (in part)—Ferrusquía-Villafranca, 1990: map (regional.).

Planiciense subprovince—Ferrusquía-Villafranca, 1990: map (regional.).

Tuxtlan subprovince—Ferrusquía-Villafranca, 1990: map (regional.).

Valle–Nacionalianan subprovince—Ferrusquía-Villafranca, 1990: map (regional.).

Gulf province—Ramírez-Pulido and Castro-Campillo, 1990: map (regional.).

Veracruz province—Ayala et al., 1996: 429 (regional.); Brown et al., 1998: 32 (veget.).

Gulf–Caribbean Slope area (in part)—Porzecanski and Cracraft, 2005: 266 (PAE).

Grijalva–Usumacinta ecoregion—Abell et al., 2008: 408 (ecoreg.).

Definition

The Veracruzan province comprises the coast of the Gulf of Mexico, in southeastern Mexico (states of Campeche, Chiapas, Hidalgo, Oaxaca, Puebla, San Luis Potosí, Tabasco, Tamaulipas, and Veracruz), Belize, and northern Guatemala (Rzedowski, 1978; Morrone, 2001e, 2006, 2014b).

Endemic and Characteristic Taxa

Morrone (2014b) provided a list of endemic and characteristic taxa. Some examples include *Zanthoxylum* sect. *Tobinia* (Rutaceae), *Troilides torquatus tolus* (Papilionidae; Figure 6.35), *Cotinis punctatostriata* and *Homophilerus quadrituberculatus* (Scarabaeidae), *Chelydra rossignonii* (Chelydridae), *Eleutherodactylus alfredi* (Eleutherodactylidae), *Sylvilagus brasiliensis truei* (Leporidae), and *Amazona belizensis* (Psittacidae).

Vegetation

There are temperate (oak and pine) forests with *Pinus caribaea*, savannas, and *palmares* (Dinerstein et al., 1995). The most typical genera include *Dialium*, *Pimenta*, *Scheelea*, and *Vochysia* (Rzedowski, 1978).

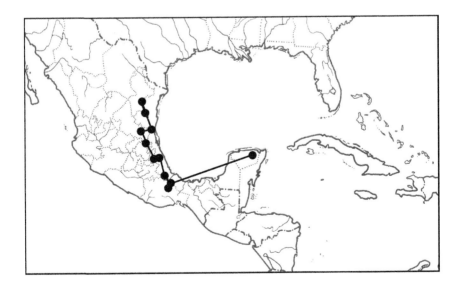

Figure 6.35   Map with the individual track of *Troilides torquatus tolus* (Papilionidae) in the Veracruzan province.

Biotic Relationships

Müller (1973) considered that several vertebrate species from this province were related to species from the Mosquito and Guatuso–Talamanca provinces. According to a parsimony analysis of endemicity, based on plant, insect, and bird taxa (Morrone et al., 1999), it is closely related to the Chiapas province. According to a track analysis based on species of Coleoptera (Morrone and Márquez, 2001), this province is related to the Puntarenas–Chiriquí, Pacific Lowlands, and Chiapas Highlands provinces.

Regionalization

Nested units identified within this province (West, 1964; Escalante et al., 1998; Campbell, 1999) were treated by Morrone (2014b) as four districts. Their preliminary delimitation is based on West's (1964) areas. The Deciduous Forest of Northern Veracruz district corresponds to the northern part of the Veracruz State (West, 1964). The Los Tuxtlas district corresponds to the area of Los Tuxtlas in southern Veracruz; some endemic taxa include *Epigomphus donnellyi* (Gomphidae), *Heteragrion azulum* (Medapodagrionidae), and *Cryptotis nelsoni* (Soricidae; González Soriano and Novelo Gutiérrez, 1996; Ceballos and Oliva, 2005; Woodman, 2005). The Valley of Chiapas district corresponds to the valley enclosed by the Chiapas Highlands (West, 1964). The Southern Veracruz–Tabasco Rainforest district corresponds to the southern part of the state of Veracruz, Tabasco, and Belize (West, 1964); endemic taxa include *Rhopalotria calonjei* (Belidae) and *Dasyprocta mexicana* (Dasyproctidae; Ceballos and Oliva, 2005; O'Brien and Tang, 2015).

YUCATÁN PENINSULA PROVINCE

Yucatán Peninsula province—Smith, 1941: 110 (regional.); Goldman and Moore, 1945: 360 (regional.); Stuart, 1964: 355 (regional.); Rzedowski, 1978: 109 (regional.); Rzedowski and Reyna-Trujillo, 1990: map (regional.); Morrone, 2001a: 43 (regional.); Morrone and Márquez, 2001: 637 (track anal.); Ibarra-Manríquez et al., 2002: 18 (cluster anal. and regional.); Morrone et al., 2002: 98 (regional.); Huber and Riina, 2003: 259 (glossary); Morrone, 2006: 479 (regional.); Contreras-Medina et al., 2007b: 408 (PAE); Mariño-Pérez et al., 2007: 79 (track anal.); Morrone and Márquez, 2008: 20 (track anal.); Maya-Martínez et al., 2011: 528 (track anal.); Cortés-Ramírez et al., 2012: 531 (cluster anal. and regional.); Duno-de Stefano et al., 2012: 1053 (PAE and cluster anal.); Morrone, 2014b: 51 (regional.).

Yucatán province—Ryan, 1963: 22 (regional.); Ayala et al., 1996: 429 (regional.); Escalante et al., 1998: 285 (cluster anal. and regional.); Morrone et al., 1999: 510 (PAE); Espinosa et al., 2000: 64 (PAE and regional.); Morrone, 2001e: 51 (regional.); Huber and Riina, 2003: 264 (glossary); Espinosa Organista et al., 2008: 62 (regional.).

Campechean province—Savage, 1966: 736 (biotic evol.); Brown et al., 1998: 31 (veget.).
Petén–Yucatán Rainforest area—West, 1964: 368 (regional.).
Eastern Lowland subregion—Savage, 1966: 736 (biotic evol.).
Yucatán center—Müller, 1973: 16 (regional.).
Yucatecan province—Udvardy, 1975: 41 (regional.); Casas-Andreu and Reyna-Trujillo, 1990: map (regional.); Ferrusquía-Villafranca, 1990: map (regional.); Ramírez-Pulido and Castro-Campillo, 1990: map (regional.).
Southeastern Coastal Plain province (in part)—Anderson and O'Brien, 1996: 332 (biotic evol.).
Gulf–Caribbean Slope area (in part)—Porzecanski and Cracraft, 2005: 266 (PAE).
Yucatán ecoregion—Abell et al., 2008: 408 (ecoreg.).
Eastern Lowlands area (in part)—Flores-Villela and Martínez-Salazar, 2009: 820 (clad. biogeogr.).

Definition

The Yucatán Peninsula province comprises the Yucatán Peninsula in southeastern Mexico (states of Campeche, Quintana Roo, and Yucatán) and the northern portions of Guatemala and Belize, below 300 meters altitude (Morrone, 2001e, 2006, 2014b; Espadas Manrique et al., 2003). Marshall and Liebherr (2000) combined the Yucatán Peninsula and Chiapas Highlands provinces into a single province, based on their shared taxa. I (Morrone, 2004a, 2006) assigned the Yucatán Peninsula province to the Antillean dominion. Duno de Stefano et al. (2012) found a close relationship with Central America, so they transferred it to the Mesoamerican dominion.

Endemic and Characteristic Taxa

Morrone (2014b) provided a list of endemic and characteristic taxa. Some examples include *Carlowrightia myriantha* (Acanthaceae), *Coccothrinax readii* and *Desmoncus quasillarius* (Arecaceae), *Bourreria pulchra* and *Cordia serratifolia* (Boraginaceae), *Acacia cedilloi* and *Caesalpinia gaumeri* (Figure 6.36a), *Calliandra belizensis*, *Pithecelobium lanceolatum*, and *P. graciliflorum* (Fabaceae), *Amblygnathus subtinctus* (Carabidae), *Priamides phanases* and *P. rogeri* (Papilionidae; Figure 6.36b), *Antrostomus badius* (Caprimulgidae), *Leptotila jamaicensis* (Columbidae), *Dasyprocta punctata yucatanica* (Dasyproctidae), *Colinus nigrogularis* (Odontophoridae), *Micronycteris schmidtorum* and *Mimon crenulatum keenani* (Phyllostomidae), *Sciurus yucatanensis* (Sciuridae), and *Rhogeessa aeneus* (Vespertilionidae).

Vegetation

There are thornbush savannas, tropical humid forests, seasonally dry tropical forests (Figure 6.37), and flooded grasslands (Müller, 1973; Dinerstein et al., 1995).

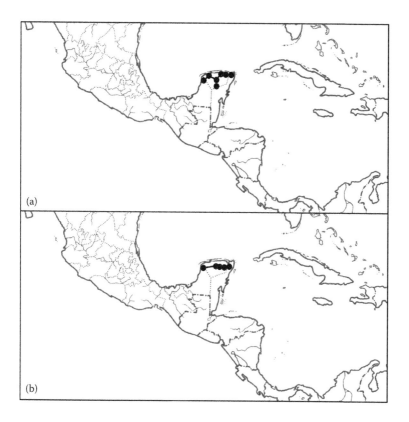

Figure 6.36 Maps with individual tracks in the Yucatán Peninsula province. (a) *Caesalpinia gaumeri* (Fabaceae); (b) *Priamides rogeri* (Papilionidae).

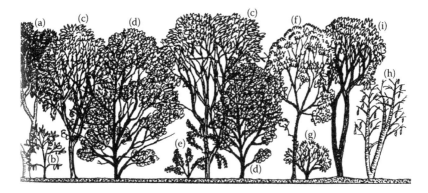

Figure 6.37 Vegetation of the Yucatán Peninsula province. (a) *Bursera simaruba*; (b) *Zexmania frutescens*; (c) *Lysiloma bahamensis*; (d) *Piscidia piscipula*; (e) *Cydista heterophylla*; (f) *Lonchocarpus rugosus*; (g) *Bixa orellana*; (h) *Pileus mexicanus*; and (i) *Cedrela odorata*. (Modified from Challenger, A., *Utilización y conservación de los ecosistemas terrestres de México: Pasado, presente y futuro*. Conabio, Mexico City, 1998.)

Biotic Relationships

Müller (1973) considered that the Yucatán Peninsula province was related to the Pacific Lowlands province. According to a parsimony analysis of endemicity, based on plant, insect, and bird taxa (Morrone et al., 1999), this province is related to the Chiapas Highlands, Veracruzan, Sierra Madre Oriental, Sierra Madre del Sur, Balsas Basin, Pacific Lowlands, and Trans–Mexican Volcanic Belt provinces.

Regionalization

Nested units identified within this province (Casas-Andreu and Reyna-Trujillo, 1990; Arriaga et al., 1997; Ibarra-Manríquez et al., 2002; Espadas Manrique et al., 2003; Ramírez-Barahona et al., 2009; Cortés-Ramírez et al., 2012) were treated by Morrone (2014b) as districts; the zones identified by Espadas Manrique et al. (2003) may be a base for the recognition of these districts (Figure 6.38). The Northern Yucatán district corresponds to the northern part of the Yucatán Peninsula (West,

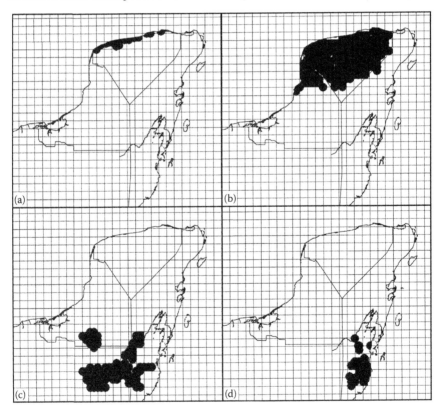

Figure 6.38   Maps with biogeographic zones recognized by Espadas Manrique et al. (2003) in the Yucatán Peninsula province. (a) Yucatán Dry; (b) Yucatán; (c) El Petén; and (d) Belize.

1964); an endemic taxon is *Reithrodontomys gracilis gracilis* (Cricetidae; Ceballos and Oliva, 2005). The Rooan district corresponds to the eastern part of the Yucatán peninsula; endemic taxa include *Reithrodontomys spectabilis* (Cricetidae) and *Procyon pygmaeus* (Procyonidae; Ceballos and Oliva, 2005). The Petén district corresponds to the southern portion of the Yucatán Peninsula, in the Petén area. The Belizean Swamp Forests district corresponds to a part of Belize.

MOSQUITO PROVINCE

Mosquito province—Ryan, 1963: 29 (regional.); Morrone, 2014b: 53 (regional.).
Caribbean Costa Rica–Panama province—Stuart, 1964: 349 (regional.).
Caribbean–Gulf Lowlands region (in part)—West, 1964: 368 (regional.); Huber and Riina, 2003: 97 (glossary).
Caribbean Rainforest Central America area—West, 1964: 368 (regional.).
Mosquito Coast area—West, 1964: 368 (regional.).
Caribbean province—Savage, 1966: 736 (biotic evol.).
Panamanian province—Savage, 1966: 736 (biotic evol.).
Coco center—Müller, 1973: 22 (regional.).
Mosquito subcenter—Müller, 1973: 24 (regional.).
Guatemala–Panama province (in part)—Samek et al., 1988: 29 (regional.); Huber and Riina, 2003: 151 (glossary).
Central American Atlantic Moist Forests ecoregion (in part)—Dinerstein et al., 1995: 87 (ecoreg.).
Miskito Pine Forests ecoregion—Dinerstein et al., 1995: 98 (ecoreg.).
Motagua Valley Thornscrub ecoregion—Dinerstein et al., 1995: 104 (ecoreg.).
Central American province (in part)—Brown et al., 1998: 31 (veget.); Huber and Riina, 2003: 99 (glossary).
Eastern Central American Atlantic Lowlands area—Flores-Villela and Goyenechea, 2001: 174 (clad. biogeogr.).
Eastern Central America province (in part)—Morrone, 2001a: 46 (regional.), 2001e: 53 (regional.); Morrone and Márquez, 2001: 637 (track anal.); Morrone, 2006: 479 (regional.); Quijano-Abril et al., 2006: 1269 (track anal.); Escalante et al., 2011: 32 (track anal.); Mercado-Salas et al., 2012: 459 (track anal.).
Gulf–Caribbean Slope area (in part)—Porzecanski and Cracraft, 2005: 266 (PAE).
Mosquitia ecoregion—Abell et al., 2008: 408 (ecoreg.).
San Juan ecoregion—Abell et al., 2008: 408 (ecoreg.).
Eastern Lowlands area (in part)—Flores-Villela and Martínez-Salazar, 2009: 820 (clad. biogeogr.).
Honduran province—Hulsey and López-Fernández, 2011: 280 (regional.).
Chortís block province (in part)—Townsend, 2014: 206 (paleogeogr.).

Definition

The Mosquito province comprises eastern Central America, between eastern Guatemala and southeastern Nicaragua (Morrone, 2014b).

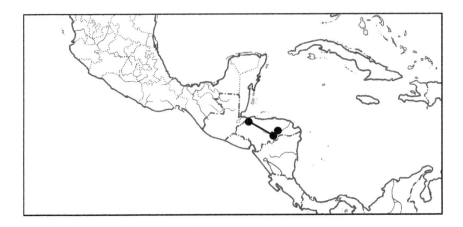

Figure 6.39   Map with the individual track of *Lithachne humilis* (Poaceae) in the Mosquito province.

Endemic and Characteristic Taxa

Morrone (2014b) provided a list of endemic and characteristic taxa. Some examples include *Lithachne humilis* (Poaceae; Figure 6.39), *Petrejoides subrecticornis* (Passalidae), *Misantlius hondurensis* (Staphylinidae), and *Marmosa alstoni* (Marmosidae).

Vegetation

There are different types of forests and grasslands with pines (*Pinus caribaea*). The Miskito pine savannas (Nicaragua and Honduras) represent the largest lowland tropical pine savannas in the Neotropics (Dinerstein et al., 1995).

Biotic Relationships

According to Müller (1973), this province is related to the Chiapas Highlands, Cuban, Jamaica, and Hispaniola provinces.

Regionalization

Hernández et al. (1992a) identified two insular districts within the Colombian portion of this province. The Providencia Island district includes the islands of Providencia and Santa Catalina, lying in the Caribbean midway between Costa Rica and Jamaica. The San Andrés Island district corresponds to the homonymous island, located in the Caribbean, 230 kilometers east of Nicaragua.

CHAPTER 7

The Chacoan Subregion

The Chacoan subregion comprises southeastern South America (Morrone, 2014b,c). It corresponds to the previously recognized Chacoan and Paraná subregions combined.

CHACOAN SUBREGION

Tropical American region (in part)—Engler, 1882: 345 (regional.).

Austral Brazilian subarea (in part)—Clarke, 1892: 381 (regional.).

South Brazilian subregion (in part)—Sclater and Sclater, 1899: 65 (regional.).

Chacoan dominion—Cabrera, 1951: 32 (regional.), 1971: 15 (regional.), 1976: 18 (regional.); Cabrera and Willink, 1973: 69 (regional.); Huber and Riina, 1997: 151 (glossary); Ojeda et al., 2002: 24 (biotic evol.).

East Brazilian province (in part)—Schmidt, 1954: 328 (regional.).

Guianan–Brazilian subregion (in part)—Ringuelet, 1961: 156 (biotic evol. and regional.); Rapoport, 1968: 72 (biotic evol. and regional.); Ringuelet, 1978: 255 (biotic evol.); Paggi, 1990: 303 (regional.).

Guianan–Brazilian region (in part)—Fittkau, 1969: 636 (regional.).

Brazilian subregion (in part)—Hershkovitz, 1969: 3 (regional.); Kuschel, 1969: 710 (regional.).

Non-Andean East area (in part)—Sick, 1969: 451 (regional.).

Caribbean Amazonian subkingdom (in part)—Rivas-Martínez and Tovar, 1983: 521 (regional.).

Central Brazilian region (in part)—Takhtajan, 1986: 251 (regional.).

Brazilian Paraná subregion (in part)—Rivas-Martínez and Navarro, 1994: map (regional.).

Eastern South America bioregion (in part)—Dinerstein et al., 1995: map 1 (ecoreg.); Huber and Riina, 1997: 37 (glossary).

Chacoan subregion—Morrone, 1999: 9 (regional.), 2000a: 52 (regional.), 2001a: 83 (regional.), 2005: 238 (regional.); López Ruf et al., 2006: 116 (track anal.); Morrone, 2006: 481 (regional.); Quijano-Abril et al., 2006: 1268 (track anal.); Nihei and de Carvalho, 2007: 497 (clad. biogeogr.); Navarro et al., 2009: 509 (endem. anal.); Morrone, 2010a: 37 (regional.); Ramos and Melo, 2010: 449 (clad. biogeogr.); Arana et al., 2011: 17 (track anal.); Moreira et al., 2011: 29 (track anal.); Pires and Marinoni,

2011: 8 (track anal.); Camardelli and Napoli, 2012: 32 (PAE); Coulleri and Ferrucci, 2012: 105 (track anal.); Lamas et al., 2014: 955 (clad. biogeogr.); Morrone, 2014b: 72 (regional.), 2014c: 207 (clad. biogeogr.); Chiapella and Demaio, 2015: 63 (biotic evol.); Klassa and Santos, 2015: 520 (endem. anal.).

Southeastern component—Nihei and de Carvalho, 2004: 271 (clad. biogeogr.).

Atlantic Forest region—Pellegrino et al., 2005: 14.

Atlantic Forest component—Sigrist and de Carvalho, 2009: 81 (clad. biogeogr.).

Chacoan–Brazilian superegion—Rivas-Martínez et al., 2011: 27 (regional.).

South American subregion (in part)—Echeverry and Morrone, 2013: 1628 (track anal.).

Endemic and Characteristic Taxa

Taxa endemic to the Chacoan subregion include many plant and animal genera and species (see e.g., Cabrera and Willink, 1973; Morrone, 2006). An example of an endemic taxon is the Violaceae genus *Anchietea* (Paula-Souza and Pirani, 2014).

Biotic Relationships

A parsimony analysis of endemicity based on primate species (Goldani et al., 2006) found a close relationship between the Atlantic Forest, Caatinga, Cerrado, and Chacoan provinces, which are considered to constitute the Chacoan subregion. In the general area cladogram of Morrone (2014c), the Chacoan and Paraná dominions are sister areas, and their sister area is the Southeastern Amazonian dominion. Their vicariance can be associated with the connection of the Parnaíba and Paraná basins, in the Paleocene (Amorim, 2001; Nihei and de Carvalho, 2004).

Regionalization

The Chacoan subregion (Figure 7.1) comprises the Southeastern Amazonian, Chacoan, and Paraná dominions (Morrone, 2014b).

SOUTHEASTERN AMAZONIAN DOMINION

Cariba, Guianan, or Amazonian center (in part)—Lane, 1943: 414 (regional.).

Amazonian dominion (in part)—Cabrera, 1971: 6 (regional.), 1976: 3 (regional.); Huber and Riina, 1997: 150 (glossary), 2003: 124 (glossary).

Amazonian province (in part)—Cabrera and Willink, 1973: 48 (regional.); Huber and Riina, 1997: 23 (glossary).

Brazilian district (in part)—Ávila-Pires, 1974b: 177 (regional.).

Madeiran province (in part)—Udvardy, 1975: 41 (regional.); Huber and Riina, 1997: 206 (glossary).

Amazonia bioregion (in part)—Dinerstein et al., 1995: map 1 (ecoreg.); Huber and Riina, 1997: 37 (glossary).

Figure 7.1   Map of the dominions and provinces of the Chacoan subregion.

Southeast Amazonia subregion (in part)—Nihei and de Carvalho, 2007: 497 (clad. biogeogr.).

Amazonian region (in part)—Rivas-Martínez et al., 2011: 26 (regional.).

Southeastern Amazonian dominion—Morrone, 2014b: 72 (regional.), 2014c: 207 (clad. biogeogr.); Daniel et al., 2016: 1167 (track anal. and regional.).

Definition

The Southeastern Amazonian dominion comprises the Amazonian forest, southeast of the Amazon River (Morrone, 2014b,c).

Endemic and Characteristic Taxa

See Xingu–Tapajós province.

Regionalization

The Southeastern Amazonian dominion comprises only the Xingu–Tapajós province (Morrone, 2014b).

Biotic Relationships

A cladistic biogeographic analysis based on plant and animal taxa (Morrone, 2014c) placed this dominion as the sister area to the Chacoan–Paraná dominions.

XINGU–TAPAJÓS PROVINCE

South sector—Rizzini, 1963: 51 (regional.).
Pará center (in part)—Cracraft, 1985: 71 (areas of endem.).
Southwestern Amazonian area (in part)—Cracraft, 1988: 223 (clad. biogeogr.).
Pantanal province (in part)—Rivas-Martínez and Navarro, 1994: map (regional.);
     Morrone, 1999: 9 (regional.), 2000b: 116 (regional.), 2006: 481 (regional.).
Xingu–Tapajós province—Rivas-Martínez and Navarro, 1994: map (regional.);
     Morrone, 2014b: 72 (regional.).
Tapajós/Xingu Moist Forests ecoregion—Dinerstein et al., 1995: 91 (ecoreg.).
Varzea Forests ecoregion (in part)—Dinerstein et al., 1995: 90 (ecoreg.).
Pará area—Silva and Oren, 1996: 430 (PAE); Bates et al., 1998: 785 (PAE); Ron, 2000:
     387 (PAE); Marks et al., 2002: 155 (biotic evol.); Racheli and Racheli, 2003: 36
     (PAE), 2004: 347 (PAE).
Tapajós–Xingu province—Morrone, 1999: 8 (regional.), 2000b: 115 (regional.), 2001a:
     78 (regional.), 2006: 481 (regional.); Daniel et al., 2016: 1169 (track anal. and
     regional.).
Varzea province (in part) Morrone, 1999: 7 (regional.), 2000b: 112 (regional.), 2001a:
     76 (regional.), 2006: 480 (regional.).
Tapajós area—Silva et al., 2005: 692 (areas of endem.); Borges, 2007: 924 (PAE).
Xingu area—Silva et al., 2005: 692 (areas of endem.); Borges, 2007: 924 (PAE).
Xingu ecoregion—Abell et al., 2008: 408 (ecoreg.).
Central Amazonian province—Rivas-Martínez et al., 2011: 27 (regional.).

Definition

The Xingu–Tapajós province comprises northwestern Brazil (Morrone, 2000b, 2006, 2014b).

Endemic and Characteristic Taxa

I (Morrone, 2014b) provided a list of endemic and characteristic taxa. Some examples include *Partamona gregaria* (Apidae), *Achia bondari* (Curculionidae), *Gypopsitta vulturina* (Psittacidae), and *Pteroglossus reichenowi* (Ramphastidae).

Vegetation

There are tropical humid forests and gallery forests (Dinerstein et al., 1995).

Biotic Relationships

According to a parsimony analysis of endemicity based on primate species (Silva and Oren, 1996) and another based on bird species (Borges, 2007), it is closely related to the Pará province. A cladistic biogeographic analysis based on plant and animal taxa (Morrone, 2014c) placed this province as the sister area to the Chacoan–Paraná dominions.

CHACOAN DOMINION

Argentinean subarea (in part)—Clarke, 1892: 381 (regional.).
Bororô province (in part)—Mello-Leitão, 1937: 246 (regional.).
Tropical district (in part)—Cabrera and Yepes, 1940: 14 (regional.).
Cariri–Bororô province (in part)—Mello-Leitão, 1943: 129 (regional.).
Chacoan dominion—Cabrera, 1951: 32 (regional.), 1971: 15 (regional.), 1976: 18
    (regional.); Cabrera and Willink, 1973: 69 (regional.); Morrone and Coscarón, 1996:
    1 (PAE); Huber and Riina, 1997: 151 (glossary); Zuloaga et al., 1999: 18 (regional.);
    Ojeda et al., 2002: 24 (biotic evol.); Morrone, 2014b: 73 (regional.), 2014c: 207
    (clad. biogeogr.); Daniel et al., 2016: 1167 (track anal. and regional.).
Nontropical East area (in part)—Sick, 1969: 457 (regional.).
Group of Chacoan regions—Rivas-Martínez and Tovar, 1983: 521 (regional.).
La Plata subregion—Procheş and Ramdhani, 2012: 263 (cluster anal. and regional.).

Definition

The Chacoan dominion occupies northern and central Argentina, southern Bolivia, western and central Paraguay, Uruguay, and central and northeastern Brazil (Fittkau, 1969; Sick, 1969; Cabrera, 1971; Cabrera and Willink, 1973; Rivas-Martínez and Tovar, 1983; Willink, 1988; Morrone and Coscarón, 1996; Morrone, 2000a, 2006, 2014b). The *savanna corridor* (Schmidt and Inger, 1951) or *diagonal of open formations* (Vanzolini, 1963) has been reviewed by Prado and Gibbs (1993), Zanella (2010), and Werneck (2011). Based on it, it was hypothesized that the Cerrado province, formerly assigned to the Amazonian dominion by Cabrera and Willink (1973), connected the Caatinga with the other Chacoan provinces, so they should be assigned to the same biogeographic unit (Morrone, 1999, 2000a, 2001a).

Endemic and Characteristic Taxa

I (Morrone, 2000a, 2001a) provided a list of endemic and characteristic taxa. Some examples include *Copernicia* and *Trithrinax* (Arecaceae), *Holocheilus* and

Panphalea (Asteraceae), *Quiabentia* (Cactaceae), *Enterolobium contortisiliquum* (Fabaceae; Figure 7.2a), *Pouteria gardneriana* (Sapotaceae), *Aspidosperma pyrifolium* (Apocynaceae), *Astronium urundeuva* (Anacardiaceae; Figure 7.2b), *Aylacostoma* (Thiaridae), *Helobdella ampullariae* (Glossiphoniidae), *Tityus trivittatus trivittatus* (Buthidae), *Procleobis patagonicus* (Ammotrechidae), *Aegla platensis* (Aeglidae), *Aramigus, Cyrtomon inhalatus, Enoplopactus*, and *Naupactus viridisquamosus* species group (Curculionidae), *Belostoma dilatatum* and *B. micantulum* (Belostomatidae), *Pompilocalus constrictus* (Pompilidae), *Cyphocharax spilotus* (Curimatidae), *Elachistocleis bicolor* (Hylidae), *Chauna torquata* (Anhimidae), *Patagioenas picazuro* (Columbidae), *Furnarius rufus* and *Lepidocolaptes angustirostris* (Furnariidae), *Molothrus rufoaxillaris* (Icteridae), *Rhea americana* (Rheidae), *Chrysocyon* and *Pseudalopex gymnocercus* (Canidae), *Monodelphis domestica* (Didelphidae), *Bolomys obscurus* and *Hololichus chacarius* (Muridae), and *Chaetophractus* spp. and *Dasypus hybridus* (Dasypodidae).

Biotic Relationships

The Chacoan dominion is closely related to the Southeastern Amazonian and Paraná dominions. The development of the *savanna corridor* during the Paleogene acted as the vicariant event that split the former continuous forest, representing an example of dynamic vicariance (Zunino, 2003). Some taxa show the Brazilian–Paraná disjunction (Amorim, 2001), whereas the other that probably evolved later are ranged in the Chacoan subregion, namely the weevil genus *Cyrtomon* (Lanteri, 1990). A comparative phylogeographic analysis based on small mammal species (Costa, 2003) has shown that central Brazil gallery and dry forests (Chacoan dominion) played (and still play) an important role as habitats for forest species. Populations of mammal species inhabiting the Chacoan dominion were revealed to have their closest relatives in the Brazilian subregion, in the Paraná dominion, or are basal to both areas.

Regionalization

The Chacoan dominion comprises the Caatinga, Cerrado, Chacoan, and Pampean provinces (Morrone, 2014b). Haffer (1985) analyzed the distributional ranges of Chacoan birds, showing their superposition in these provinces (Figure 7.3).

CAATINGA PROVINCE

Gê province—Mello-Leitão, 1937: 246 (regional.).
Northeastern subprovince—Rizzini, 1963: 46 (regional.), 1997: 622 (regional.).
Cariri province—Fittkau, 1969: 642 (regional.).
Caatinga province—Cabrera and Willink, 1973: 70 (regional.); Udvardy, 1975: 41 (regional.); Rivas-Martínez and Navarro, 1994: map (regional.); Morrone, 1999: 10 (regional.), 2000a: 54 (regional.), 2001a: 86 (regional.), 2006: 481 (regional.); Zanella, 2010: 202 (biotic evol.); Moreira et al., 2011: 29 (track anal.); Camardelli

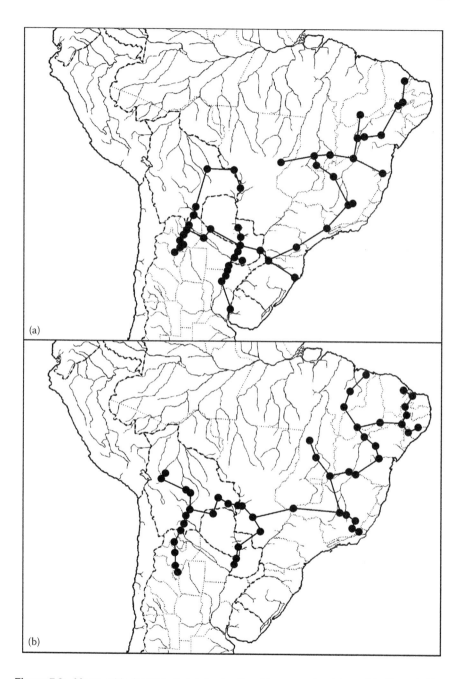

Figure 7.2 Maps with individual tracks in the Chacoan dominion. (a) *Enterolobium contortisiliquum* (Fabaceae); (b) *Astronium urundeuva* (Anacardiaceae).

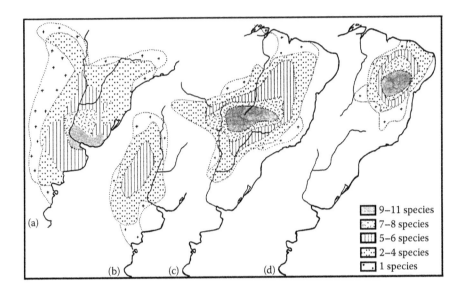

Figure 7.3 Maps showing the superposition of distributional areas of Chacoan bird species.
(a) Pampean province; (b) Chaco province; (c) Cerrado province; and (d) Caatinga
province. (Modified from Haffer, J., *Ornithological Monographs*, 36: 13–145, 1985.)

and Napoli, 2012: 32 (PAE); Morrone, 2014b: 73 (regional.); Daniel et al., 2016:
1169 (track anal. and regional.).
Caatinga center—Müller, 1973: 115 (regional.); Cracraft, 1985: 75 (areas of endem.).
Northeastern Brazil province—Ringuelet, 1975: 107 (regional.).
Caatingas dominion—Ab'Sáber, 1977: map (climate).
Caatinga region—Rivas-Martínez and Tovar, 1983: 521 (regional.).
Caatingas province—Fernandes and Bezerra, 1990: 159 (regional.); Fernandes, 2006:
131 (regional.).
Caatinga ecoregion—Dinerstein et al., 1995: 105 (ecoreg.).
Northeastern Brazil Restingas ecoregion—Dinerstein et al., 1995: 106 (ecoreg.).
Northeastern Caatinga and Coastal Drainages ecoregion—Abell et al., 2008: 408 (ecoreg.).
Caatinga dominion—Fiaschi and Pirani, 2009: 485 (regional.).
Catingan [sic] province—Rivas-Martínez et al., 2011: 27 (regional.).

Definition

The Caatinga province comprises northeastern Brazil in the states of Alagoas,
Bahia, Ceará, Minas Gerais, Paraíba, Pernambuco, Piaui, Rio Grande do Norte, and
Sergipe (Cabrera and Willink, 1973; Cracraft, 1985; Morrone, 2000a, 2006, 2014b).
Its limits are not well marked, but roughly coincide with the Paranaíba River to the
west and northwest, the lowlands of coastal Brazil, and areas beyond the *caatinga*
vegetation type (Cracraft, 1985). The Caatinga province is characterized by semiarid to arid climates with long dry seasons, irregular rainfall (concentrated in the

summer), average annual precipitation of 400–600 millimeters, high annual temperatures, and a large temperature range (Camardelli and Napoli, 2012). More humid environments are found on the mountain sides facing the winds (Sampaio, 1995).

Endemic and Characteristic Taxa

I (Morrone, 2014b) provided a list of endemic and characteristic taxa. Some examples include *Selaginella convoluta* (Selaginellaceae), *Chionolaena jeffreyi* (Asteraceae), *Cereus variabilis* (Cactaceae), *Amburana cearensis, Caesalpinia bracteosa, Dalbergia cearensis, Mimosa caesalpiniifolia, Moldenhawera acuminata,* and *Senna obtusifolia* (Fabaceae), *Bothriurus asper* (Bothriuridae), *Rhinacloa fernandoana* (Miridae), *Stygnus polyacanthus* (Stygnidae; Figure 7.4), *Tantilla marcovani* (Colubridae), *Penelope jacucaca* (Cracidae), *Spinus yarrellii* (Fringillidae), *Epictia borapeliotes* (Leptotyphlopidae), *Aratinga cactorum* and *Forpus xanthopterygius flavissimus* (Psittacidae), and *Tropidurus semitaeniatus* species group (Tropiduridae).

Vegetation

The *caatinga* is comprised of several types of tropical thorn scrub, including tall scrub forests and savannas with cacti and thorny plants (Cabrera and Willink, 1973; Dinerstein et al., 1995; Rizzini, 1997; Camardelli and Napoli, 2012). On some hills over 500 meters altitude, there are moist forests known as *brejos* (Dinerstein et al., 1995).

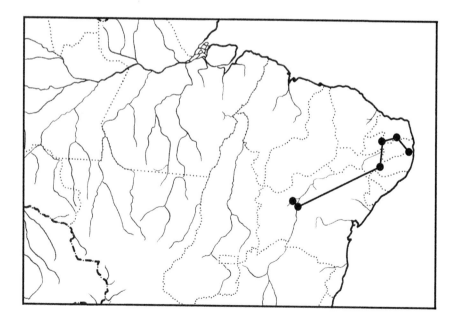

Figure 7.4   Map with the individual track of *Stygnus polyacanthus* (Stygnidae) in the Caatinga province.

Dominant plant species include *Apuleia leiocarpa, Aspidosperma pyrifolium, Astronium urundeuva, Bursera leptophloeos, Caesalpinia leiostachya, Ceiba erianthos, Cephalocereus dybowskii, Cereus jamaracu, Combretum leprosum, Copernicia alba, C. cerifera, Croton* spp., *Dalbergia variabilis, Erythrina velutina, Geoffroea superba, Mimosa caesalpiifolia, Myroxylon balsamum, Orbignia speciosa, Parkinsonia aculeata, Patagonula bahiana, Pilocereus gounellei, Schinopsis brasiliensis, Selaginella convoluta, Spondias tuberosa, Syagrus* spp., *Tabebuia avellanedae, T. caraiba, T. serratifolia, Torresea cearensis, Zizyphus joazeiro*, and *Zollernia ilicifolia* (Cabrera and Willink, 1973; Fernandes and Bezerra, 1990).

Biotic Relationships

Müller (1973) suggested a close relationship of the Caatinga province with the Cerrado and Chaco provinces. Cabrera and Willink (1973) considered that its fauna is related to the Cerrado province. A parsimony analysis of endemicity based on bird taxa (Porzecanski and Cracraft, 2005) also postulated this relationship. A phylogeographic analysis of two species of four-eyed frogs of the genus *Pleurodema* (Thomé et al., 2016) suggested that taxa endemic to the Caatinga were affected by past forest interactions, when rainforests partially replaced the *caatinga*, in recurrent pulses of moister climate. *Caatinga* vegetation isolates within the Cerrado and rainforests isolates embedded in the *caatinga* are evidence of the dynamic relationship between these biomes.

Regionalization

Fernandes and Bezerra (1990) and Fernandes (2006) identified two sectors, which were treated by Morrone (2014b) as the Sertão and Agreste districts (Figure 7.5); the Fernando de Noronha Island is added herein as a third district. The Sertão district corresponds to most of the Caatinga province, with some hills called *morros* that can reach altitudes of up to 500 meters (Fernandes and Bezerra, 1990; Fernandes, 2006). The Agreste district corresponds to the eastern part of the Caatinga province, situated between the Sertão district and the Atlantic province, in the states of Pernambuco, Paraíba, Sergipe, and Alagoas (Fernandes and Bezerra, 1990; Fernandes, 2006). The Fernando de Noronha Island district is an archipelago situated in the Atlantic Ocean, 354 kilometers offshore from the Brazilian coast, comprising 21 islands (Fernando de Noronha, Ilha da Rata, Ilha do meio, ilha Rasa, Sela Gineta, São Jose, Cuscuz, Ilha Viuvinha, Chapéu do Nordeste, Morro de Fora, Dois Irmãos, Morro da Viúva, Morro do Leão, Chapeu da Sueste, Ilha Cabeluda, Ilha dos Ovos, Ilha do Frade, and some unnamed rocklets).

CERRADO PROVINCE

Central Plateau, Brazilian Plateau, Central Brazilian Plateau, and Chapada or Bororô
    center (in part)—Lane, 1943: 414 (regional.).
Central province—Rizzini, 1963: 48 (regional.), 1997: 623 (regional.).

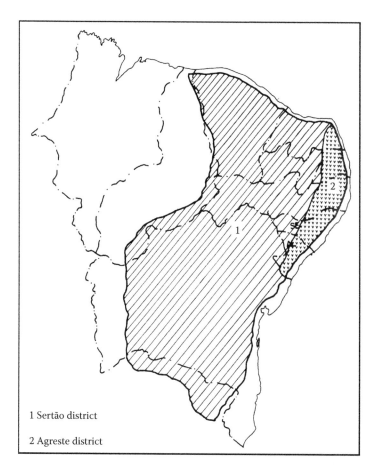

1 Sertão district

2 Agreste district

Figure 7.5   Map of the districts of the Caatinga province. (Modified from Fernandes, A. and P. Bezerra, *Estudo fitogeográfico do Brasil*. Stylus Comunicaçoes, Fortaleza, 1990.)

Bororô province (in part)—Fittkau, 1969: 642 (regional.).

Cerrado province—Cabrera and Willink, 1973: 56 (regional.); Rivas-Martínez and Navarro, 1994: map (regional.); Huber and Riina, 1997: 272 (glossary); Morrone, 1999: 10 (regional.), 2000a: 55 (regional.), 2001a: 87 (regional.), 2006: 481 (regional.); Zanella, 2010: 202 (biotic evol.); Moreira et al., 2011: 29 (track anal.); Morrone, 2014b: 74 (regional.); del Río et al., 2015: 1294 (track anal. and clad. biogeogr.); Daniel et al., 2016: 1160 (track anal. and regional.).

Campo Cerrado center—Müller, 1973: 120 (regional.); Cracraft, 1985: 75 (areas of endem.).

Campos Cerrados province—Udvardy, 1975: 42 (regional.).

Cerrados dominion—Ab'Sáber, 1977: map (climate).

Cerrados province—Fernandes and Bezerra, 1990: 125 (regional.); Fernandes, 2006: 95 (regional.).

Cerrado subregion—Rivas-Martínez and Navarro, 1994: map (regional.).

Tocantins province—Rivas-Martínez and Navarro, 1994: map (regional.).
Cerrado area—Coscarón and Coscarón-Arias, 1995: 726 (areas of endem.).
Cerrado ecoregion—Dinerstein et al., 1995: 99 (ecoreg.).
Central South America area—Porzecanski and Cracraft, 2005: 266 (PAE).
Sao Francisco ecoregion—Abell et al., 2008: 408 (ecoreg.).
Tocantins–Araguaia ecoregion—Abell et al., 2008: 408 (ecoreg.).
Cerrado dominion—Fiaschi and Pirani, 2009: 482 (regional.).
East Cerrado province—Rivas-Martínez et al., 2011: 27 (regional.).
Tocantins province (in part)—Rivas-Martínez et al., 2011: 27 (regional.).
West Cerrado province—Rivas-Martínez et al., 2011: 27 (regional.).

Definition

The Cerrado province comprises south central Brazil (Goiás, Maranhao, Mato Grosso, Minas Gerais, Paraná, Piaui, and São Paulo states), northeastern Paraguay, and Bolivia (Cabrera and Willink, 1973; Cracraft, 1985; Morrone, 2000a, 2006, 2014b). Its approximate boundaries include the Paraná and Paranaíba rivers to the south, the Chaco to the east, the tropical lowlands to the north and northwest, and the *caatinga* vegetation to the east (Cracraft, 1985). The Cerrado biome is basically confined to the Central Brazil Plateau as a result of climatic, topographic, and edaphic interactions (Werneck, 2011).

Endemic and Characteristic Taxa

I (Morrone, 2014b) provided a list of endemic and characteristic taxa. Some examples include *Cecropia saxatilis* (Cecropiaceae), *Actinocladum* (Poaceae), *Parascopas chapadensis*, *Propedies auricularis*, and *P. lobipennis* (Acrididae), *Ananteris mariaterezae*, *Tityus blaseri*, and *T. mattogrossensis* (Buthidae), *Sirthenea peruviana gracilis* (Reduviidae), *Stygnus multispinosus* (Stygnidae; Figure 7.6a), *Acanthoscurria* aff. *gomesiana* (Figure 7.6b), *Avicularia taunayi* (Theraphosidae), *Liotyphlops schubarti* and *L. wilderi* (Anomalepididae), *Pseudalopex vetulus* (Canidae), *Proechimys roberti* (Echimyidae), *Gymnodactylus amarali* (Gekkonidae), *Hoplocercus spinosus* (Hoplocercidae), *Colaptes campestris campestris* (Picidae), *Cnemidophorus jalapensis* and *Tupinambis quadrilineatus* (Teiidae), and *Bothropoides marmoratus*, *B. mattogrossensis*, and *Rhinocerophis itapetiningae* (Viperidae).

Vegetation

The *cerrado*, or *campo cerrado*, represents a large savanna and seasonally dry tropical forest complex, with open forests containing low trees of 8–12 meters height, shrubs and herbs (especially Poaceae and Fabaceae), savannas, and gallery forests along the rivers (Cabrera and Willink, 1973; Rizzini, 1997; Graham, 2004; Marinho-Filho et al., 2010; Werneck, 2011). It represents one of the largest savanna–forest complexes in the world, containing a diverse mosaic of habitat types and natural communities (Dinerstein et al., 1995). Dominant plant species include *Acosmium*

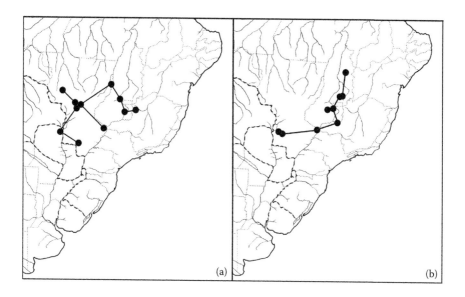

Figure 7.6 Maps with individual tracks in the Cerrado province. (a) *Stygnus multispinosus* (Stygnidae); (b) *Acanthoscurria* aff. *gomesiana* (Theraphosidae).

dasycarpum, Agonandra brasiliensis, Anacardium microcarpum, A. humilis, Andira laurifolia, Annona coriacea, Astronium fraxinofolium, Bowdichia virgilioides, Byrsonina coccolobifolia, Caryocar brasiliense, Connarus suberosus, Curatella americana, Dalbergia violacea, Eriocema coriaceum, Harpalyce brasiliana, Hirtella racemosa, Jacaranda brasiliana, Kielmeyera coriacea, Krameria argentea, K. tomentosa, Lippia fruticosa, Machaerium opacum, Pterodon pubescens, Qualea parviflora, Salvertia convallariodora, Sclerobium paniculatum, Simarouba versi-color, Stryphnodendron coriaceum, Tabebuia alba, Terminalia argentea, T. fagifolia, and *Vochysia thyrsoidea* (Cabrera and Willink, 1973; Fernandes and Bezerra, 1990). Based on the proportion of plant growth forms, four different types have been recognized (Marinho-Filho et al., 2010): Campo limpo, dominated by Poaceae, with few bushes and lacking trees; Campo sujo, dominated by Poaceae and bushes, with some trees; Cerrado in the strict sense, low dominance of Poaceae and bushes, and medium presence of trees; and Cerradão, forests with a herbaceous layer lacking Poaceae, dominated by seedlings, and with an important presence of trees.

Biotic Relationships

Cabrera and Willink (1973) assigned the Cerrado province to the Amazonian dominion, mostly taking into account its flora. Müller (1973) and Prado and Gibbs (1993) suggested instead a close relationship with the Chaco and Caatinga provinces. A parsimony analysis of endemicity based on bird taxa (Porzecanski and Cracraft, 2005) and another based on lizards (Colli, 2005) supported a closer relationship between the Cerrado and Chaco province than with the Caatinga. A track analysis based on beetle taxa (Daniel et al., 2016) supported relationships of this province with the Paraná, Chaco, *Araucaria* Forest, Atlantic, Caatinga, Xingu–Tapajós, Madeira, and Western Ecuador provinces. Daniel et al. (2016) considered that the forest intrusions in the savanna formations represented remnants of a paleoflora that was fragmented during the climatic fluctuations of the Quaternary.

Werneck et al. (2012) modeled the historical distribution of the Cerrado vegetation under different time projections. They postulated that the diagonal of open formations was a composite area, defined by shared species as a result of dispersal, and that different geomorphological events shaped the origin and diversification of the open diagonal biotas (e.g., marine introgressions in the Chaco, global savanna expansion in the Cerrado, and regional geological surface influences in the Caatinga). They suggested that the biogeographic counterparts to the Cerrado are, in fact, the disjunct savannas located on the Guyana shield (Llanos and Gran Sabana).

Regionalization

Fernandes and Bezerra (1990) recognized three sectors, which were treated by Morrone (2014b) as districts (Figure 7.7). The Parnaiba Basin district corresponds to the basin of the Parnaiba River in the states of Maranhão and Piauí (Fernandes and Bezerra, 1990). The Planalto district includes most of the Cerrado province (2,000,000 km²) in the states of Goiás, Tocantins, Mato Grosso, Mato Grosso do Sul, Minas Gerais, São Paulo, and Bahia (Fernandes and Bezerra, 1990); some

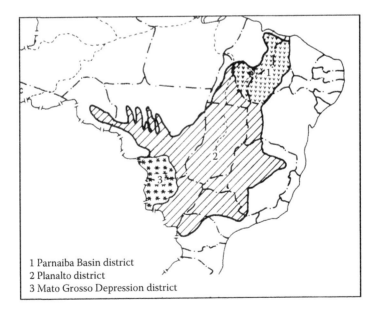

1 Parnaiba Basin district
2 Planalto district
3 Mato Grosso Depression district

Figure 7.7 Map of the districts of the Cerrado province. (Modified from Fernandes, A. and P. Bezerra, *Estudo fitogeográfico do Brasil*. Stylus Comunicaçoes, Fortaleza, 1990.)

endemic taxa include *Apostolepis cerradoensis* (Dipsadidae), *Bachia didactyla* and *Rhachisaurus brachylepis* (Gymnophthalmidae), *Gymnodactylus guttulatus* (Phyllodactylidae), *Cnemidophorus mumbuca* (Teiidae), *Eurolophosaurus nanuzae*, *Tropidurus callathelys*, and *T. chromatops* (Tropiduridae; Nogueira et al., 2011). The Mato Grosso Depression district corresponds to the southeastern part of the Cerrado province (Fernandes and Bezerra, 1990); some endemic taxa include *Rhinoleucophenga matogrossensis* and *R. nigrescens* (Drosophilidae), *Bronia bedai* (Amphisbaenidae), and *Apostolepis intermedia* (Colubridae; Chaves et al., 2010; Nogueira et al., 2011). Proença et al. (2010) made a preliminary analysis based on angiosperm species, but considered that the available information was not sufficient to identify diversity centers accurately. Nogueira et al. (2011) found 10 areas of endemism based on reptiles, but they did not name them formally.

CHACO PROVINCE

Chaco formation—Holmberg, 1898: 447 (regional.).
Subtropical Forests and Savannas area—Hauman, 1920: 46 (regional.).
Chaco Forests and Savannas province—Hauman, 1931: 59 (regional.).
Gran Chaco region—Shannon, 1927: 5 (regional.).
Chacoan Park area—Parodi, 1934: 171, 1945: 128 (regional.).
Subtropical district—Cabrera and Yepes, 1940: 15 (regional.).
Chacoan province—Castellanos and Pérez-Moreau, 1944: 79 (regional.); Cabrera, 1951: 32 (regional.), 1953: 107 (regional.), 1958: 200 (regional.); Morello, 1958: 131 (regional.); Cabrera, 1971: 15 (regional.); Cabrera and Willink, 1973: 72 (regional.); Cabrera, 1976: 18 (regional.); Huber and Riina, 1997: 124 (glossary); Morrone, 1999: 9 (regional.); Zuloaga et al., 1999: 38 (regional.); Morrone, 2000a: 56 (regional.); Ojeda et al., 2002: 24 (biotic evol.); Arzamendia and Giraudo, 2009: 1741 (track anal.); del Río et al., 2015: 1294 (track anal. and clad. biogeogr.).
Chacoan Forest area—Bölcke, 1957: 2 (regional.).
Chacoan region—Hueck, 1957: 40 (regional.); Rivas-Martínez and Navarro, 1994: map (regional.); Rivas-Martínez et al., 2011: 27 (regional.).
Subtropical dominion—Ringuelet, 1961: 160 (biotic evol. and regional.).
Gran Chaco area—Sick, 1969: 452 (regional.).
Chaco center—Müller, 1973: 143 (regional.); Cracraft, 1985: 75 (areas of endem.).
Gran Chaco province—Udvardy, 1975: 41 (regional.); Huber and Riina, 1997: 184 (glossary).
Central Chaco dominion—Ab'Sáber, 1977: map (climate).
Chaco area—Haffer, 1985: 126 (refugia).
Chaco province—Morrone, 2001a: 88 (regional.); López Ruf et al., 2006: 116 (track anal.); Morrone, 2006: 481 (regional.); Aagesen et al., 2009: 310 (endem. anal.); Zanella, 2010: 202 (biotic evol.); Arana et al., 2011: 17 (track anal.); Moreira et al., 2011: 29 (track anal.); Nori et al., 2011: 1008 (cluster anal. and endem. anal.); Ferretti et al., 2012: 1 (track anal.); Ferro, 2013: 324 (biotic evol.); Sandoval and Barquez, 2013: 77 (endem. anal.); Ferretti et al., 2014: 1089; Morrone, 2014b: 75 (regional.); Chiapella and Demaio, 2015: 63 (biotic evol.); Daniel et al., 2016: 1169 (track anal. and regional.).

Chaco ecoregion—Salazar Bravo et al., 2002: 78 (ecoreg.); Abell et al., 2008: 408
    (ecoreg.).
Iguassu ecoregion—Abell et al., 2008: 408 (ecoreg.).
Paraguay ecoregion—Abell et al., 2008: 408 (ecoreg.).
Great Rivers province—López et al., 2008: 1574 (PAE).

Definition

The Chaco province comprises the plains of southeastern Bolivia, western
Paraguay, southwestern Brazil (Mato Grosso do Sul state), and north central Argentina
(Morello, 1958; Sick, 1969; Cabrera and Willink, 1973; Müller, 1973; Cracraft, 1985;
Willink, 1988; Prado, 1993a,b; Morrone, 2000a, 2006, 2014b; Werneck, 2011). It is
bordered on the east by the Paraguay River, on the north by more mesic vegetation
of central Bolivia, the Andes on the west, and the more temperate dry forests of
northern central Argentina (Cracraft, 1985). The Chacoan biome is likely a Pliocene
or Early Pleistocene relict, established over the salty soils left after a withdrawal of
the sea formed by the Andean uplift during the Oligocene (Werneck, 2011). During
the Quaternary interglacial periods, it may have been more humid than today (Nores,
1992). It contains a great complexity of habitat types (Dinerstein et al., 1995).

Endemic and Characteristic Taxa

I (Morrone, 2014b) provided a list of endemic and characteristic taxa. Some
examples include *Trixis antimenorhoea* var. *discolor* (Asteraceae), *Chlorus borellii*,
Pseudoscopas nigrigena, *P. viridis*, and *Scotussa brachyptera* (Acrididae), *Cloeodes
irvingi* (Baetidae), *Tytius confluens* (Buthidae; Figure 7.8a), *Austrochthonius
argentinae* (Chthoniidae), *Echemoides giganteus* (Gnaphosidae), *Azelia neotropica*
and *Helina argentina* (Muscidae), *Simulium chaquense* (Simuliidae), *Callonetta
leucophrys* (Anatidae), *Chunga burmeisteri* (Cariamidae), *Dolichotis salinicola*
(Caviidae), *Alouatta caraya* (Cebidae), *Chlamyphorus retusus* (Dasypodidae),
Furnarius cristatus (Furnariidae; Figure 7.8b), and *Forpus xanthopterygius flaves-
cens* and *Myiopsitta monachus* (Psittacidae).

Vegetation

There are xerophytic scrublands/chaparral, xerophytic forests, and savannas,
with a stratum of herbs, cacti, and terrestrial Bromeliaceae (Cabrera and Willink,
1973; Cabrera, 1976; Burkart et al., 1999; Dinerstein et al., 1995; Graham, 2004).
Dominant plant species include *Acacia aroma, A. caven, Aspidosperma quebracho-
blanco, Astronium balansae, Baccharis salicifolia, Bromelia hieronymi,
Bulnesia sarmientoi, Caesalpinia paraguariensis, Celtis tala, Cercidium praecox, Chorisia insignis, Copernicia australis, Cyperus giganteus, Deinacanthum
urbanianum, Elionurus muticus, Geoffroea decorticans, Gleditsia amorphoides,
Leptochloa chloridiformis, Mimosa pigra, Opuntia quimilo, Paspalum interme-
dium, Patagonula americana, Pisonia zapallo, Polylepis australis, Prosopis alba,*

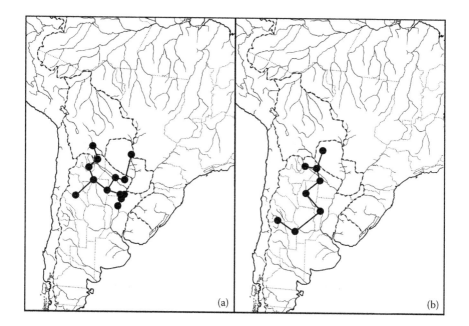

Figure 7.8   Maps with individual tracks in the Chaco province. (a) *Tytius confluens* (Buthidae); (b) *Furnarius cristatus* (Furnariidae).

P. algarobilla, P. kuntzei, P. nigra, P. ruscifolia, Salix humboldtiana, Schinopsis balansae, Schinus longifolia, Tabebuia avellanedae, T. ipe, Trithrinax campestris, and *Zizyphus mistol* (Cabrera and Willink, 1973; Sandoval and Barquez, 2013). Ramella and Spichiger (1989), Spichiger et al. (1991), Galán de Mera and Navarro (1992), and Prado (1993a,b) presented detailed vegetational studies (Figure 7.9).

Biotic Relationships

Müller (1973) suggested a close relationship of this province with the Cerrado and Caatinga provinces. Cabrera (1976) considered that the Chaco province is related to the Pampa and Cuyan High Andean provinces. In the Western Chacoan district, there is a premontane forest that is transitional with the Yungas (Ayarde, 1995). A parsimony analysis of endemicity based on bird taxa (Porzecanski and Cracraft, 2005) and another based on lizards (Colli, 2005) suggested a close relationship between the Cerrado and Chaco province, and both with the Caatinga. According to a cladistic biogeographic analysis based on beetle and plant taxa (Morrone, 1993), this province is closely related to the Monte province.

Regionalization

Some authors (Cabrera, 1971, 1976; Cabrera and Willink, 1973; Rivas-Martínez and Navarro, 1994; Burkart et al., 1999; Werneck, 2011) have recognized nested

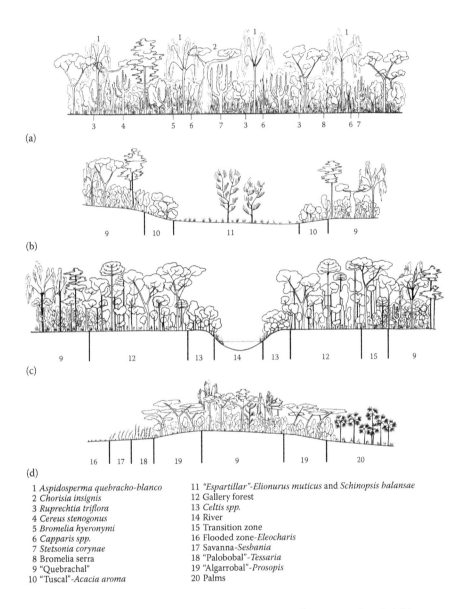

(a)

(b)

(c)

(d)

1 *Aspidosperma quebracho-blanco*	11 *"Espartillar"-Elionurus muticus* and *Schinopsis balansae*
2 *Chorisia insignis*	12 Gallery forest
3 *Ruprechtia triflora*	13 *Celtis spp.*
4 *Cereus stenogonus*	14 River
5 *Bromelia hyeronymi*	15 Transition zone
6 *Capparis spp.*	16 Flooded zone-*Eleocharis*
7 *Stetsonia corynae*	17 Savanna-*Sesbania*
8 Bromelia serra	18 *"Palobobal"-Tessaria*
9 "Quebrachal"	19 *"Algarrobal"-Prosopis*
10 *"Tuscal"-Acacia aroma*	20 Palms

Figure 7.9   Vegetation types in the Chaco province. (a) Xeric forest or *quebrachal*; (b) savanna or *espartillar*; (c) gallery forest; and (d) vegetation sequence in a humid area. (Modified from Ramella, L. and R. Spichiger, *Candollea*, 44: 639–680, 1989.)

units within the Chaco province. I (Morrone, 2014b) considered the existence of only two districts (Figure 7.10). The Western Chacoan district corresponds to the western part of the Chacoan province; endemic taxa include *Acanthocalycium* (Cactaceae) and *Parodiodendron* (Euphorbiaceae; Zuloaga et al., 1999). The Eastern Chacoan district corresponds to the eastern part of the Chacoan province; it has a more humid

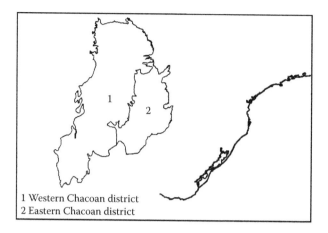

Figure 7.10 Map of the districts of the Chaco province. (Modified from Sandoval, M. L. and R. M. Barquez, *Revista Chilena de Historia Natural*, 86: 75–94, 2013.)

climate, with rainfall increasing from west to east (Cabrera, 1971). Daniel et al. (2016) recently suggested treating the Pantanal area (herein considered within the Rondônia province) as another district of the Chaco province.

PAMPEAN PROVINCE

Pampean subregion—Blyth, 1871: 428 (regional.).

Pampa formation—Holmberg, 1898: 403 (regional.).

Pampean Prairie area—Hauman, 1920: 62 (regional.).

Pampas region—Shannon, 1927: 5 (regional.); Good, 1947: 237 (regional.); Takhtajan, 1986: 251 (regional.).

Pampean Grassland province—Hauman, 1931: 59 (regional.); Bölcke, 1957: 2 (regional.).

Pampean Steppe area—Parodi, 1934: 171 (regional.), 1945: 130 (regional.).

Guaraní province (in part)—Mello-Leitão, 1937: 246 (regional.).

Pampasic district—Cabrera and Yepes, 1940: 15 (regional.).

Pampas area—Hueck, 1957: 40 (regional.); Sick, 1969: 452 (regional.); Haffer, 1985: 128 (refugia); Coscarón and Coscarón-Arias, 1995: 726 (areas of endem.); Porzecanski and Cracraft, 2005: 266 (PAE).

Pampasic dominion—Ringuelet, 1961: 160 (biotic evol. and regional.).

Pampa area—Hueck, 1966: 3 (regional.).

Pampa province—Fittkau, 1969: 642 (regional.); Morrone, 2000a: 58 (regional.), 2001a: 91 (regional.), 2006: 481 (regional.); Moreira et al., 2011: 29 (track anal.); Ferretti et al., 2012: 1 (track anal.), 2014: 1089.

Pampean province—Cabrera, 1951: 42 (regional.), 1953: 107 (regional.), 1958: 200 (regional.), 1971: 24 (regional.); Cabrera and Willink, 1973: 79 (regional.); Cabrera, 1976: 42 (regional.); Huber and Riina, 1997: 273 (glossary); Morrone, 1999: 10 (regional.); Zuloaga et al., 1999: 35 (regional.); Ojeda et al., 2002: 24 (biotic evol.); López et al., 2008: 1575 (PAE); Arzamendia and Giraudo, 2009: 1741 (track anal.);

Morrone, 2014b: 77 (regional.); del Río et al., 2015: 1294 (track anal. and clad. biogeogr.); Daniel et al., 2016: 1169 (track anal. and regional.); Dos Santos et al., 2016: 362.

Pampean region—Rivas-Martínez and Navarro, 1994: map (regional.); Rivas-Martínez et al., 2011: 27 (regional.).

Pampas ecoregion—Dinerstein et al., 1995: 99 (ecoreg.).

Pampa ecoregion—Huber and Riina, 1997: 244 (glossary); Burkart et al., 1999: 30 (ecoreg.).

Pampas province—Nori et al., 2011: 1009 (cluster anal. and endem. anal.).

Definition

The Pampean province comprises central-western Argentina between the latitudes 30° and 39° south, Uruguay, and the southern part of the Brazilian state of Rio Grande do Sul (Cabrera and Willink, 1973; Takhtajan, 1986; Willink, 1988; Rivas-Martínez and Navarro, 1994; Dinerstein et al., 1995; Morrone, 2000a, 2006, 2014b).

Endemic and Characteristic Taxa

I (Morrone, 2014b) provided a list of endemic and characteristic taxa. Some examples include *Ephedra tweediana* (Ephedraceae), *Haylockia* (Amaryllidaceae), *Criscia* (Figure 7.11), *Panphalea bupleurifolia*, *Schlechtendalia*, and *Sommerfeltia* (Asteraceae), *Rivudiva coveloe* (Baetidae), *Belostoma martini* (Belostomatidae), *Pachipterinella* (Cercopidae), *Atrichonotus marginatus*, *Listroderes elegans*, *Naupactus dissimulator*, *Priocyphus bosqi*, and *Trichonaupactus densius* (Curculionidae), *Hyalella bonaeriensis* and *H. warmingii* (Hyalellidae), *Tallium buonoae* (Mutillidae), *Dichotomius haroldi* and *Xenocanthon sericans* (Scarabaeidae), *Mitragenius nudus* and *Nyctelia saundersi* (Tenebrionidae), *Phrynops hillarii* (Chelidae), *Akodon kempi*, *Calomys musculinus*, and *Scapteromys tumidus* (Cricetidae), *Cyphocharax spiluropsis* (Curimatidae), and *Asthenes hudsoni* and *Spartonoica maluroides* (Furnariidae).

Vegetation

There are savannas, with species of Poaceae (that can reach 1 m height), herbs, and shrubs. There are also seasonally dry tropical forests similar to those of the Chaco province, but impoverished; flooded savannas; and gallery forests along the rivers (Cabrera and Willink, 1973; Dinerstein et al., 1995). In the savannas, dominant plant species belong to the genera *Aristida*, *Bothriochloa*, *Briza*, *Bromus*, *Eragrostis*, *Melica*, *Panicum*, *Paspalum*, *Piptochaetium*, *Schizachirium*, and *Stipa* (Cabrera, 1971; Cabrera and Willink, 1973; Burkart et al., 1999). Dominant plant species include *Acacia caven*, *Allophyllus edulis*, *Andropogon lateralis*, *Aspidosperma quebracho-blanco*, *Butia yatay*, *Celtis iguanea*, *C. spinosa*, *Cyclolepis genistoides*, *Elionurus muticus*, *Fagara hiemalis*, *Geoffroea decorticans*, *Heterostachys ritteriana*, *Melica magra*, *Panicum racemosum*, *Paspalum almum*, *P. notatum*,

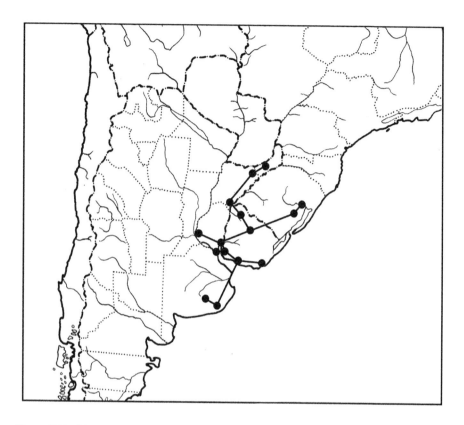

Figure 7.11   Map with the individual track of *Criscia* (Asteraceae) in the Pampean province.

Prosopis alba, P. algarobillo, P. caldenia, P. flexuosa, P. nigra, and *Schinus longifolia* (Cabrera, 1971, 1976; Cabrera and Willink, 1973).

Biotic Relationships

The Pampean province is related to the Chaco and Monte provinces, where a sequence of biotic impoverishment from the Chaco to the Pampean was hypothesized by Cabrera (1976), with the Monte province being intermediate between both (Figure 7.12). Amazonian, Patagonian, and Subantarctic biotic affinities have been also postulated (Ringuelet, 1955, 1981; Ribichich, 2002; Ferretti et al., 2012). In particular, biotic affinities with the Paraná dominion have suggested a past extension of a tropical biota that gradually contracted (Ringuelet, 1978, 1981; Morrone and Lopretto, 1994; Crisci et al., 2001; Goldani and de Carvalho, 2003).

Regionalization

Cabrera (1971, 1976) and Cabrera and Willink (1973) have identified four districts, to which I (Morrone, 2014b) added the Espinal. The Western Pampean district

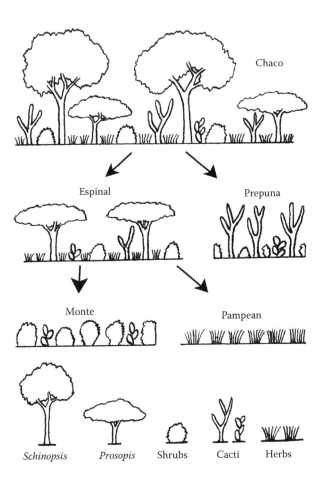

Figure 7.12 Sequence of biotic impoverishment from the Chaco to the Espinal, Prepuna, Monte, and Pampean provinces. (Modified from Cabrera, A. L., "Regiones fitogeográficas argentinas." In: Kugler, W. F. (ed.), *Enciclopedia Argentina de Agricultura y Jardinería*. II, ACME, Buenos Aires, pp. 1–85, 1976.)

corresponds to southern Córdoba, northeastern La Pampa, and northwestern Buenos Aires (Cabrera, 1971; Cabrera and Willink, 1973). The Austral Pampean district corresponds to the southern part of Buenos Aires province (Cabrera, 1971; Cabrera and Willink, 1973). The Eastern Pampean district corresponds to the northern and eastern parts of Buenos Aires province (Cabrera, 1971; Cabrera and Willink, 1973). The Uruguayan district corresponds to Uruguay, the southern part of the Brazilian state of Rio Grande do Sul, and southern Entre Ríos and Santa Fe in Argentina (Cabrera, 1971; Cabrera and Willink, 1973); some endemic taxa include *Amphisbaena kingii* (Amphisbaenidae), *Ctenomys pearsoni* (Ctenomyidae), and *Pleurodema bibroni* (Leptodactylidae; Ortega Baes et al., 2002). The Espinal district is an irregular arch surrounding the rest of the Pampean province, from central Corrientes and northern Entre Ríos to central Santa Fe and Córdoba, San Luis, central La Pampa, and southern Buenos Aires (Cabrera, 1971; Burkart et al., 1999); two endemic taxa include *Parachernes* sp. (Chernetidae) and *Ketianthidium* (Megachilidae; Ceballos and Rosso de Ferradás, 2008; Durante et al., 2008).

A recent analysis (Dos Santos et al., 2016) based on Ephemeroptera suggested that the Uruguayan district is not part of the Pampean province. The authors concluded that it represents the southern limit of tropical affinities and should be treated as a distinct province, assigned to the Paraná dominion. Until more evidence is gathered, I prefer to keep this district as part of the Pampean province.

PARANÁ DOMINION

Guaraní province (in part)—Mello-Leitão, 1937: 246 (regional.).
Tupí district (in part)—Cabrera and Yepes, 1940: 15 (regional.).
Tupí, Araucariland or Tupí–Guaraní center (in part)—Lane, 1943: 414 (regional.).
Southeastern Brazil Subtropical Forests region—Hueck, 1957: 40 (regional.).
Atlantic province—Rizzini, 1963: 46 (regional.), 1997: 622 (regional.).
Atlantic–Paraná subregion—Rivas-Martínez and Navarro, 1994: map (regional.).
Southern Brazil Mountains area—Coscarón and Coscarón-Arias, 1995: 726 (areas of endem.).
Paraná subregion—Morrone, 1999: 10 (regional.), 2001a: 97 (regional.), 2005: 238 (regional.); López Ruf et al., 2006: 116 (track anal.); Morrone, 2006: 482 (regional.); Quijano-Abril et al., 2006: 1268 (track anal.); Nihei and de Carvalho, 2007: 497 (clad. biogeogr.); Navarro et al., 2009: 509 (endem. anal.); Morrone, 2010a: 37 (regional.), 2010b: 1 (biotic evol.); Moreira et al., 2011: 29 (track anal.); Pires and Marinoni, 2011: 8 (track anal.); Coulleri and Ferrucci, 2012: 105 (track anal.); Campos-Soldini et al., 2013: 16 (track anal.); Lamas et al., 2014: 955 (clad. biogeogr.).
Atlantic Forest dominion—Fiaschi and Pirani, 2009: 483 (regional.).
Atlantic Forest area—Bertoncello et al., 2011: 3414 (cluster anal. and regional.).
Brazilian–Paranense region—Rivas-Martínez et al., 2011: 27 (regional.).
Paraná dominion—Morrone, 2014b: 80 (regional.), 2014c: 207 (clad. biogeogr.); Daniel et al., 2016: 1167 (track anal. and regional.).

Definition

The Paraná dominion comprises northeastern Argentina, eastern Paraguay, southern Brazil (west of the Serra do Mar), and toward central Rio Grande do Sul and eastern Brazil, between the latitudes 7° and 32° south (Cabrera and Willink, 1973; Willink, 1988; Rivas-Martínez and Navarro, 1994; Morrone, 2001b, 2006, 2014b). This dominion, which corresponds to all the forests east of the Andes and south of the Amazon, originally covered an area of nearly 1,300,000 square kilometers, ranging between the states of Rio Grande do Norte and Rio Grande do Sul, and from the eastern Atlantic coast to the Central Brazilian plateau and parts of Paraguay and Argentina (Bertoncello et al., 2011). It is among the most threatened biomes on earth, with about 11.4–16 percent of the original area of vegetation cover remaining (Ribeiro et al., 2009). It is highly fragmented and reduced, with its larger and best preserved remains confined to the high-elevation areas along the Brazilian southern and southeastern coastal regions, mostly on the Serra do Mar mountain ranges (Bertoncello et al., 2011).

Endemic or Characteristic Taxa

Taxa endemic to this dominion (Morrone, 2001a,b) include *Lithachne horizontalis* (Poaceae), *Podocarpus lambertii* (Podocarpaceae), *Castalia undosa* (Hyriidae), *Trachelopachys gracilis* (Clubionidae), *Homalocerus* (Belidae; Figure 7.13), *Cyrtoneurina biseta*, *Dolicophaonia noctiluca*, *Limnophora vittata*, and *Pseudotilolepis chrysella* (Muscidae), *Alchisme frontomaculata* and *A. obscura* (Membracidae), *Acanthagrion ascendens* (Coenagrionidae), *Oligosarcus* (Characidae), *Clibanornis dendrocolaptoides* (Furnariidae), *Amazona pretrei*, *Pionopsitta pileata*, and *Pyrrurha frontalis chiripepe* (Psittacidae), and *Pygoderma bilabiatum* (Phyllostomidae).

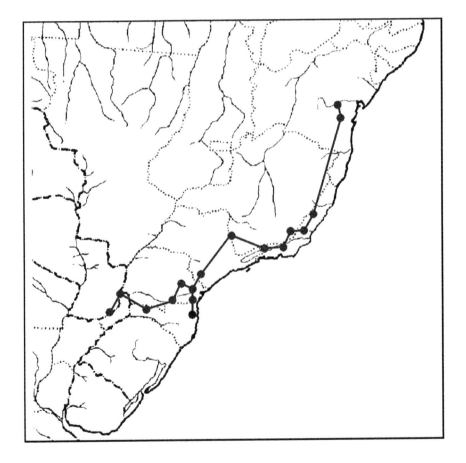

Figure 7.13 Map with the individual track of *Homalocerus* (Belidae) in the Paraná dominion.

Biotic Relationships

The inclusion of the Atlantic and Paraná forests within the Amazonian dominion (Cabrera and Willink, 1973) has been questioned by Costa (2003), who undertook a comparative phylogeographic analysis of forest-dwelling small mammals distributed between and within the Amazon and Atlantic forests, finding that sequence similarity was often greater between samples from the Atlantic Forest and either Amazon or central Brazilian forests than it was within each of both forest areas. The Atlantic Forest clades were either not reciprocally monophyletic or were the sister group to all the other clades, and the central Brazilian area did not behave as a separate region but as complementary to either the Amazon or Atlantic forests.

Several authors have discussed the relationships of this dominion with the Subantarctic subregion (Kuschel, 1960; Cabrera, 1976; Vanin, 1976; Morrone and Lopretto, 1994; Maury et al., 1996). According to paleontological, paleoclimatological, and geological evidence, a temperate climate prevailed in southern South America during the Paleogene, allowing the existence of a continuous forest that extended further south than today. A trend to cooling and aridification began in Oligocene and Miocene, and later the forest fragmented simultaneously with the climatic changes induced by the uplift of the Andes and the expansion of the Chacoan biota (Kuschel, 1969). The disjunct distributions of the weevil families Belidae (Vanin, 1976) and Nemonychidae (Kuschel and May, 1997), in the Sub-Antarctic subregion and the Paraná dominion, support this relationship. According to a parsimony analysis of endemicity based on species of anurans (Ron, 2000), the Paraná dominion is more closely related to the Amazonian subregion.

Regionalization

The Paraná dominion comprises the Atlantic, Paraná, and *Araucaria* Forest provinces (Morrone, 2014b).

ATLANTIC PROVINCE

Tupí province—Mello-Leitão, 1937: 246; Fittkau, 1969: 642 (regional.).
Austro-Oriental subprovince—Rizzini, 1963: 47 (regional.), 1997: 623 (regional.).
Littoral sector—Rizzini, 1963: 47 (regional.), 1997: 623 (regional.).
Beach subsector—Rizzini, 1963: 47 (regional.).
Low Slope subsector—Rizzini, 1963: 47 (regional.).
Maritime subsector—Rizzini, 1963: 47 (regional.).
Restinga subsector—Rizzini, 1963: 47 (regional.).
Tabuleiros district—Rizzini, 1963: 47 (regional.).
Atlantic province—Cabrera and Willink, 1973: 64 (regional.); Ávila-Pires, 1974b: 169 (regional.); Rivas-Martínez and Navarro, 1994: map (regional.); Fernandes and Bezerra, 1990: 99 (regional.); Fernandes, 2006: 67 (regional.); Arzamendia and Giraudo, 2009: 1740 (track anal.); Morrone, 2014b: 80 (regional.); del Río et al.,

2015: 1294 (track anal. and clad. biogeogr.); Daniel et al., 2016: 1169 (track anal. and regional.).

Serra do Mar center—Müller, 1973: 125 (regional.); Cracraft, 1985: 72 (areas of endem.).

Serra do Mar province—Udvardy, 1975: 41 (regional.).

Atlantic Tropical dominion—Ab'Sáber, 1977: map (climate).

Southeastern Brazil area—Cracraft, 1988: 223 (clad. biogeogr.).

Littoral or Coastal subprovince—Fernandes and Bezerra, 1990: 114 (regional.); Fernandes, 2006: 84 (regional.).

Brazilian Atlantic Coast Restingas ecoregion—Dinerstein et al., 1995: 106 (ecoreg.).

Brazilian Coastal Atlantic Forests ecoregion—Dinerstein et al., 1995: 93 (ecoreg.).

Maritime Cordillera sector—Rizzini, 1997: 623 (regional.).

Tabuleiros sector—Rizzini, 1997: 623 (regional.).

Brazilian Atlantic Forest province—Morrone, 1999: 11 (regional.), 2001a: 98 (regional.); López Ruf et al., 2006: 116 (track anal.); Morrone, 2006: 482 (regional.); Quijano-Abril et al., 2006: 1270 (track anal.); Moreira et al., 2011: 29 (track anal.).

Brazilian Atlantic Coast province—Morrone, 2001b: 2 (regional.).

Atlantic Forest area—Porzecanski and Cracraft, 2005: 266 (PAE).

Brazilian Atlantic province—Rivas-Martínez et al., 2011: 27 (regional.).

Definition

The Atlantic province comprises a narrow strip of tropical rainforest along the Brazilian Atlantic Coast east of the coastal cordillera, between the latitudes 7° and 32° south (Cabrera and Willink, 1973; Müller, 1973; Fernandes and Bezerra, 1990; Morrone, 2001b, 2006, 2014b). It is basically located from the São Francisco River in the north (Pernambuco) to the southern end of the rainforest in Santa Catarina, with its boundaries presumably determined by autoecological factors relating to habitat and physiological tolerance of more arid conditions (Cracraft, 1985).

Endemic and Characteristic Taxa

I (Morrone, 2014b) provided a list of endemic and characteristic taxa. Some examples include *Crinodendron brasiliense* (Elaeocarpaceae), *Arachosia bifasciata* (Anyphaenidae), *Ericydeus bahiensis*, *Prosicoderus crassipes*, *Sicoderus analis*, and *S. subcoronatus* (Curculionidae), *Nicomia interrupta* and *N. monticola* (Membracidae), *Curicta bilobata* and *C. longimanus* (Nepidae), *Coprophaneus bellicosus* (Scarabaeidae), *Rhynchosciara americana* (Sciaridae), *Neolindus schubarti* (Staphylinidae), *Leucopternis lacernulata* (Figure 7.14) and *L. polionota* (Accipitridae), *Bradypus torquatus* (Bradypodidae), *Tijuca atra* (Cotingidae), *Monodelphis iheringi* (Didelphidae), *Ilicura militaris* and *Muscipipra vetula* (Pipridae), and *Batara cinerea*, *Biatas nigropectus*, and *Rhopornis ardesiaca* (Thamnophilidae).

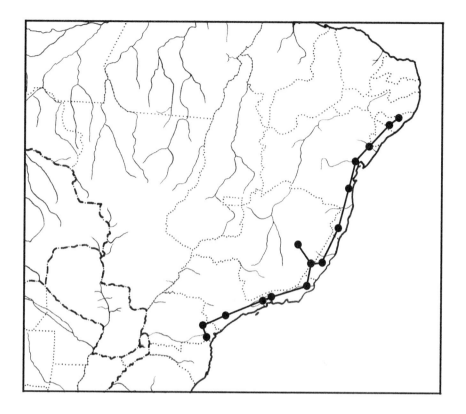

Figure 7.14 Map with the individual track of *Leucopternis lacernulata* (Accipitridae) in the Atlantic province.

Vegetation

The Atlantic province comprises two types of tropical humid forests: the rainforest that includes tall trees (30–40 m height), a stratum of palms, many lianas and epiphytes (mainly Bromeliaceae), Orchidaceae, and ferns, which is situated on the eastern slope of the mountain chain; and the semideciduous forest that extend across the plateau farther inland (Pinto-da-Rocha and Silva, 2005; Perret et al., 2006). In the coast, there are mangroves and dunas or *restingas* (Cabrera and Willink, 1973; Dinerstein et al., 1995). Dominant plant species include *Alchornea triplinervia, Cedrela velloziana, Cecropia candida, Euterpe edulis, Meriania excelsa, Ocotea catharinensis, Pouteria laurifolia, P. venosa, Sickingia glaziovii, Sideroxylon crassipedicellatum, Sloanea guianensis, Tabebuia umbellata*, and *Torrubia olfersiana* (Cabrera and Willink, 1973).

Biotic Relationships

According to a parsimony analysis of endemicity based on Heteroptera of the family Reduviidae (Morrone and Coscarón, 1996), this province is related to the Paraná province. Another parsimony analysis of endemicity, based on bird taxa (Porzecanski and Cracraft, 2005), postulated a close relationship of this province with the Caatinga, Cerrado, and Chacoan provinces. A cladistic biogeographic analysis based on bird taxa (Cracraft and Prum, 1988) found a close relationship of this province with the Pará province.

Regionalization

Müller (1973), Silva et al. (2004), and DaSilva and Pinto-da-Rocha (2010) identified nested units, which were treated by Morrone (2014b) as three districts. The Pernambuco district corresponds to the northernmost part of the province, situated between Bahia and Recife (Müller, 1973); some endemic taxa include *Philydor novaesi* and *Synallaxis infuscata* (Furnariidae), *Myrmotherula snowi* and *Terenura sicki* (Thamnophilidae), *Tangara fastuosa* (Thraupidae), *Tinamus solitarius pernambucensis* (Tinamidae), *Phylloscartes ceciliae* (Tyrannidae), and *Bothrops leucurus* (Viperidae; Müller, 1973; Silva et al., 2004). The Bahia district is situated south of the Pernambuco district (Müller, 1973); some endemic taxa include *Alvimia, Anomochloa, Criciuma, Eremocaulon*, and *Sucrea monophylla* (Poaceae), *Goniosoma modestum* (Gonyleptidae), *Synallaxis cinerea* (Furnariidae), *Formicivora iheringi* and *Rhopornis ardesiaca* (Thamnophilidae), and *Phylloscartes beckeri* (Tyrannidae; Soderstrom et al., 1988; Silva et al., 2004; Pinto-da-Rocha and Silva, 2005). The Paulista district corresponds to the southeastern part of the Brazilian coast from Cabo Frio in the north to Florianópolis in the south (Müller, 1973); some endemic taxa include *Sucrea maculata* and *S. sampaiana* (Poaceae), *Acutisoma banhadoae, Ampheres fuscopunctatus, Gonyleptoides acanthoscelis, Heliella singularis, Iguapeia melanocephala, Moreiranula mamillata, Zortalia bicalcarata*, and *Z. leprevosti* (Gonyleptidae), *Eidmanacris bidentata* and *E. tridentata* (Phalangopsidae), *Pteroglossus bailloni* (Ramphastidae), *Myrmotherula gularis* (Thamnophilidae), and *Dacnis nigripes, Orthogonys chloricterus*, and *Tangara desmaresti* (Thraupidae; Silva et al., 2004; Pinto-da-Rocha and Silva, 2005; Yamaguti and Pinto-da-Rocha, 2009; Souza-Dias et al., 2015).

PARANÁ PROVINCE

Misiones formation—Holmberg, 1898: 451 (regional.).
Austral Brazil Forests and Savannas province—Hauman, 1931: 59 (regional.).
Misiones Subtropical Forest area—Parodi, 1934: 171 (regional.).
Misiones–Brazilian Forest area—Castellanos and Pérez-Moreau, 1941: 378 (regional.).
Misiones province—Castellanos and Pérez-Moreau, 1944: 90 (regional.).
Misiones Forest area—Parodi, 1945: 127 (regional.).

Eastern Subtropical province—Cabrera, 1951: 28 (regional.), 1953: 114 (regional.);
   Huber and Riina, 1997: 275 (glossary).
Maritime Cordillera sector—Rizzini, 1963: 47 (regional.).
Paraná province—Cabrera, 1971: 11 (regional.), 1976: 10 (regional.); Cabrera and
   Willink, 1973: 60 (regional.); Rivas-Martínez and Navarro, 1994: map (regional.);
   Huber and Riina, 1997: 273 (glossary); Ojeda et al., 2002: 24 (biotic evol.);
   Arzamendia and Giraudo, 2009: 1740 (track anal.); Morrone, 2014b: 81 (regional.);
   del Río et al., 2015: 1294 (track anal. and clad. biogeogr.); Daniel et al., 2016: 1169
   (track anal. and regional.).
Brazilian Rainforest province—Udvardy, 1975: 41 (regional.).
Montane subprovince—Fernandes and Bezerra, 1990: 114 (regional.); Fernandes,
   2006: 72 (regional.).
Meridional Planalto sector (in part)—Fernandes and Bezerra, 1990: 114 (regional.);
   Fernandes, 2006: 79 (regional.).
Paraná Forest ecoregion—Burkart et al., 1999: 21 (ecoreg.).
Forests province—Morrone, 1999: 10 (regional.).
Brazilian Interior Atlantic Forests ecoregion—Dinerstein et al., 1995: 93 (ecoreg.);
   Huber and Riina, 1997: 42 (glossary).
Paraná Forest province—Morrone, 2001a: 99 (regional.), 2001b: 2 (regional.); López
   Ruf et al., 2006: 116 (track anal.); Morrone, 2006: 482 (regional.); Moreira et al.,
   2011: 29 (track anal.); Ferretti et al., 2012: 1 (track anal.), 2014: 1089.
Paranense province (in part)—Rivas-Martínez et al., 2011: 27 (regional.); Campos-
   Soldini et al., 2013: 16 (track anal.).

Definition

The Paraná province comprises southeastern Brazil west of the Serra do Mar to central Rio Grande do Sul, northeastern Argentina, and eastern Paraguay (Cabrera, 1971; Cabrera and Willink, 1973; Morrone, 2001b, 2006, 2014b).

Endemic and Characteristic Taxa

I (Morrone, 2014b) provided a list of endemic and characteristic taxa. Some examples include *Chionolaena arbuscula*, *Holocheilus illustris*, *Jungia floribunda*, and *Panphalea missionum* (Asteraceae), *Fuchsia bracelinae*, *F. coccinea*, and *F. regia* subsp. *regia* (Onagraceae; Figure 7.15a), *Gomphochernes savignyi* (Chernetidae), *Schendylops demartini*, *S. paulistus*, and *S. sublaevis* (Schendylidae), *Neobisnius brasilianus* (Staphylinidae), *Melanolestes lugens* (Reduviidae), *Rhynchosciara hollaenderi* (Sciaridae), and *Cyphocharax modestus* (Curimatidae; Figure 7.15b).

Vegetation

There are tropical humid forests, with trees of 20–30 meters in height, Bambusoideae, and tree ferns, as also savannas and isolated trees (Cabrera and Willink, 1973; Dinerstein et al., 1995). Dominant plant species include *Andropogon lateralis*, *Aspidosperma polyneuron*, *Astronium balansae*, *Balfourodendron*

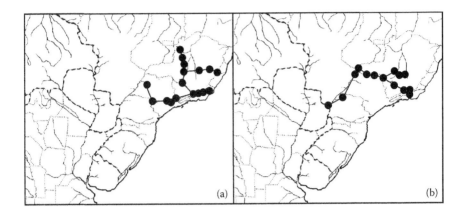

Figure 7.15   Maps with individual tracks in the Paraná province. (a) *Fuchsia regia* subsp. *regia* (Onagraceae); (b) *Cyphocharax modestus* (Curimatidae).

riedelianum, Cabralea oblongifolia, Cedrela fissilis, Chusquea pinifolia, Cortaderia modesta, Diatenopteryx sorbifolia, Elionurus muticus, Escallonia montevidensis, E. organensis, Euterpe edulis, Halocalyx balansae, Ilex paraguariensis, Lonchocarpus leucanthus, Nectandra saligna, Parapiptadenia rigida, Paspalum guaraniticum, Peltophorum dubium, Podostemon aguirensis, P. comata, Senecio cuneifolius, Syagrus romanzzoffianus, and *Tabebuia pulcherrima* (Cabrera, 1971, 1976; Cabrera and Willink, 1973).

Biotic Relationships

According to a parsimony analysis of endemicity based on Heteroptera of the family Reduviidae (Morrone and Coscarón, 1996), the Paraná province is closely related to the Atlantic province.

Regionalization

Three districts have been identified in Argentina: Campos, Mixed Forests, and Montane (Cabrera, 1971, 1976; Cabrera and Willink, 1973; Morrone, 2014b). It seems that there is some correspondence with subsectors and districts previously recognized by Rizzini (1963), but this should be investigated more thoroughly. The Campos district corresponds to southern Misiones and northeastern Corrientes (Cabrera, 1971, 1976). The Mixed Forests district corresponds to the low forest that reaches 1,800 meters in height (Rizzini, 1963), occupying almost all the extension of the province of Misiones (Cabrera, 1971); two endemic taxa are *Iguazua lilloana* and *Passaliolla eugastrica* (Scarabaeidae; Ocampo and Ruiz Manzano, 2008). The Montane district includes the most elevated areas of the Paraná province, above 1,800 meters in altitude (Cabrera and Willink, 1973).

ARAUCARIA FOREST PROVINCE

Araucaria Forest area—Hueck, 1953: 16 (regional.).
Southeastern Brazil *Araucaria* Forests region—Hueck, 1957: 40.
Araucaria angustifolia Forest province—Aubreville, 1962: 46 (veget.); Morrone,
 2001a: 101 (regional.), 2001b: 5 (regional.), 2006: 482 (regional.); Moreira et al.,
 2011: 29 (track anal.); Ferretti et al., 2012: 1 (track anal.), 2014: 1089.
Meridional Planalto sector—Rizzini, 1963: 48 (regional.); Fernandes and Bezerra,
 1990: 114 (regional.); Rizzini, 1997: 623 (regional.); Fernandes, 2006: 79 (regional.).
Pine forests district—Cabrera and Willink, 1973: 61 (regional.).
Paraná center—Müller, 1973: 138 (regional.); Cracraft, 1985: 73 (areas of endem.).
Brazilian Planalto province—Udvardy, 1975: 41 (regional.).
Araucaria Planaltos dominion—Ab'Sáber, 1977: map (climate).
Araucaria subsector—Fernandes and Bezerra, 1990: 111 (regional.).
Brazilian *Araucaria* Forests ecoregion—Dinerstein et al., 1995: 98 (ecoreg.).
Pine forests province—Morrone, 1999: 11 (regional.).
Paranense province (in part)—Rivas-Martínez et al., 2011: 27 (regional.).
Araucaria Forest province—Morrone, 2014b: 82 (regional.); del Río et al., 2015: 1294
 (track anal. and clad. biogeogr.); Daniel et al., 2016: 1169 (track anal. and regional.).

Definition

The *Araucaria* Forest province comprises southern Brazil and northeastern
Argentina, between 600 and 1,800 meters altitude (Cabrera and Willink, 1973;
Morrone, 2001b, 2006, 2014b). Its northern boundary lies at about the level of São
Paulo, extends to the west to the Paraná River, and has its southern limits at the Jacuí
River (Cracraft, 1985).

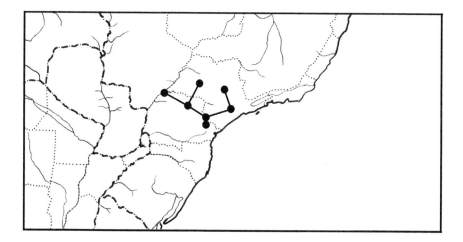

Figure 7.16 Map with the individual track of *Cyphocharax nagelii* (Curimatidae) in the
Araucaria forest province.

Endemic and Characteristic Taxa

I (Morrone, 2014b) provided a list of endemic and characteristic taxa. Some examples include *Araucaria angustifolia* (Araucariaceae), *Fuchsia hatschbachii* and *F. regia* subsp. *reitzii* (Onagraceae), *Taphroderes sahlbergi* (Brentidae), *Araucarius brasiliensis*, *A. crassipunctatus*, *Corthylus praealtus*, *Pandeleteius colatus*, *Phloeotribus argentinensis*, *P. cylindricus*, *Platypus araucariae*, and *Spermologus rufus* (Curculionidae), *Sepedonea trichotypa* (Sciomyzidae), *Cyphocharax nagelii* (Curimatidae; Figure 7.16), and *Cinclodes pabsti*, *Clibanornis dendrocolaptoides*, and *Leptasthenura setaria* (Furnariidae).

Vegetation

There are *Araucaria angustifolia* forests, typically associated with dense tropical moist forests (with *Drimys brasiliensis*, *Ilex paraguariensis*, *Podocarpus lambertii*, and several species of Myrtaceae and Lauraceae), although pure stands also occur (Cabrera and Willink, 1973; Dinerstein et al., 1995).

CHAPTER 8

The South American Transition Zone

The South American transition zone comprises the highlands of the Andes between western Venezuela and Chile, desert areas of coastal Peru and northern Chile, and central western Argentina (Morrone, 2004b, 2006, 2014b). It spans from sea level to 4,500 meters. It has been shaped by the uplift of the Andes in the Neogene. This rise progressed from south to north and from west to east, with two major events: one in the Middle Miocene and another at the beginning of the Pliocene (Amarilla et al., 2015).

SOUTH AMERICAN TRANSITION ZONE

Peruvian subregion (in part)—Blyth, 1871: 428 (regional.).
Tropical Andean subarea (in part)—Clarke, 1892: 381 (regional.).
Argentinean subarea (in part)—Clarke, 1892: 381 (regional.).
Patagonian subregion (in part)—Sclater and Sclater, 1899: 65 (regional.); Kuschel, 1964: 447 (regional.); Hershkovitz, 1969: 8 (regional.); Kuschel, 1969: 712 (regional.); Ojeda et al., 2002: 23 (biotic evol.).
Andean region (in part)—Shannon, 1927: 3 (regional.); Good, 1947: 236 (regional.); O'Brien, 1971: 198 (regional.); Morain, 1984: 178 (text); Rivas-Martínez and Navarro, 1994: map (regional.); Huber and Riina, 1997: 24 (glossary); Morrone, 2001a: 103 (regional.), 2001d: 70 (regional.), 2006: 483 (regional.); Quijano-Abril et al., 2006: 1268 (track anal.); López et al., 2008: 1564 (PAE); Löwenberg-Neto and de Carvalho, 2009: 1751 (PAE); Procheş and Ramdhani, 2012: 263 (cluster anal. and regional.).
Andean dominion (in part)—Hauman, 1931: 60 (regional.); Cabrera, 1951: 48 (regional.); Orfila, 1941: 86 (regional.); Cabrera, 1957: 335 (regional.).
Andean–Patagonian subregion (in part)—Mello-Leitão, 1937: 232 (regional.), 1943: 129 (regional.); Ringuelet, 1961: 156 (biotic evol. and regional.); Rapoport, 1968: 75 (biotic evol. and regional.); Ringuelet, 1978: 255 (biotic evol.); Fittkau, 1969: 639 (regional.); Paggi, 1990: 303 (regional.).
Andean district (in part)—Cabrera and Yepes, 1940: 16 (regional.).
Subandean district—Cabrera and Yepes, 1940: 15 (regional.).

High Andean province—Cabrera, 1951: 49 (regional.), 1953: 107 (regional.), 1957: 337
    (regional.), 1958: 200 (regional.); Cabrera and Willink, 1973: 84 (regional.); Huber
    and Riina, 1997: 270 (glossary); Ojeda et al., 2002: 24 (biotic evol.).
Andean province (in part)—Fittkau, 1969: 642 (regional.).
Subandean province (in part)—Fittkau, 1969: 642 (regional.).
Central Andes area—Sick, 1969: 463 (regional.).
Andean–Patagonian dominion (in part)—Cabrera, 1971: 29 (regional.), 1976: 50
    (regional.); Cabrera and Willink, 1973: 83 (regional.); Huber and Riina, 1997: 150
    (glossary); Zuloaga et al., 1999: 18 (regional.); Ojeda et al., 2002: 24 (biotic evol.).
Austral subregion (in part)—Ringuelet, 1978: 255 (biotic evol.); Almirón et al., 1997:
    23 (regional.).
Andean subkingdom (in part)—Rivas-Martínez and Tovar, 1983: 516 (regional.);
    Huber and Riina, 1997: 332 (glossary).
Argentine subregion (in part)—Smith, 1983: 462 (cluster anal. and regional.).
Andean subregion (in part)—Morrone, 1994b: 190 (PAE); Morrone, 1996: 105
    (regional.); Posadas et al., 1997: 2 (PAE).
Austro–American subkingdom (in part)—Rivas-Martínez and Navarro, 1994: map
    (regional.); Rivas-Martínez et al., 2011: 27 (regional.).
Central Andes bioregion—Dinerstein et al., 1995: map 1 (ecoreg.); Huber and Riina,
    1997: 37 (glossary).
Neotemperate region (in part)—Amorim and Pires, 1996: 187 (regional.).
Páramo–Punan subregion—Morrone, 1999: 11 (regional.), 2001a: 106 (regional.),
    2001c: 1 (regional.); Quijano-Abril et al., 2006: 1268 (track anal.).
South American transition zone—Morrone, 2004a: 158 (track anal.), 2004b: 42
    (regional.), 2006: 482 (regional.); Quijano-Abril et al., 2006: 1271 (track anal.);
    Couri and de Carvalho, 2008: 2677 (biotic evol.); Roig-Juñent et al., 2008: 23 (biotic
    evol.); Roig et al., 2009: 164 (regional.); de Carvalho and Couri, 2010: 295 (track
    anal.); Morrone, 2010a: 37 (regional.); Urtubey et al., 2010: 505 (track anal. and
    clad. biogeogr.); Hechem et al., 2011: 46 (track anal.); Löwenberg-Neto et al., 2011:
    1942 (macroecol.); Luebert, 2011: 109 (biotic evol.); Coulleri and Ferrucci, 2012:
    105 (track anal.); Ferretti et al., 2012: 1 (track anal.); Mercado-Salas et al., 2012: 459
    (track anal.); Campos-Soldini et al., 2013: 16 (track anal.); Ferro, 2013: 323 (biotic
    evol.); Lamas et al., 2014: 955 (clad. biogeogr.); Morrone, 2014b: 82 (regional.),
    2014c: 203 (clad. biogeogr.); Klassa and Santos, 2015: 520 (endem. anal.); Amarilla
    et al., 2015: 1 (biotic evol.); Morrone, 2015d: 210 (regional.); Aagesen and del Valle
    Elías, 2016: 161 (endem. anal.).
Tropical South Andean superegion—Rivas-Martínez et al., 2011: 27 (regional.).
Tropical South Andean region—Rivas-Martínez et al., 2011: 27 (regional.).

Endemic and Characteristic Taxa

I (Morrone, 2001c) provided a list of endemic and characteristic taxa. Some examples include *Ephedra breana*, *E. multiflora*, and *E. rupestris* (Ephedraceae), *Azorella
compacta* (Apiaceae), *Adaetobdella cryptica* (Glossiphoniidae), *Trachelopachys bicolor*
(Clubionidae), *Amblygnathus gilvipes peruanus* and *Notiobia aquilalorum* (Carabidae),
Belostoma dallasi (Belostomatidae), *Crites*, *Incacris*, and *Punacris* (Tristiridae), *Buteo
poecilochrous* (Accipitridae), *Phoenicopterus andinus* and *P. jamesi* (Phoenicopteridae),

Diuca speculifera, Poospiza garleppi, Sicalis lutea, and *S. uropygialis* (Thraupidae), *Picumnus dorbignyanus* and *Veniliornis frontalis* (Picidae), *Hippocamelus* (Cervidae), *Felis jacobita* and *Lynchailurus pajerus garleppi* (Felidae), *Sturnira bogotensis* (Phyllostomidae), and *Chaetophractus nationi* (Dasypodidae).

Biotic Relationships

Urtubey et al. (2010) undertook a cladistic biogeographic analysis of the provinces of this transition zone and the subregions of the Neotropical and Andean regions, based on Asteraceae. They obtained a general area cladogram (Figure 8.1), based on the area cladograms of *Barnadesia, Chuquiraga, Dasyphyllum, Hypochaeris,* and the *Lucilia* group, which indicates a basic separation between the Atacaman, Monte, and Cuyan High Andean provinces closely related to the Andean region and the Páramo, Desert, and Puna provinces closely related to the Neotropical region. These results corroborate the transitional character of the provinces assigned to the South American transition zone.

Regionalization

The South American transition zone comprises the Páramo, Desert, Puna, Atacama, Cuyan High Andean, and Monte provinces (Figure 8.2).

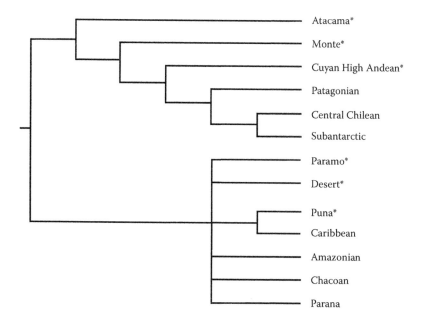

Figure 8.1 General area cladogram obtained by Urtubey et al. (2010). Asterisks indicate the provinces of the South American transition zone.

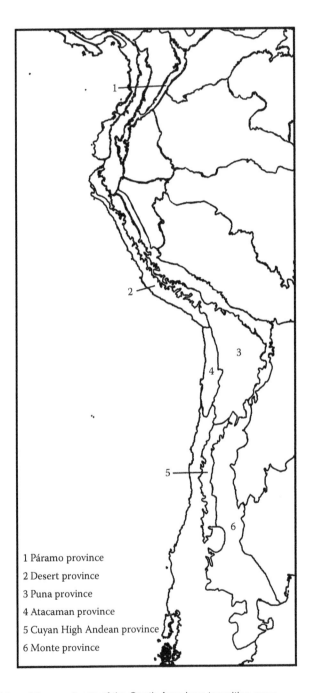

1 Páramo province
2 Desert province
3 Puna province
4 Atacaman province
5 Cuyan High Andean province
6 Monte province

Figure 8.2 Map of the provinces of the South American transition zone.

Cenocrons

Roig-Juñent et al. (2008) distinguished four cenocrons based on insect taxa. The Pangeic cenocron consists of old taxa that were present before the breakup of Pangaea, including two varieties: taxa found in eremic (xeric) environments and taxa from more humid habitats. The Gondwanic cenocron consists of taxa that were present before the breakup of Gondwana, including five varieties: Gondwanic, Peripampasic Arc, Patagonian, autochthonous, and Neotropical. The Holarctic cenocron consists of taxa that dispersed from the Nearctic region. The synanthropic cenocron includes taxa that dispersed to the areas within the last 500 years, associated with humankind.

PÁRAMO PROVINCE

Incasic province (in part)—Mello-Leitão, 1943: 130 (regional.); Fittkau, 1969: 642 (regional.).

Páramo province—Cabrera, 1957: 335 (regional.); Cabrera and Willink, 1973: 66 (regional.); Morrone, 1994b: 190 (PAE); Rivas-Martínez and Navarro, 1994: map (regional.); Morrone, 1996: 108 (regional.); Katinas et al., 1997: 112 (track anal.); Posadas et al., 1997: 2 (PAE); Morrone, 1999: 13 (regional.), 2014b: 83 (regional.); del Río et al., 2015: 1294 (track anal. and clad. biogeogr.); Morrone, 2015d: 210 (regional.).

Northern Andes area (in part)—Sick, 1969: 461 (regional.); Porzecanski and Cracraft, 2005: 266 (PAE).

North Andean center—Müller, 1973: 45 (regional.); Cracraft, 1985: 62 (areas of endem.).

Bogotá subcenter—Müller, 1973: 46 (regional.).

Peruvian Andes subcenter—Müller, 1973: 46 (regional.).

Colombian Montane province—Udvardy, 1975: 42 (regional.); Huber and Riina, 1997: 130 (glossary).

Páramo region—Rivas-Martínez and Tovar, 1983: 516 (regional.); Huber and Riina, 1997: 284 (glossary), 2003: 275 (glossary).

Peruvian Andean center—Cracraft, 1985: 64 (areas of endem.).

East Peruvian Andean subcenter—Cracraft, 1985: 64 (areas of endem.).

South Peruvian Andean subcenter—Cracraft, 1985: 64 (areas of endem.).

West Peruvian Andean subcenter—Cracraft, 1985: 64 (areas of endem.).

North Andean area—Coscarón and Coscarón-Arias, 1995: 726 (areas of endem.).

Cordillera Central Páramo ecoregion—Dinerstein et al., 1995: 102 (ecoreg.); Huber and Riina, 1997: 248 (glossary).

Cordillera de Mérida Páramo ecoregion—Dinerstein et al., 1995: 102 (ecoreg.); Huber and Riina, 1997: 248. (glossary)

Northern Andean Páramo ecoregion—Dinerstein et al., 1995: 102 (ecoreg.); Huber and Riina, 1997: 249 (glossary).

Santa Marta Páramo ecoregion—Dinerstein et al., 1995: 101 (ecoreg.).

Páramo ecoregion—Huber and Riina, 1997: 154 (glossary).

North Andean Páramo province—Morrone, 2001a: 107 (regional.), 2001c: 3 (regional.), 2004b: 45 (regional.), 2006: 483 (regional.); Urtubey et al., 2010: 506 (track anal.

and clad. biogeogr.); Escalante et al., 2011: 32 (track anal.); Hechem et al., 2011: 46 (track anal.); Mercado-Salas et al., 2012: 459 (track anal.).
Norandean Páramo province—Quijano-Abril et al., 2006: 1270 (track anal.).

Definition

The Páramo province comprises the high mountain peaks of the Andean cordillera of Venezuela, Colombia, and Ecuador, from the upper forest line at 3,000–3,500 meters upwards and below the permanent snowline at ca. 5,000 meters (Cabrera and Willink, 1973; Müller, 1973; Ringuelet, 1975; Rivas-Martínez and Tovar, 1983; Luteyn, 1992; Posadas et al., 1997; Gradstein, 1998; Rangel, 2000a; Morrone, 2001c, 2006, 2014b; Londoño et al., 2014). This high-elevation biome has been influenced by glaciations, being the landscape irregular, from very rough to flat and stretches from 3,000 meters to the perennial snowline at altitudes of 4,800–5,000 meters (Luteyn, 1992). It has been debated whether this biome is natural or manmade. Luteyn (1992) considered that it is clear that páramos of the present are more extensive than in earlier times, beginning at much lower elevations because of the anthropogenic destruction of the forest. This province is part of the tropical Andean biodiversity hotspot and is characterized by its high endemism, which reaches 60 percent of the species (Hughes and Eastwood, 2006).

Endemic and Characteristic Taxa

I (Morrone, 2014b) provided a list of endemic and characteristic taxa. Some examples include *Bomarea angustipetala*, *B. bredemeyerana*, and *B. holtoni* (Alstroemeriaceae), *Blakiella*, *Espeletia* complex, *Floscaldasia*, and *Westoniella* (Asteraceae), *Passiflora truxillensis* (Passifloraceae), *Aragoa* (Scrophulariaceae), *Bogotacris*, *Chibchacris*, and *Timotes* (Acrididae), Strengerianini (Pseudothelphusidae; Figure 8.3a), *Gigantodax cervicornis* (Figure 8.3b) and *G. siberianus* (Simuliidae), *Atelopus tamaense* (Bufonidae), *Cavia porcellus anolaimae* (Caviidae), *Mazama rufina* (Cervidae), *Momotus momota olivaresii* (Momotidae), and *Crypturellus kerriae* and *C. saltuarius* (Tinamidae).

Vegetation

It consists of different types of vegetation, including moorlands, xerophytic scrublands, grasslands, and peat bogs (Cleef, 1978, 1981; van der Hammen and Cleef, 1986; Rangel et al., 1997; Gradstein, 1998). There are steppes of *Festuca* and *Deyeuxia* (Poaceae), with the typical *frailejones* (Asteraceae: Espeletiinae) and forests on the higher areas (Cabrera and Willink, 1973; Cuatrecasas, 1986; Monasterio, 1986; Sturm, 1990). Dominant plant species belong to the genera *Calamagrostis*, *Chusquea*, *Deyeuxia*, *Diplostephium*, *Cynoxys*, *Espeletia*, *Espeletiopsis*, *Festuca*, *Gentiana*, *Gunnera*, *Hypericum*, *Lupinus*, *Miconia*, *Paepalanthus*, and *Rubus* (Cabrera, 1957; Cabrera and Willink, 1973; Cleef, 1978, 1981; Monasterio, 1986; Hernández and Sánchez, 1992; Hernández et al., 1992c). The páramo has been

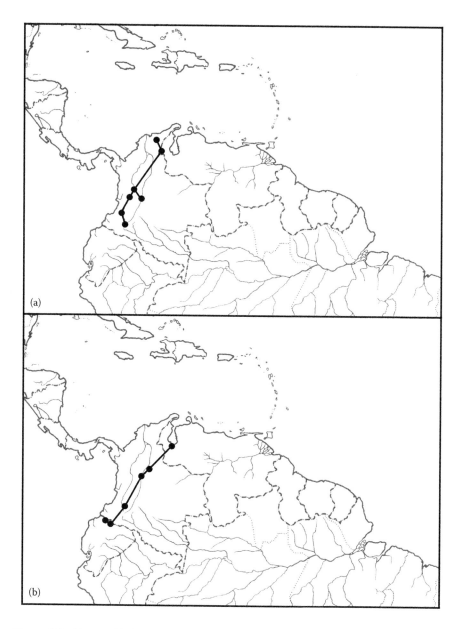

Figure 8.3 Maps with individual tracks in the Páramo province. (a) Strengerianini (Pseudothelphusidae); (b) *Gigantodax cervicornis* (Simuliidae).

divided into three altitudinal zones (Figure 8.4a; van der Hammen and Cleef, 1986; Luteyn, 1992; Huber and Riina, 1997; Rangel, 2000b):

1. Subpáramo or ceja andina: Shrubby transition zone from 3,000 to 3,500 meters altitude, made up from forest elements from the lower slopes and grass páramo above. Shrubs belong to genera *Gynoxys*, *Vaccinium*, *Befaria*, *Maclenia*, *Brachyotum*, *Miconia*, and *Hesperomeles*.

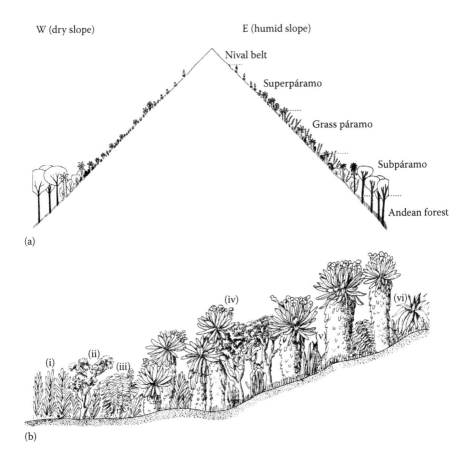

Figure 8.4 Páramo vegetation. (a) Schematic representation of the vegetational zonation. (Modified from van der Hammen, T. and A. M. Cleef, "Development of the High Andean páramo flora and vegetation." In: Vuilleumier, F. and M. Monasterio (eds.), *High altitude tropical biogeography*. Oxford University Press and American Museum of Natural History, New York and Oxford, pp. 153–201, 1986.); (b) detail of the grass páramo. (Rangel, J. O., et al., "Tipos de vegetación en Colombia: Una aproximación al conocimiento de la terminología fitosociológica, fitoecológica y de uso común," in: Rangel, J. O., P. D. Lowy, and M. Aguilar (eds.), *Colombia Diversidad Biótica II: Tipos de vegetación en Colombia*, Instituto de Ciencias Naturales, Universidad Nacional de Colombia, Santafé de Bogotá, pp. 89–381, 1997.) (i) *Chusquea tessellata*; (ii) *Diplostephium schultzii*; (iii) *Blechnum loxense*; (iv) *Espeletia hartwegiana*; (v) *Neurolepis cf. aperta*; and (vi) *Puya* sp.

2. Grass páramo, páramo proper or páramo in the strict sense (Figure 8.4b): From 3,500 to 4,100 meters, it has a markedly xeromorphic vegetation. It is composed of buch grasslands dominated by species of *Calamagostris* and *Festuca*, and dwarfed bamboos (*Chusquea*) on the wet slopes. It is also rich in shrubs (*Hypericum, Diplostephium, Pentacalia, Pernettya*, and *Valeriana*), acaulescent rosette plants (*Acaena*) and cushion plants (*Werneria*). The most characteristic plants are the species of *Espeletia* and *Espeletiopsis*, with their columnar, wooly, rosette plant growth form.

3. Superpáramo: Narrow zone from 4,100 to 4,800 meters altitude, with sandy soils, situated between the grass páramo and the snowline that are subject to regular snowfalls. It has scattered plants belonging to the genera *Senecio, Draba, Ephedra*, and *Lupinus*.

Van der Hammen and Cleef (1986) recognized in each of these altitudinal belts a lower and an upper zone, and also described them for each side (west or east) of the cordillera.

Restrepo and Duque (1992) analyzed the vegetational types in Llano de Paletara in the Colombian Central Cordillera, and distinguished eight different types: *frailejones* with *Espeletia hartwegiana, Blechnum loxense*, and *Sphagnum sancto-josephense*; *chuscal* with *Chusquea tessellata* and *Sphagnum sancto-josephense*; swamp with *Carex jamesonii* and *Arthoxanthum odoratum*; *pajonal* with *Calamagrostis intermedia*; scrubland with *Hypericum lancioides* and *Blechnum loxense*; scrubland with *Ageratina tinifolia*; scrubland with *Diplostephium cf. cinerascens*; and forest with *Escallonia myrtilloides*.

Biotic Relationships

The preeminence given to some tropical elements made Cabrera and Willink (1973) assign the Páramo province to the Neotropical region, whereas their close relationships with other Andean provinces (Rivas-Martínez and Tovar, 1983; Fjeldså, 1982) led other authors (Morrone, 1994a,b, 1996b; Posadas et al., 1997) to place it in the Andean region. Additionally, this province contains numerous north temperate plant genera, such as *Alnus, Draba, Lupinus, Quercus, Salix, Sanbucus, Valeriana*, and *Viburnum*, which arrived after the uplift of the northern Andes (Hughes and Eastwood, 2006). The placement of the Páramo province in the South American transition zone (Morrone, 2014b) reflects its conflicting relationships.

Vuilleumier (1986) suggested a close relationship with the Puna province, based on bird taxa. A parsimony analysis of endemicity based on bird taxa (Porzecanski and Cracraft, 2005) postulated a close relationship of this province with the Guatuso–Talamanca, Pantepui, and Puna provinces. Track, parsimony, and cladistic biogeographic analyses based on insect and plant taxa (Morrone, 1994a, 1994b; Posadas et al., 1997) suggest that the Páramo province is closely related to the Puna province.

Regionalization

Hernández et al. (1992a) identified 43 districts in the Colombian portion of this province: Alto Cauca Highland, Alto Patía, Alto Patía Subandean, Andalucía, Awa, Cañón Chicamocha, Cañón del Cauca, Cañón del Dagua, Cauca and Valle Western

Cordillera Andean Forest, Cauca Pacific Slope Subandean Forest, Catatumbo Mountains, Cauca–Huila Eastern Subandean Forests, Cauca–Huila–Valle–Tolima Andean Forests, Cauca–Huila–Valle–Tolima Páramos, Cauca–Valle Cordillera Subandean Forests, Central Cordillera Southeastern Subandean, Citara, Dabeiba, Eastern Andean, Eastern Cordillera Cloud Forests, Eastern Cordillera Páramos, Eastern Nariño Andean Forests, Farallones de Cali, Frontino, Andean Forest, Nariño–Putumayo Páramos, Paramillo del Sinú, Perijá, Perijá Páramos, Quindío Andean Forests, Quindío–Antioquia Central Cordillera Subandean Forests, Quindío Páramo, San Agustín, San Juan Cloud Forest, San Lucas Mountains, Southern Perijá, Tachira, Tolima, Tolima Central Cordillera Subandean Forests, Western Cordillera Eastern Subandean Forests, Western Cordillera Northern Andean Forests, Western Cordillera Northern Subandean Forests, and Western Nariño Andean Forests.

Cenocrons

It has been postulated that the biota of this province is derived from different biotic sources (Cleef, 1978, 1981; Vuilleumier, 1986; Sklenář et al., 2011). The most probable scenario suggests an initial pre-páramo flora developing on savanna-like hilltops from Neotropical elements already adapted to a climate with marked pluvial seasonality and/or special edaphic conditions. During, or shortly after, the final upheaval of the northern Andes, the early páramo vegetation or proto-páramo (typical páramo taxa, but floristically poorer than present-day páramo) was invaded gradually by Subantarctic elements that migrated northward along the Andes and by Nearctic elements which crossed the Panama Isthmus southwards since ca. 3 mya, finding temperate-like conditions in a tropical setting (Londoño et al., 2014). Additionally, there are numerous Holarctic plant genera, such as *Alnus, Draba, Lupinus, Quercus, Salix, Sambucus, Valeriana,* and *Viburnum*, which might have arrived before the existence of the Panama Isthmus and after the uplift of the northern Andes (Antonelli et al., 2009). During the Pleistocene, páramo vegetation was exposed to at least 20 events of glacial/interglacial periods from the recurrent upward and downward displacement of the Andean vegetation zones (van der Hammen, 1974). Sklenář et al. (2011) analyzed distributional and phylogenetic evidence available on several plant genera, finding that half of the Páramo species were derived through dispersal from temperate areas, being those from the southern Andes—the initial source of immigrants. Taxa of northern temperate origin, however, show currently a higher number of species in the Páramo, including several genera that underwent adaptive radiations in the area, namely *Gentianella, Draba, Valeriana, Cerastium,* and *Lupinus*.

DESERT PROVINCE

Desert province—Cabrera and Willink, 1973: 89 (regional.); Huber and Riina, 1997: 272 (glossary); Morrone, 1999: 12 (regional.); Donato, 2006: 422 (clad. biogeogr.); Morrone, 2014b: 86 (regional.); del Río et al., 2015: 1294 (track anal. and clad. biogeogr.); Morrone, 2015d: 210 (regional.); Daniel et al., 2016: 1169 (track anal. and regional.).

Andean Pacific center—Müller, 1973: 100 (regional.).
Peruvian subcenter—Müller, 1973: 101 (regional.).
Salares zone—Artigas, 1975: 20 (regional.).
Pacific Desert province—Udvardy, 1975: 41 (regional.); Huber and Riina, 1997: 238 (glossary).
Pacific Coastal Deserts dominion—Ab'Sáber, 1977: map (climate).
Arequipa unit—Lamas, 1982: 353 (regional.).
Callao unit—Lamas, 1982: 352 (regional.).
Mollendo unit—Lamas, 1982: 353 (regional.).
Porculla unit—Lamas, 1982: 353 (regional.).
Surco unit—Lamas, 1982: 353 (regional.).
Pacific Desert region—Rivas-Martínez and Tovar, 1983: 516 (regional.).
Peruvian Arid Coastal center—Cracraft, 1985: 68 (areas of endem.).
Peruvian Pacific Desert region—Rivas-Martínez and Navarro, 1994: map (regional.).
Desert area (in part)—Coscarón and Coscarón-Arias, 1995: 726 (areas of endem.).
Sechura Desert ecoregion—Dinerstein et al., 1995: 105 (ecoreg.).
Pacific Desert ecoregion—Huber and Riina, 1997: 154 (glossary).
Coastal Peruvian Desert province—Morrone, 2001a: 110 (regional.), 2001c: 4 (regional.), 2004b: 46 (regional.), 2006: 483 (regional.); Vidal et al., 2009: 161 (endem. anal.); Urtubey et al., 2010: 506 (track anal. and clad. biogeogr.); Hechem et al., 2011: 46 (track anal.).
Desertic Peruvian–Ecuadorean province—Rivas-Martínez et al., 2011: 27 (regional.).
Hyperdesertic North Peruvian province—Rivas-Martínez et al., 2011: 27 (regional.).
Hyperdesertic Tropical Pacific region—Rivas-Martínez et al., 2011: 27 (regional.).
Hyperdesertic Tropical Chilean–Arequipan province (in part)—Rivas-Martínez et al., 2011: 27 (regional.).

Definition

The Desert province comprises a narrow strip along the Pacific Ocean coast, from northern Peru to northern Chile (Cabrera and Willink, 1973; Morrone, 2001c, 2006, 2014b). Its extreme aridity is maintained by several factors, including the South Pacific anticyclone, the Humboldt Current, and a rain shadow created by the Andes (Dillon et al., 2009).

Endemic and Characteristic Taxa

I (Morrone, 2014b) provided a list of endemic and characteristic taxa. Some examples include *Nolana* spp. (Solanaceae), *Notiobia moffetti* (Carabidae), *Echemoides aguilari* (Gnaphosidae), *Amorphochilus schnablii* (Furipteridae), *Conepatus rex inca* (Mephitidae; Figure 8.5), and *Rhodopis* (Trochilidae).

Vegetation

Scarce, permanent vegetation is abundant only near the rivers and the sea; between 1,500 and 3,000 meters tree-like cacti are abundant, among them shrubs and herbs grow when it rains (Cabrera and Willink, 1973). Dominant plant species include

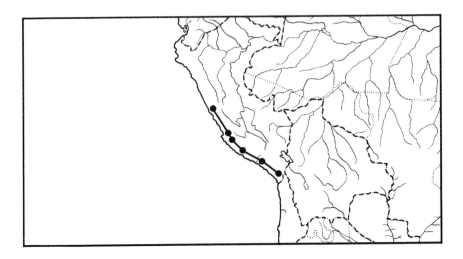

Figure 8.5 Map with the individual track of *Conepatus rex inca* (Mephitidae) in the Desert province.

Acacia macrantha, Caesalpinia tinctoria, Diplostephium tacorense, Franseria fruticosa, Inga feuillei, Kaegeneckia lanceolata, Lemaireocereus cartwrightianus, Paspalum vaginatum, Prosopis chilensis, P. limensis, Salicornia ambigua, Schinus areira, Tillandsia latifolia, T. purpurea, T. straminea, and *Trichocereus peruvianus* (Cabrera and Willink, 1973; Rivas-Martínez and Tovar, 1983).

Biotic Relationships

Rivas-Martínez and Tovar (1983) consider that the Desert province is related to the Chaco province. The Desert province, assigned in the past to the Neotropical region, was assigned to the South American transition zone, based on its close biotic links with the Puna and Páramo (Fjeldså, 1992; Morrone and Urtubey, 1997; Posadas et al., 1997).

Regionalization

Cabrera and Willink (1973) have delimited two districts. The Coastal Desert district corresponds to the driest portions, situated in the coast of the Desert province (Cabrera and Willink, 1973). The Cardonales district corresponds to the western slopes of the Andes, between 1,500 and 3,000 meters.

Cenocrons

Dillon et al. (2009) analyzed the diversification of *Nolana* (Solanaceae) in the Desert province and northern Chile. Based on a molecular phylogenetic analysis, they found that the genus diverged from its sister genus 8.5 mya. The crown group of *Nolana*, dated at 4 mya, corresponds to the time when western South America was suffering increasing aridity. The authors considered that the species of the genus from the Desert province were derived from northern Chilean ancestors, which dispersed into the province in several episodes, and then radiated there.

PUNA PROVINCE

Puna formation—Holmberg, 1898: 433 (regional.).
Puna province—Cabrera, 1951: 52 (regional.), 1953: 107 (regional.), 1957: 336 (regional.), 1958: 200 (regional.), 1971: 32 (regional.); Cabrera and Willink, 1973: 87 (regional.); Udvardy, 1975: 42 (regional.); Cabrera, 1976: 59 (regional.); Morrone, 1994b: 190 (PAE); Huber and Riina, 1997: 271 (glossary); Posadas et al., 1997: 2 (PAE); Morrone, 1996: 108 (regional.); Katinas et al., 1997: 112 (track anal.); Morrone, 2001a: 111 (regional.), 2001c: 5 (regional.); Ojeda et al., 2002: 24 (biotic evol.); Roig-Juñent et al., 2003: 275 (track anal. and clad. biogeogr.); Morrone, 2004b: 46 (regional.); Donato, 2006: 422 (clad. biogeogr.); Morrone, 2006: 483 (regional.); Aagesen et al., 2009: 309 (endem. anal.); Urtubey et al., 2010: 506 (track anal. and clad. biogeogr.); Hechem et al., 2011: 46 (track anal.); Ferretti et al., 2012: 2 (track anal.); Campos-Soldini et al., 2013: 16 (track anal.); Ferro, 2013: 324 (biotic evol.); Morrone, 2014b: 87 (regional.); Amarilla et al., 2015: 8 (biotic evol.); del Río et al., 2015: 1294 (track anal. and clad. biogeogr.); Morrone, 2015d: 210 (regional.).
Northern Andean Cordillera region—Peña, 1966a: 5 (regional.), 1966b: 213 (regional.).
High Andean province (in part)—Cabrera, 1971: 30 (regional.); Huber and Riina, 1997: 270 (glossary); Ojeda et al., 2002: 24 (biotic evol.); Aagesen et al., 2009: 309 (endem. anal.).
Quechua High Andean district—Cabrera, 1971: 30 (regional.), 1976: 52 (regional.).
Puna center—Müller, 1973: 92 (regional.).
Altiplanic zone—Artigas, 1975: 20 (regional.).
Puna zone—Artigas, 1975: 20 (regional.).
Cuyan Subandean province—Ringuelet, 1975: 107 (regional.).
Titicaca province—Ringuelet, 1975: 107 (regional.).
Lake Titicaca province—Udvardy, 1975: 42 (regional.).
Punas dominion—Ab'Sáber, 1977: map (climate).
Puna region—Rivas-Martínez and Tovar, 1983: 516 (regional.); Huber and Riina, 1997: 284 (glossary).
Austral Andean center—Cracraft, 1985: 65 (areas of endem.).
Punan subregion—Rivas-Martínez and Navarro, 1994: map (regional.).
Bolivian province—Rivas-Martínez and Navarro, 1994: map (regional.).
Peruvian province—Rivas-Martínez and Navarro, 1994: map (regional.).
Puna area—Coscarón and Coscarón-Arias, 1995: 726 (areas of endem.).
Central Andean Dry Puna ecoregion—Dinerstein et al., 1995: 102 (ecoreg.).
Central Andean Puna ecoregion—Dinerstein et al., 1995: 102 (ecoreg.).
Central Andean Wet Puna ecoregion—Dinerstein et al., 1995: 102 (ecoreg.).
Puna ecoregion—Huber and Riina, 1997: 154 (glossary); Burkart et al., 1999: 11 (ecoreg.).
Arid Puna province—Morrone, 1999: 12 (regional.).
Central Puna province—Morrone, 1999: 12 (regional.).

Humid Puna province—Morrone, 1999: 12 (regional.).
Altiplano ecoregion—Salazar Bravo et al., 2002: 78 (ecoreg.).
Central Andes area—Porzecanski and Cracraft, 2005: 266 (PAE).
Punean province—Donato, 2006: 422 (clad. biogeogr.).
Andean Cuyan province (in part)—López et al., 2008: 1572 (PAE).
Aymara province—López et al., 2008: 1574 (PAE).
Mesophytic Punenian province—Rivas-Martínez et al., 2011: 27 (regional.).
Xerophytic Punenian province—Rivas-Martínez et al., 2011: 27 (regional.).
Bolivian–Tucumanan province—Rivas-Martínez et al., 2011: 27 (regional.).

Definition

The Puna province corresponds to the plateau at altitudes of approximately 3,800–4,500 meters, situated from southern Peru to northwestern Argentina between the latitudes 15° and 27° south (Peña, 1966a; Cabrera and Willink, 1973; Müller, 1973; Willink, 1988; Rivas-Martínez and Navarro, 1994; Posadas et al., 1997; Burkart et al., 1999; Morrone, 2001c, 2006, 2014b; Palma et al., 2005). The climate is cold and dry. Mean annual precipitation is 150–230 millimeters (Palma et al., 2005), with the rain falling predominately during summer (Aagesen et al., 2009). The High Andean and Mesoandean floors were lumped together, by Rivas-Martínez and Tovar (1983).

Endemic and Characteristic Taxa

I (Morrone, 2014b) provided a list of endemic and characteristic taxa. Some examples include *Chuquiraga atacamensis* and *C. kuscheli* (Asteraceae), *Epilobium pedicellare* (Figure 8.6), *Fuchsia austromontana* and *F. tincta* species group (Onagraceae), *Piper carrapanum*, *P. edurumglaberimicaule*, and *P. tardum*

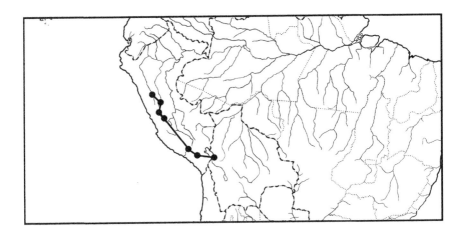

Figure 8.6   Map with the individual track of *Epilobium pedicelare* (Onagraceae) in the Puna province.

(Piperaceae), *Trachelopachys bidentatus* and *T. tarma* (Clubionidae), *Curicta peruviana* (Nepidae), *Atacamacris* and *Punacris* (Tristiridae), *Akodon lutescens* and *Punomys* spp. (Cricetidae), *Leopardus jacobitus* (Felidae), *Grallaria andicola* (Formicaridae), and *Rhea pennata garleppi* (Rheidae).

Vegetation

It consists of bush steppes, montane grasslands, low scrubland, trees, and herbaceous plants (Cabrera, 1957; Cabrera and Willink, 1973; Dinerstein et al., 1995; Martínez Carretero, 1995; Roig and Martínez Carretero, 1998; Roig-Juñent et al., 2003; Palma et al., 2005). The open bush steppes are mainly formed by species of Solanaceae, Fabaceae, and Asteraceae, accompanied by grass species such as *Jarava leptostachya*, *Pennisetum chilense*, *Deyeuxia rigida*, and *Aristida antoniana* (Aagesen et al., 2009). Steppe grasses include species of *Festuca* and *Stipa* (Palma et al., 2005).

Biotic Relationships

Vuilleumier (1986) suggested a close relationship with the Páramo province, based on bird taxa. This is supported by track, parsimony, and cladistic biogeographic analyses based on insect and plant taxa (Morrone, 1994a, 1994b; Posadas et al., 1997). A parsimony analysis of endemicity based on bird taxa (Porzecanski and Cracraft, 2005) postulated a close relationship of this province with the Guatuso–Talamanca, Pantepui, and Páramo provinces.

Regionalization

Martínez Carretero (1995) identified four districts within the Argentinean portion of this province: Bolivian, Jujuyan, Central, and Cuyan. Five units identified by Lamas (1982) within the Peruvian portion of this province may be treated as districts: Cajamarca, Huancapata, Pasco, Shimbe, and Vilcanota.

ATACAMA PROVINCE

High Plateau region—Peña, 1966a: 4 (regional.), 1966b: 212 (regional.).
Atacaman subregion—O'Brien, 1971: 199 (regional.).
Coquimban district (in part)—Cabrera and Willink, 1973: 91 (regional.); Huber and Riina, 1997: 146 (glossary).
Chilean subcenter—Müller, 1973: 101 (regional.).
Atacaman area (in part)—Artigas, 1975: 20 (regional.).
Argentinean–Atacaman province (in part)—Rivas-Martínez and Navarro, 1994: map (regional.).
Atacama Hyperdesert province—Rivas-Martínez and Navarro, 1994: map (regional.).
Desert area (in part)—Coscarón and Coscarón-Arias, 1995: 726 (areas of endem.).

Atacama Desert ecoregion—Dinerstein et al., 1995: 105 (ecoreg.); Huber and Riina, 1997: 141 (glossary).
Atacama province—Morrone, 1999: 12 (regional.), 2001a: 112 (regional.), 2001c: 6 (regional.), 2004b: 46 (regional.), 2006: 483 (regional.); Vidal et al., 2009: 161 (endem. anal.); Urtubey et al., 2010: 506 (track anal. and clad. biogeogr.); Hechem et al., 2011: 46 (track anal.); Morrone, 2014b: 88 (regional.); del Río et al., 2015: 1294 (track anal. and clad. biogeogr.); Morrone, 2015d: 210 (regional.).
Subtropical Pacific area (in part)—Porzecanski and Cracraft, 2005: 266 (PAE).
Atacama ecoregion—Abell et al., 2008: 408 (ecoreg.).
Hyperdesertic Tropical Chilean–Arequipan province (in part)—Rivas-Martínez et al., 2011: 27 (regional.).

Definition

The Atacama province is situated in northern Chile, between the latitudes 18° and 28° south (Peña, 1966a; O'Brien, 1971; Cabrera and Willink, 1973; Müller, 1973; Artigas, 1975; Rivas-Martínez and Navarro, 1994; Dinerstein et al., 1995; Morrone, 2001c, 2006, 2014b; Luebert, 2011). It is characterized by arid (≤ 50 mm/year) to hyperarid (≤ 5 mm/year) climates (Guerrero et al., 2013). Some of the most diverse communities of the Atacama province occur in *lomas* supported by winter fogs, which form over cool Pacific Ocean currents (Dinerstein et al., 1995).

Endemic and Characteristic Taxa

I (Morrone, 2014b) provided a list of endemic and characteristic taxa. Some examples include *Cleome chilensis* (Cleomaceae), *Heliotropium* sect. *Cochranea* (Heliotropiaceae), *Malesherbia arequipana* and *M. tocopillana* (Passifloraceae), *Zephyra* (Tecophilaeaceae), *Chileotracha* (Ammotrechidae), *Listroderes robustior* (Curculionidae; Figure 8.7), *Gigantodax cortesi*, *Simulium hectorvargasi*, and *S. tenuipes* (Simuliidae), *Liolaemus atacamensis* and *L. poconchilensis* (Iguanidae), and *Octodon degus* (Octodontidae).

Vegetation

There are desert areas with scarce vegetation and coastal and desert scrubland (Figure 8.8). There are rich communities in the *lomas*, due to the fog that the currents of the Pacific Ocean form during winter (Dinerstein et al., 1995). Some species rich taxa include *Copiapoa* (Cactaceae), *Heliotropium* sect. *Cochranea* (Heliotropiaceae), *Nolana* and *Solanum* sect. *Regmandra* (Solanaceae), and *Oxalis* sect. *Carnosae* (Oxalidaceae; Luebert, 2011). Most abundant species in the coastal desert include *Adesmia argentea*, *Atriplex clivicola*, *Bulnesia chilensis*, *Ephedra breana*, *Eulychnia iquiquensis*, *E. morromorenoensis*, *Euphorbia lactiflua*, *Frankenia chilensis*, *Gypothamnium pinifolium*, *Heliotropium eremogenum*, *H. floridum*, *H. pychnophyllum*, *H. stenophyllum*, *Nolana adansonii*, *N. lycioides*, and *Oxalis gigantea*. In the desert scrub, the most abundant species are *Atriplex atacamensis*, *A. deserticola*, *Flourensia thurifera*, *Gymnophyton foliosum*, *Heliotropium*

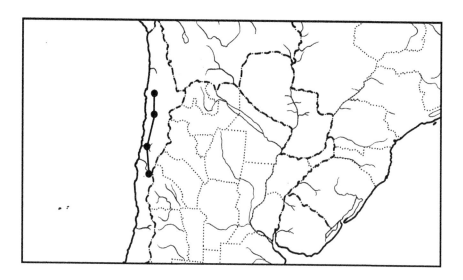

Figure 8.7 Map with the individual track of *Listroderes robustior* (Curculionidae) in the Atacama province.

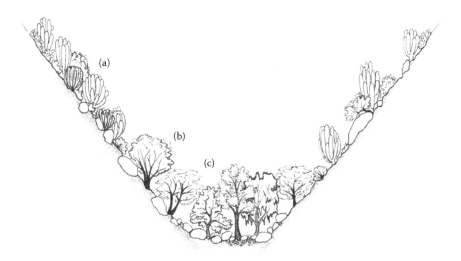

Figure 8.8 Vegetation in the scrub desert of the Atacama province. (a) Scrub; (b) xeric forest with *Acacia caven*; and (c) ravine forest with *Salix humboldtiana*. (Modified from Luebert, F. and P. Pliscoff, *Sinopsis bioclimática y vegetacional de Chile*. Editorial Universitaria, Santiago de Chile, 2006.)

stenophyllum, Huidobria chilensis, Nolana leptophylla, Oxyphyllum ulicinum, Skytanthus acutus, and *Tessaria absinthioides* (Luebert and Pliscoff, 2006).

Biotic Relationships

O'Brien (1971) considered that the entomofauna of the Atacama province is related to the Coquimbo province (Central Chilean subregion, Andean region).

Regionalization

Peña (1966a,b), Di Castri (1968), and Artigas (1975) identified five nested units, which were treated by Morrone (2014b) as the Northern Coast, Interior Desert, Tamarugal, Northern Precordilleran, and Northern Andean districts. Their preliminary delimitation is based on Artigas' (1975) map.

Cenocrons

Luebert (2011) distinguished four floristic elements, which may be assimilated to cenocrons: Neotropical (basically from the Desert province), Central Chilean, Transandean (basically from the Chacoan province), and Antitropical (basically from the Nearctic region). Guerrero et al. (2013) analyzed the time of colonization of three plant genera (*Chaetanthera, Nolana,* and *Malesherbia*) and one lizard genus (*Liolaemus*) in the Atacama province and other arid areas in western South America. Based on their time-calibrated phylogenetic hypotheses, the authors reconstructed the timing of the shifts in climate and the invasion of these taxa to desert areas. *Chaetanthera, Malesherbia,* and *Liolaemus* might have invaded the region 10 mya, some 20 mya after the initial onset of regional aridity. *Nolana,* the most diverse of these taxa, colonized it 2 mya, suggesting that some lineages may require longer timescales to adapt to climate change and aridity.

CUYAN HIGH ANDEAN PROVINCE

Subandean province (in part)—Fittkau, 1969: 642 (regional.).
Cuyan High Andean district—Cabrera, 1971: 31 (regional.), 1976: 55 (regional.); Huber and Riina, 1997: 145 (glossary).
Southern Andean province (in part)—Udvardy, 1975: 42 (regional.); Morrone, 1996: 108 (regional.); Huber and Riina, 1997: 328 (glossary).
Argentinean–Atacaman province (in part)—Rivas-Martínez and Navarro, 1994: map (regional.).
Southern Andean Steppe ecoregion—Dinerstein et al., 1995: 102 (ecoreg.); Olson et al., 2001: 935 (ecoreg.).
Andean Cuyan province (in part)—López et al., 2008: 1572 (PAE).
Prepuna province—Morrone, 1999: 12 (*non* Cabrera, 1951; regionalization), 2001a: 112 (regional.), 2006: 483 (regional.), 2014b: 89 (regional.), 2015d: 210 (regional.).
Cuyan High Andean province—Morrone and Ezcurra, 2016: 287 (regional.).

Definition

The Cuyan High Andean province comprises the southern Andes of Catamarca, La Rioja, San Juan, Mendoza, and northern Neuquén provinces in Argentina and limiting areas in central Chile, between the latitudes 27° and 38° south, from altitudes of aproximately 2,200 to 4,500 meters (Cabrera 1971; Ezcurra 2001; Morrone and Ezcurra, 2016).

Endemic and Characteristic Taxa

Morrone and Ezcurra (2016) provided a list of endemic and characteristic taxa. Some examples include *Adesmia pinifolia* (Fabaceae), *Chuquiraga echegarayi*, *Huarpea andina*, *Nassauvia cumingii*, and *Senecio uspallatensis* (Asteraceae), *Azorella cryptantha* (Figure 8.9a) and *Laretia acaulis* (Apiaceae), *Stipa ruiz-lealii* (Poaceae), *Jaborosa laciniata* (Solanaceae), *Oxychloe bisexualis* (Juncaceae), *Geositta isabellina* (Furnariidae), *Sicalis auriventris* (Emberizidae), *Euneomys chinchilloides noei* (Cricetidae), and *Cyanoliseus patagonus andinus* (Psittacidae; Figure 8.9b).

Vegetation

There are bush steppes with columnar cacti (Cabrera, 1971, 1976; Cabrera and Willink, 1973). Dominant plant species include *Abromeitiella brevifolia*, *Acacia visco*, *Adesmia inflexa*, *Baccharis salicifolia*, *Caesalpinia trichocarpa*, *Cassia crassirostris*, *Cercidium andicola*, *Deuterocohnia strobilifera*, *Dickia spp.*, *Gochnatia glutinosa*, *Prosopis ferox*, *Puya dickioides*, *Schinus areira*, *Tillandsia gilliensi*, *Trichocereus pasacana*, *T. terscheckii*, and *Zucagnia punctata* (Cabrera, 1971, 1976; Cabrera and Willink, 1973; Zuloaga et al., 1999). Bushes belonging to

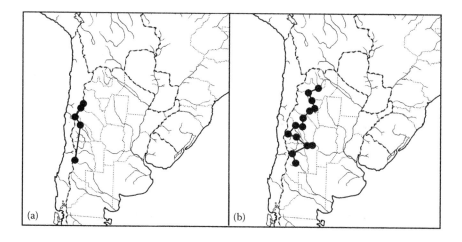

Figure 8.9 Maps with individual tracks in the Cuyan High Andean province. (a) *Azorella cryptantha* (Apiaceae); (b) *Cyanoliseus patagonus andinus* (Psittacidae).

families Asteraceae and Fabaceae and several species of Cactaceae are characteristic. Cushion-forming species of Bromeliaceae and the columnar cactus *Trichocereus atacamensis* are rather common (Aagesen et al., 2009, 2012).

Biotic Relationships

The Cuyan High Andean province, assigned in the past to the Neotropical region, was assigned to the South American transition zone, based on its close biotic links with the Puna and Páramo provinces (Fjeldså, 1992; Morrone and Urtubey, 1997; Posadas et al., 1997).

MONTE PROVINCE

Monte formation—Holmberg, 1898: 419 (regional.).
Monte area—Hauman, 1920: 54 (regional.); Hueck, 1957: 40 (regional.), 1966: 3 (regional.); Roig-Juñent, 1994: 183 (clad. biogeogr.); Coscarón and Coscarón-Arias, 1995: 726 (areas of endem.).
Monte province—Hauman, 1931: 60 (regional.); Soriano, 1950: 33 (veget.); Cabrera, 1951: 41 (regional.), 1953: 107 (regional.); Morello, 1955: 386 (regional.); Cabrera, 1958: 200 (regional.); Morello, 1958: 131 (regional.); Cabrera, 1971: 22 (regional.); Cabrera and Willink, 1973: 77 (regional.); Cabrera, 1976: 36 (regional.); Udvardy, 1975: 41 (regional.); Stange et al., 1976: 78 (biotic evol.); Rivas-Martínez and Navarro, 1994: map (regional.); Huber and Riina, 1997: 228 (glossary); Roig, 1998: 136 (regional.); Morrone, 1999: 9 (regional.); Zuloaga et al., 1999: 37 (regional.); Marino et al., 2001: 115 (PAE); Morrone, 2000a: 61 (regional.), 2001a: 94 (regional.); Roig-Juñent et al., 2001: 78 (areas of endem.); Ojeda et al., 2002: 24 (biotic evol.); Roig-Juñent et al., 2003: 275 (track anal. and clad. biogeogr.); Morrone, 2004b: 46 (regional.), 2006: 483 (regional.); Abraham et al., 2009: 145 (geography); Roig et al., 2009: 164 (regional.); Urtubey et al., 2010: 506 (track anal. and clad. biogeogr.); Hechem et al., 2011: 46 (track anal.); Ferretti et al., 2012: 2 (track anal.); Campos-Soldini et al., 2013: 16 (track anal.); Ferretti et al., 2014: 1089; Morrone, 2014b: 90 (regional.); Amarilla et al., 2015: 8 (biotic evol.); del Río et al., 2015: 1294 (track anal. and clad. biogeogr.); Morrone, 2015d: 210 (regional.); Aagesen and del Valle Elías, 2016: 161 (endem. anal.).
Xerophyllous Forests area—Parodi, 1934: 171 (regional.).
Central Xerophyllous Forest area—Castellanos and Pérez-Moreau, 1941: 382 (regional.).
Western Monte area—Parodi, 1945: 130 (regional.); Bölcke, 1957: 2 (regional.).
Central province—Soriano, 1949: 198 (regional.).
Central or Subandean dominion—Ringuelet, 1961: 160 (biotic evol. and regional.).
Monte center—Müller, 1973: 146 (regional.).
Monte with Cactaceae dominion—Ab'Sáber, 1977: map (climate).
Argentina Monte ecoregion—Dinerstein et al., 1995: 99 (ecoreg.).
Monte de Sierras y Bolsones ecoregion—Burkart et al., 1999: 13 (ecoreg.).
Mar Chiquita–Salinas Grandes ecoregion—Abell et al., 2008: 408 (ecoreg.).

Definition

The Monte province comprises central Argentina, between the latitudes 24° and 44° south, from Jujuy to northeastern Chubut, and southern Bolivia (Morello, 1955, 1958; Cabrera, 1958, 1971; Cabrera and Willink, 1973; Müller, 1973; Stange et al., 1976; Willink, 1988; Roig-Juñent, 1994; Coscarón and Coscarón-Arias, 1995; Morrone, 2000a, 2006, 2014b; Roig-Juñent et al., 2001; Abraham et al., 2009; Roig et al., 2009; Aagesen and del Valle Elías, 2016). Attitudinally, it ranges from sea level to 3,500 meters, depending on latitude (Aagesen and del Valle Elías, 2016). Based on geomorphological features, it has been divided into the High Monte (mountainous areas in the north) and the Low Monte (piedmonts, hills, and desert valleys in the central and southern areas; Olson et al., 2001).

Endemic and Characteristic Taxa

I (Morrone, 2014b) and Aagesen and del Valle Elías (2016) provided lists of endemic taxa. Some examples include *Ephedra boelckei* (Ephedraceae), *Hickenia* (Asclepiadaceae), *Denmoza* and *Tephrocactus* (Cactaceae), *Halophytum* (Halophytaceae), *Monttea aphylla* (Scrophullariaceae), *Larrea divaricata*, *L. cuneifolia*, and *L. nitida* (Zygophyllaceae), *Urophonius brachycentrus* (Bothriuridae), *Bradynobaenus subandinus* (Bradynobaenidae), *Amblycerus caryoboriformis* (Chrysomelidae), *Cyrtomon hirsutus*, *Enoplopactus lizeri* (Figure 8.10a), *Listroderes bruchi*, *Mendozella*, and *Pantomorus luteipes* (Curculionidae), *Catadacus* (Ichneumonidae), *Scaptodactyla* (Mutillidae), *Papipappus* (Ommexechidae), *Burmeisteriellus* and *Vulcanocanthon seminulum* (Scarabaeidae), *Heterolopuhus*

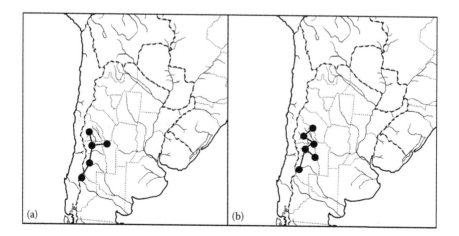

Figure 8.10 Maps with individual tracks in the Monte province. (a) *Enoplopactus lizeri* (Curculionidae); (b) *Bothrops ammodytoides* (Viperidae).

guttier (Tridenchthoniidae), *Pseudasthenes patagonica* (Furnariidae), *Phrygilus carbonarius, Poospiza ornata,* and *Sicalis mendozae* (Thraupidae), and *Bothrops ammodytoides* (Viperidae; Figure 8.10b).

Vegetation

There are open scrubland, with species of the genera *Aloysia, Parkinsonia, Capparis,* and the Zygophyllaceae genera *Larrea, Bulnesia,* and *Plectrocarpa* (Cabrera and Willink, 1973; Roig-Juñent et al., 2003; Roig et al., 2009; Aagesen and del Valle Elías, 2016). Dominant plant species include *Allenrolfea vaginata, Atriplex lampa, Baccharis salicifolia, Bougainvillea spinosa, Bulnesia retama, B. schikendantzii, Cassia aphylla, C. rigida, Cercidium praecox, Fabiana patagonica, Larrea cuneifolia, L. divaricata, L. nitida, Mimosa ephedroides, Monttea aphylla, Plectrocarpa rougesii, Portulaca spp., Prosopis alpataco, P. chilensis, P. flexuosa, P. torquata, Proustia cuneifolia, Tessaria dodoneaefolia,* and *Zuccagnia punctata* (Morello, 1955; Cabrera, 1971, 1976; Cabrera and Willink, 1973; Roig, 1998; Roig-Juñent et al., 2001). Plants similar to those in the North American dry areas belong to the genera *Acacia, Caesalpinia, Cassia, Cercidium, Larrea, Mimosa,* and *Prosopis* (Graham, 2004).

Biotic Relationships

It has been suggested that the Monte represents a biotically "impoverished" Chaco (Willink, 1991). According to a cladistic biogeographic analysis based on beetle and plant taxa (Morrone, 1993), the Monte province is closely related to the Chaco province. In addition to its historical relationship with the Chacoan provinces, the Monte province shows biotic similarities with the Sonora province of the Nearctic region (Clark, 1979; Willink, 1988; Morello, 1984; Wen and Ickert-Bond, 2009). The hypotheses postulated to explain that these similarities have been reviewed by Solbrig et al. (1977) and Wen and Ickert-Bond (2009), usually involving long-distance dispersal, although vicariance and parallel evolution due to adaptations to arid environments from closely related ancestors have been also postulated. Examples of the latter include Bruchinae (Coleoptera: Chrysomelidae) associated with Fabaceae shrubs in both areas (Stange et al., 1976) and weevils (Coleoptera: Curculionidae) of the *Sibinia sulcifer* species group (Clark, 1979).

Regionalization

Roig et al. (2009) identified three districts within this province, to which Morrone and Ezcurra (2016) added the Prepuna district. The Prepuna district extends from southern Bolivia to northwestern Argentina (Jujuy to La Rioja provinces), approximately from 1,000 to 3,500 meters. The Northern district encompasses the largest extension of the province, from Salta and Tucumán to southern Mendoza (Roig et al., 2009); endemic taxa include *Ephedra boelckei* (Ephedraceae), *Heliotropium curassavicum* var. *fruticulosum* (Boraginaceae), *Prosopis alpataco* (Fabaceae), *Mimodromius proseni* (Carabidae), *Cylydrorhinus oblongus, Enoplopactus catamarcensis,* and *E. sanjuaninus* (Curculionidae), *Glyphoderus monticola* (Scarabaeidae), *Entomoderes infernalis, Nyctelia alutacea,* and *Thylacoderes sphaericus* (Tenebrionidae), *Liolaemus anomalus* and *L. scapularis* (Iguanidae), and *Leiosaurus catamarcencis* (Leiosauridae; Roig-Juñent et al., 2001). The Eremean district is a narrow strip along the length of the high pre-Andean valleys of Mendoza and San Juan (Roig et al., 2009); endemic taxa include *Maihueniopsis clavarioides* (Cactaceae), *Cistanthe densiflora* (Montiaceae), *Larrea divaricata subsp. monticellii* (Zygophyllaceae), and *Homonota andicola* (Phyllodactylidae Roig et al., 2009). The Southern district lies south of the Colorado river (Roig et al., 2009); endemic taxa include *Aylacophora deserticola, Chuquiraga avellanedae,* and *C. rosulata* (Asteraceae), *Maihuenia patagonica* (Cactaceae), *Tetraglochin caespitosum* (Rosaceae), *Barypus dentipennis, B. schajovskoyi, Mimodromius fleissi,* and *M. straneoi* (Carabidae), *Emmallodera crenatocostata* and *Scotobius casicus* (Tenebrionidae), *Amphisbaena angustifrons* (Amphisbaenidae), and *Liolaemus donosobarrosi* and *L. melanops* (Iguanidae; Roig et al., 2009; Roig-Juñent et al., 2001; Ocampo and Ruiz Manzano, 2008).

Cenocrons

The entomofauna of the Monte has basically Chacoan affinities, showing also Patagonian, Cuyan High Andean, and Subantarctic elements (Willink, 1988; Ellenrieder, 2001; Roig-Juñent et al., 2001, 2003; Marino et al., 2001). Roig et al. (2009) identified four cenocrons in the Monte province:

1. Paleorelict taxa with related groups in other continents: Taxa originated before the breakup of Gondwana. They include Zygophyllaceae, *Prosopis*, and Daesiidae (Solifuga).
2. Taxa with American related groups: Taxa that evolved after the breakup of Gondwana. They include Cactaceae and *Larrea* (Zygophyllaceae).
3. Endemic desert taxa: Taxa with their sister taxa in mesic or humid tropical areas of South America. Examples include the beetle tribes Cnemalobini (Carabidae) and Eucranini (Scarabaeidae).
4. Neoendemic desert taxa: Taxa that underwent rapid speciation in the area. One example is the lizard genus *Pristidactylus*, which originated in the Subantarctic Chile.

Epilogue

The time is out of joint.

Hamlet, Scene 5, 188

The state of biogeographic regionalization in the twenty-first century is quite paradoxical. On one hand, biogeographic scenarios for particular taxa (based on molecular phylogenetic analyses and parametric event-based approaches) are prevalent, while on the other hand, analyses searching for biotic patterns based on the congruence among distributional data of different taxa seem to be at the periphery. Biogeographic regionalizations are a result of the latter, and thus are increasingly less common, although they may occassionally find space in the pages of prestigious scientific journals like *Science*. Why, then, are regionalizations still valid? Simple! Communication. Whenever biogeographers communicate their ideas they need to refer to a regionalization, usually in the form of a map. Ecologists, systematists, and conservationists, among others, also use regionalizations as general reference systems. I hope that the biogeographic units recognized in this book are useful for people undertaking different types of biogeographic analyses in the Neotropics or for communicating the results of their research to others.

During the past few decades, several authors have noted that biogeography lacks the conceptual coherence of other biological disciplines, with different approaches and methods coexisting and competing (e.g., Morrone and Crisci, 1995; Ebach and Humphries, 2003; Morrone, 2009). I believe that this should not be seen as a problem, but should be seen as an opportunity for communication between different biogeographic disciplines. Biogeographic regionalizations are, by their nature, synthetic and may help biogeographers communicate more effectively. When communicating between themselves, biogeographers of different viewpoints may discover opportunites to work on common problems. In the case of the Neotropical region, there are decades of biogeographic work done on different taxa, using different approaches that are synthesized into a regionalization. This vast amount of knowledge may be profitably used as background knowledge for other studies dealing with biogeography, ecology, biodiversity conservation, and evolutionary biology.

The biogeographic regionalization of the Neotropical region, presented in this book, is open to scrutiny and further revision. Future efforts may be focused on the following tasks:

- Analyze a diverse group of taxa, such as invertebrates, for which published distributional data are scarce and phylogenetic relationships are largely unknown.
- Use a variety of approaches and methods (as well as develop new ones) to discover new distributional patterns and provide novel interpretations of known taxa.
- Better characterize known biogeographic units by finding more endemic taxa that help diagnose them.
- Test recognized biogeographic units and propose others, especially districts, which are lacking for many Neotropical provinces.

- Refine identified cenocrons and propose new ones, through dating plant and animal lineages with molecular clocks and fossil data.
- Integrate geological data to infer more accurate geobiotic scenarios, in order to refine the existing explanations for the biotic evolution of the Neotropics.

The biogeographic regionalization of the Neotropical region is in its early years. Hopefully, many biogeographers will feel confident enough to contribute to its growth!

References

Aagesen, L., M. J. Bena, S. Nomdedeu, A. Panizza, R. P. López, and F. O. Zuloaga. 2012. "Areas of endemism in the southern central Andes." *Darwiniana*, 50: 218–251.

Aagesen, L. and G. del Valle Elías. 2016. "Areas of vascular plants endemism in the Monte desert (Argentina)." *Phytotaxa*, 266: 161–251.

Aagesen, L., C. A. Szumik, F. O. Zuloaga, and O. Morrone. 2009. "Quantitative biogeography in the South America highlands—Recognizing the Altoandina, Puna and Prepuna through the study of Poaceae." *Cladistics*, 25: 295–210.

Abell, R., M. L. Thieme, C. Revenga, M. Bryer, M. Kottelat, N. Bogutskaya, B. Coad, N. Mandrak, S. Contreras Balderas, W. Bussing, M. L. L. Stiassny, P. Skelton, G. R. Allen, P. Unmack, A. Naseka, R. Ng, N. Sinforf, J. Robertson, E. Armijo, J. V. Higgins, T. J. Heibel, E. Wikramanayake, D. Olson, H. L. López, R. E. Reis, J. G. Lundberg, M. H. Sabaj Pérez, and P. Petry. 2008. "Freshwater ecoregions of the world: A new map of biogeographic units for freshwater biodiversity conservation." *BioScience*, 58: 493–414.

Abraham, E., H. F. del Valle, F. Roig, L. Torres, J. O. Ares, F. Coronato, and R. Godagnone. 2009. "Overview of the geography of the Monte Desert biome (Argentina)." *Journal of Arid Environments*, 73: 144–153.

Abrahamovich, A., N. B. Díaz, and J. J. Morrone. 2004. "Distributional patterns of the Neotropical and Andean species of the genus *Bombus* (Hymenoptera: Apidae)." *Acta Zoológica Mexicana (nueva serie)*, 20: 99–117.

Ab'Sáber, A. N. 1977. "Os domínios morfoclimáticos na América do Sul. Primeira aproximação." *Geomorfología*, 52: 1–21.

Acevedo-Rodríguez, P. and M. T. Strong. 2008. "Floristic richness and affinities in the West Indies." *Botanical Review*, 74: 5–36.

Acosta, L. E. 2002. "Patrones zoogeográficos de los opiliones argentinos (Arachnida: Opiliones)." *Revista Ibérica de Aracnología*, 6: 69–84.

Albert, J. S., H. L. Bart, and R. E. Reis. 2011. "Species richness and cladal diversity." In: Albert, J. S. and R. E. Reis (eds.), *Historical biogeography of Neotropical freshwater fishes*. University of California Press, Berkeley and Los Angeles, pp. 89–104.

Ali, J. R. 2012. "Colonizing the Caribbean: Is the GAARlandia land-bridge hypothesis gaining a foothold?" *Journal of Biogeography*, 39: 431–433.

Ali, J. R. and J. C. Aitchison. 2014. "Exploring the combined role of eustasy and oceanic island thermal subsidence in shaping biodiversity on the Galapagos." *Journal of Biogeography*, 41: 1227–1241.

Allen, G. 1911. "Mammals of the West Indies." *Bulletin of the American Museum of Natural History*, 54: 175–263.

Allen, J. A. 1892. "The geographical distribution of North American mammals." *Bulletin of the American Museum of Natural History*, 4: 199–243.

Almendra, A. L., D. S. Rogers, and F. X. González-Cózatl. 2014. "Molecular phylogenetics of the *Handleyomis chapmani* complex in Mesoamerica." *Journal of Mammalogy*, 95: 26–40.

Almirón, A., M. Azpelicueta, J. Casciotta, and A. López Cazorla. 1997. "Ichthyogeographic boundary between the Brazilian and Austral subregions in South America, Argentina." *Biogeographica*, 73: 23–30.

Alonso, R., A. J. Carawford, and E. Bermingham. 2011. "Molecular phylogeny of an endemic radiation of Cuban toads (Bufonidae: *Peltophryne*) based on mitochondrial and nuclear genes." *Journal of Biogeography*, 39: 434–451.

Amarilla, L. D., A. M. Anton, J. O. Chiapella, M. M. Manifesto, D. F. Angulo, and V. Sosa. 2015. "*Munroa argentina*, a grass of the South American transition zone, survived the Andean uplift, aridification and glaciations of the Quaternary." *PLOS ONE*, 10: 1–21.

Amorim, D. S. 2001. "Dos Amazonias." In: Llorente, J. and J. J. Morrone (eds.), *Introducción a la biogeografía en Latinoamérica: Conceptos, teorías, métodos y aplicaciones, Las Prensas de Ciencias*. UNAM, Mexico City, pp. 245–255.

Amorim, D. S. and M. R. S. Pires. 1996. "Neotropical biogeography and a method for maximum biodiversity estimation." In: Bicudo, C. E. M. and N. A. Menezes (eds.), *Biodiversity in Brazil: A first approach*. CNPq, São Paulo, pp. 183–219.

Amorim, D. S., C. M. D. Santos, and S. S. de Oliveira. 2009. "Allochronic taxa as an alternative model to explain circumantarctic distributions." *Systematic Entomology*, 34: 2–9.

Amorim, D. S. and H. S. Tozoni. 1994. "Phylogenetic and biogeographic analysis of the Anisopodoidea (Diptera, Bibionomorpha), with an area cladogram for intercontinental relationships." *Revista Brasileira de Entomologia*, 38: 517–543.

Anderson, R. S. and C. W. O'Brien. 1996. "Curculionidae (Coleoptera)." In: Llorente Bousquets, J., A. N. García Aldrete, and E. González Soriano (eds.), *Biodiversidad, taxonomía y biogeografía de artrópodos de México: Hacia una síntesis de su conocimiento*. Universidad Nacional Autónoma de México, Mexico City, pp. 329–351.

Andrés Hernández, A. R., J. J. Morrone, T. Terrazas, and L. López Mata. 2006. "Análisis de trazos de las especies mexicanas de *Rhus* subgénero *Lobadium* (Angiospermae: Anacardiaceae)." *Interciencia*, 31: 900–904.

Antonelli, A., J. A. Nylander, C. Person, and I. Sanmartín. 2009. "Tracing the impact of the Andean uplift on Neotropical plant evolution." *Proceedings of the National Academy of Sciences*, 106: 9749–9754.

Arana, M. D., J. J. Morrone, and A. J. Oggero. 2011. "Licofitas (Tracheophyta: Lycopodiophyta) de las Sierras Centrales de Argentina: Un enfoque panbiogeográfico." *Gayana Botánica*, 68: 16–21.

Arbeláez-Cortés, E., A. S. Nyari, and A. G. Navarro-Sigüenza. 2010. "The differential effect of lowlands on the phylogeographic pattern of a Mesoamerican montane species (*Lepidocolaptes affinis*, Aves: Furnariidae)." *Molecular Phylogenetics and Evolution*, 57: 658–668.

Arbogast, B. S., S. V. Drovetski, R. L. Curry, P. T. Boag, G. Seutin, P. R. Grant, B. R. Grant, and D. J. Anderson. 2006. "The origin and diversification of the Galapagos mockingbirds." *Evolution*, 60: 370–382.

Arellano, E., F. X. González-Cozátl, and D. S. Rogers. 2005. "Molecular systematics of Middle American harvest mice *Reithrodontomys* (Muridae), estimated from mitochondrial cytochrome b gene sequences." *Molecular Phylogenetics and Evolution*, 37: 529–540.

Arriaga, L., C. Aguilar, D. Espinosa, and R. Jiménez. (eds.). 1997. *Regionalización ecológica y biogeográfica de México*. Taller de la Comisión Nacional para el Conocimiento y Uso de la Biodiversidad (Conabio), November 1997, Mexico City.

Arroyo-Cabrales, J., O. J. Polaco, C. Laurito, E. Johnson, M. T. Alberdi, and A. L. Valerio-Zamora. 2007. "The proboscideans (Mammalia) from Mesoamerica." *Quaternary International*, 169–170: 17–23.

Artigas, J. N. 1975. "Introducción al estudio por computación de las áreas zoogeográficas de Chile continental basado en la distribución de 903 especies de animales terrestres." *Gayana, miscelánea*, 4: 1–25.

Arzamendia, V. and A. R. Giraudo. 2009. "Influence of large South American rivers of the Plata Basin on distributional patterns of tropical snakes: A panbiogeographical analysis." *Journal of Biogeography*, 36: 1739–1749.

Asiain, J. and J. Márquez. 2003. "Revisión sistemática y análisis filogenético del género *Misantlius* Sharp, 1885 (Coleoptera: Staphylinidae: Staphylinini)." *Folia Entomológica Mexicana*, 42: 37–64.

Asiain, J., J. Márquez, and J. J. Morrone. 2010. "Track analysis of the species of *Agrodes* and *Plochionocerus* (Coleoptera: Staphylinidae)." *Revista Mexicana de Biodiversidad*, 81: 177–181.

Aubreville, A. 1962. "Savanization tropicale et glaciations quaternaires." *Adansonia (Nouvelle Serie)*, 2: 16–84.

Ávila-Pires, F. D. 1974a. "Caracterizaçao zoogeográfica da provincia Amazônica. I. Expediçoes científicas na Amazônia brasileira." *Anais da Academia Brasileira de Ciencias*, 46: 133–158.

Ávila-Pires, F. D. 1974b. "Caracterizaçao zoogeográfica da provincia Amazônica. II. A familia Callithricidae e a zoogeografía Amazônica." *Anais da Academia Brasileira de Ciencias*, 46: 159–181.

Avise, J. C. 2000. *Phylogeography: The history and formation of species.* Harvard University Press, Cambridge, Massachusetts.

Ayala, R., T. L. Griswold, and D. Yanega. 1996. "Apoidea (Hymenoptera)." In: Llorente Bousquets, J., A. N. García Aldrete, and E. Gónzalez Soriano (eds.), *Biodiversidad, taxonomía y biogeografía de artrópodos de México: Hacia una síntesis de su conocimiento.* Universidad Nacional Autónoma de México, Mexico City, pp. 423–464.

Ayarde, H. R. 1995. "Estructura de un sector de selva pedemontana: Reserva Fiscal Parque La Florida, Tucumán (Argentina)." In: Brown, A. D. and H. R. Grau (eds.), *Investigación, conservación y desarrollo en selvas subtropicales de montaña.* LIEY, San Miguel de Tucumán, pp. 69–78.

Bagley, J. C. and J. B. Johnson. 2014. "Phylogeography and biogeography of the lower Central American Neotropics: Diversification between two continents and between two seas." *Biological Review*, 89: 767–790.

Bailey, R. G. 1998. *Ecoregions: The ecosystem geography of the oceans and continents.* Springer-Verlag, New York.

Bănărescu, P. and N. Boşcaiu. 1978. *Biogeographie: Fauna und Flora der Erde und ihre geschichtliche Entwicklung.* Veb Gustav Fischer Verlag, Jena.

Barbour, T. 1914. "A contribution to the zoogeography of the West Indies, with especial reference to amphibians and reptiles." *Memoirs of the Museum of Comparative Zoology*, 44: 209–359.

Barker, F. K., K. J. Burns, J. Klicka, S. M. Lanyon, and I. J. Lovette. 2013. "Going to extremes: Contrasting rates of diversification in a recent radiation of New World passerine birds." *Systematic Biology*, 62: 298–320.

Barrera, A. 1962. "La península de Yucatán como provincia biótica." *Revista de la Sociedad Mexicana de Historia Natural*, 23: 71–105.

Barrera-Moreno, O., T. Escalante, T., and G. Rodríguez. 2011. "Panbiogeografía y modelos digitales de elevación: Un caso de estudio con roedores en la Faja Volcánica Transmexicana." *Revista de Geografía Norte Grande*, 48: 11–25.

Bartholomew, J. G., W. E. Clark, and P. H. Grimshaw. 1911. *Atlas of zoogeography: A series of maps illustrating the distribution of over seven hundred families, genera, and species of existing animals.* Edinburgh Geographical Institute, Edinburgh.

Bates, J. M., S. J. Hackett, and J. Cracraft. 1998. "Area-relationships in the Neotropical lowlands: An hypothesis based on raw distributions of Passerine birds." *Journal of Biogeography*, 25: 783–793.

Becerra, J. X. 2005. "Timing the origin and expansion of the Mexican dry forest." *Proceedings of the National Academy of Sciences*, 102: 10919–10923.

Beierkuhnlein, C. 2007. *Biogeographie: Die räumliche Organisation des Lebens in einer sich verändernden Welt*. Verlag Eugen Ulmer, Stuttgart.

Benzing, D. H. 2000. *Bromeliaceae: Profile of an adaptive radiation*. Cambridge University Press, Cambridge.

Bertoncello, R., K. Yamamoto, L. D. Meireles, and G. J. Shepherd. 2011. "A phytogeographic analysis of cloud forests and other forest subtypes amidst the Atlantic forests in south and southeast Brazil." *Biodiversity and Conservation*, 20: 3413–3433.

Beven, S., E. Connor, and K. Beven. 1984. "Avian biogeography in the Amazon basin and the biological model of diversification." *Journal of Biogeography*, 11: 383–399.

Bisconti, M., W. Landini, G. Bianucci, G. Cantalamessa, G. Carnevale, L. Ragaini, and G. Valleri. 2001. "Biogeographic relationships of the Galapagos terrestrial biota: Parsimony analyses of endemicity based on reptiles, land birds and *Scalesia* land plants." *Journal of Biogeography*, 28: 495–510.

Blancas-Calva, E., A. G. Navarro-Sigüenza, and J. J. Morrone. 2010. "Patrones biogeográficos de la avifauna de la Sierra Madre del Sur." *Revista Mexicana de Biodiversidad*, 81: 561–568.

Blanford, W. T. 1890. "Anniversary address to the Geological Society." *Proceedings of the Geological Society of London*, 1890: 43–110.

Blyth, E. 1871. "A suggested new division of the Earth into zoological regions." *Nature*, 3: 427–429.

Bölcke, O. 1957. "La situación forrajera argentina." *IDIA*, 113: 1–36.

Bonaccorso, E. and J. M. Guayasamin. 2013. "On the origin of Pantepui montane biotas: A perspective based on the phylogeny of *Aulacorhynchus* toucanets." *PLOS ONE*, 8: 1–10.

Borges, S. H. 2007. "Análise biogeográfica da avifauna da região oeste do baixo Rio Negro, Amazônia brasileira." *Revista Brasileira de Zoologia*, 24: 919–940.

Borhidi, A. 1996. *Phytogeography and vegetation ecology of Cuba. 2a ed*. Akademiai Kiado, Budapest.

Borhidi, A. and O. Muñiz. 1986. "The phytogeographic survey of Cuba. II. Floristic relationships and phytogeographic subdivision." *Acta Botanica Hungarica*, 32: 3–48.

Bragagnolo, C., M. R. Hara, and R. Pinto-da-Rocha. 2015. "A new family of Gonyleptoidea from South America (Opiliones, Laniatores)." *Zoological Journal of the Linnean Society*, 173: 296–319.

Brattstrom, B. H. 1990. "Biogeography of the Islas Revillagigedo, Mexico." *Journal of Biogeography*, 17: 177–183.

Briggs, J. C. 1994. "The genesis of Central America: Biology versus geophysics." *Global Ecology and Biogeography Letters*, 4: 169–172.

Brooks, D. R. 2004. "Reticulations in historical biogeography: The triumph of time over space in evolution." In: Lomolino, M. V. and L. R. Heaney (eds.), *Frontiers of biogeography: New directions in the geography of nature*. Sinauer Associates Inc., Sunderland, Massachusetts, pp. 125–144.

Brower, A. V. Z. 1994. "Phylogeny of *Heliconius* butterflies inferred from mitochondrial DNA sequences (Lepidoptera: Nymphalidae)." *Molecular Phylogenetics and Evolution*, 3: 159–174.

Brown, D. E., F. Reichenbacher, and S. E. Franson. 1998. *A classification of North American biotic communities*. The University of Utah Press, Salt Lake City.

Burkart, R., N. O. Bárbaro, R. O. Sánchez, and D. A. Gómez. 1999. *Eco-regiones de la Argentina*. Administración de Parques Nacionales, Buenos Aires.

Bush, M. B. 1994. "Amazonian speciation: A necessarily complex model." *Journal of Biogeography*, 21: 5–17.

Bush, M. B. and P. A. Colinvaux. 1990. "A pollen record of a complete glacial cycle from lowland Panama." *Journal of Vegetation Science*, 1: 105–118.

Byrne, H., A. B. Rylands, J. C. Carneiro, J. W. Lynch Alfaro, F. Bertuol, M. N. F. da Silva, M. Messias, C. P. Groves, R. A. Mittermeier, I. Farias, T. Hrbek, H. Schneider, I. Sampaio, and J. P. Boubli. 2016. "Phylogenetic relationships of the New World titi monkeys (*Callicebus*): First appraisal of taxonomy based on molecular evidence." *Frontiers in Zoology*, 13: 1–25.

Cabrera, A. and J. Yepes. 1940. *Mamíferos sud-americanos (vida, costumbres y descripción)*. Historia Natural Ediar, Buenos Aires.

Cabrera, A. L. 1951. "Territorios fitogeográficos de la República Argentina." *Boletín de la Sociedad Argentina de Botánica*, 4: 21–65.

Cabrera, A. L. 1953. "Esquema fitogeográfico de la República Argentina." *Revista del Museo de la Ciudad Eva Perón, Botánica*, 8: 87–168.

Cabrera, A. L. 1957. "La vegetación de la Puna argentina." *Revista de Investigaciones Agrícolas*, 11: 317–412.

Cabrera, A. L. 1958. "Fitogeografía en la Argentina." *Suma de Geografía*, 3: 101–207.

Cabrera, A. L. 1971. "Fitogeografía de la República Argentina." *Boletín de la Sociedad Argentina de Botánica*, 14: 1–42.

Cabrera, A. L. 1976. "Regiones fitogeográficas argentinas." In: Kugler, W. F. (ed.), *Enciclopedia Argentina de Agricultura y Jardinería*. *II*, ACME, Buenos Aires, pp. 1–85.

Cabrera, A. L. and A. Willink. 1973. *Biogeografía de América Latina*. Monografía 13, Serie de Biología, OEA, Washington, D.C.

Cadle, J. E. 1982. "The Neotropical colubrid snake fauna (Serpentes: Colubridae): Lineage component and biogeography." *Systematic Zoology*, 34: 1–65.

Calvillo-Canadell, L. and S. R. S. Cevallos-Ferris. 2005. "Diverse assemblage of Eocene and Oligocene Leguminosae from Mexico." *International Journal of Plant Sciences*, 166: 671–692.

Camardelli, M. and M. F. Napoli. 2012. "Amphibian conservation in the Caatinga biome and semiarid region of Brazil." *Herpetologica*, 68: 31–47.

Camargo, J. M. F. 1996. *Meliponini neotropicais (Apinae, Apidae, Hymenoptera): Biogeografía histórica*. Anais do Encontro sobre Abelhas de Ribeirão Preto, São Paulo, pp. 107–121.

Camargo, J. M. F. and J. S. Moure. 1996. "Meliponini neotropicais: O gênero *Geotrigona* Moure, 1943 (Apinae, Apidae, Hymenoptera), com especial referência à filogenia e biogeografia." *Arquivos de Zoologia, São Paulo*, 33: 95–161.

Camargo, J. M. F. and S. R. M. Pedro. 2003. "Meliponini neotropicais: O gênero *Partamona* Schwarz, 1939 (Hymenoptera, Apidae, Apinae)—Bionomía e biogeografia." *Revista Brasileira de Entomologia*, 47: 311–372.

Campbell, J. A. 1999. "Distribution patterns of amphibians in Middle America." In: Duellman, W. E. (ed.), *Patterns of distribution of amphibians: A global prespective*. The Johns Hopkins University Press, Baltimore and London, pp. 111–210.

Campos-Soldini, M. P., M. G. Del Río, and S. A. Roig-Juñent. 2013. "Análisis panbiogeográfico de las especies de *Epicauta* Dejean, 1834 (Coleoptera: Meloidae) en América del Sur austral." *Revista de la Sociedad Entomológica Argentina*, 72: 15–25.

Cano, E., A. V. Ramírez, A. Cano-Ortiz, and F. J. E. Esteban Ruiz. 2009. "Distribution of Central American Melastomataceae: Biogeographical analysis of the Caribbean islands." *Acta Botanica Gallica*, 156: 527–557.

Cano-Ortiz, A., C. M. Musarella, J. C. Piñar Fuentes, C. J. Pinto Gomes, and E. Cano. 2016. "Distribution patterns of endemic flora to define hotspots on Hispaniola." *Systematics and Biodiversity*, 14: 261–275.

Carleton, M. D., O. Sánchez, and G. Urbano-Vidales. 2002. "A new species of *Habromys* (Muroidea: Neotominae) from Mexico, with a generic review of species definitions and remarks on diversity patterns among Mesoamerican small mammals restricted to humid montane forests." *Proceedings of the Biological Society of Washington*, 115: 488–533.

Carvalho, C. J. B. de, and M. S. Couri. 2010. "Biogeografia de Muscidae (Insecta, Diptera) da América do Sul." In: Carvalho, C. J. B. and E. A. B. Almeida (eds.), *Biogeografia da América do Sul: Padroes e processos*. Editora Roca Limitada, São Paulo, pp. 277–298.

Casas-Andreu, G. and T. Reyna-Trujillo. 1990. "Herpetofauna (anfibios y reptiles). Mapa IV. 8. 6." In: *Atlas Nacional de México. Vol. III*. Instituto de Geografía, UNAM, Mexico City, map.

Castellanos, A. and R. A. Pérez-Moreau. 1941. "Carta fitogeográfica de la República Argentina, en contribución a la bibliografía botánica argentina." *Lilloa*, 7: 1–497.

Castellanos, A. and R. A. Pérez-Moreau. 1944. "Los tipos de vegetación de la República Argentina." *Monografías del Instituto de Estudios Geográficos, Universidad Nacional de Tucumán*, 4: 1–154.

Castillo, C. and P. Reyes-Castillo. 1984. "Biosistemática del género *Petrejoides* Kuwert (Coleoptera, Lamellicornia, Passalidae)." *Acta Zoológica Mexicana (nueva serie)*, 4: 1–84.

Cavers, S., C. Navarro, and A. J. Lowe. 2003. "Chloroplast DNA phylogeography reveals colonization history of a Neotropical tree, *Cedrela odorata* L., in Mesoamerica." *Molecular Ecology*, 12: 1451–1460.

Ceballos, A. and B. Rosso de Ferradás. 2008. "Pseudoscorpiones." In: Claps, L. E., G. Debandi, and S. Roig-Juñent (eds.), *Biodiversidad de artrópodos argentinos, Vol. 2*. Sociedad Entomológica Argentina, San Miguel de Tucumán, pp. 105–116.

Ceballos, G., J. Arroyo-Cabrales, and R. A. Medellín. 2002. "Mamíferos de México." In: G. Ceballos and J. A. Simonetti (eds.), *Diversidad y conservación de mamíferos neotropicales*. Conabio and UNAM, Mexico City, pp. 377–413.

Ceballos, G., L. Martínez, A. García, E. Espinoza, J. B. Creel, and R. Dirzo (eds.). 2010. *Diversidad, amenazas y áreas prioritarias para la conservación de las selvas secas del Pacífico de México*. Fondo de Cultura Económica, Mexico City.

Ceballos, G. and G. Oliva. 2005. *Los mamíferos silvestres de México*. Conabio and Fondo de Cultura Económica, Mexico City.

Cecca, F., J. J. Morrone, and M. C. Ebach. 2011. "Biogeographical convergence and time-slicing: Concepts and methods in cladistic biogeography." In: Upchurch, P., A. McGowan, and C. Slater (eds.), *Palaeogeography and palaeobiogeography: Biodiversity in space and time, Systematics Association Special Volume*. Taylor & Francis, CRC Press, Boca Raton, Florida, pp. 1–12.

Cerón, C., W. Palacios, R. Valencia, and R. Sierra. 1999. "Las formaciones naturales de la costa del Ecuador." In: Sierra, R. (ed.), *Propuesta preliminar de un sistema de clasificación de vegetación para el Ecuador continental*. Proyecto INEFAN/GEF-BIRF and EcoCiencia, Quito, pp. 55–78.

Challenger, A. 1998. *Utilización y conservación de los ecosistemas terrestres de México: Pasado, presente y futuro*. Conabio, Mexico City.

Chaves, N. B., F. Roque, and R. Tidon. 2010. "Aspectos biogeográficos dos drosofilídeos (Insecta: Diptera) do bioma Cerrado." In: Diniz, I. R., J. M. Filho, R. B. Machado, and R. B. Cavalcanti (eds.), *Cerrado: Conhecimiento científico quantitativo como subsídio para açoes de conservação*. Universidade de Brasília, Brasília, pp. 203–222.

Chiapella, J. O. and P. H. Demaio. 2015. "Plant endemism in the Sierras of Córdoba and San Luis (Argentina): Understanding links between phylogeny and regional biogeographical patterns." *PhytoKeys*, 47: 57–96.

Cione, A. L., G. M. Gasparini, E. Soibelzon, L. H. Soibelzon, and E. P. Tonni. 2015. *The Great American biotic interchange: A South American perspective*. Springer Briefs in Earth System Sciences, Dordrecht.

Clark, W. E. 1979. "Taxonomy and biogeography of weevils of the genus *Sibinia* Germar (Coleoptera: Curculionidae) associated with *Prosopis* (Leguminosae: Mimosoideae) in Argentina." *Proceedings of the Entomological Society of Washington*, 81: 153–170.

Clarke, C. B. 1892. "On biologic regions and tabulation areas." *Philosophical Transactions of the Royal Society of London*, 183: 371–387.

Cleef, A. M. 1978. "Characteristics of Neotropical páramo vegetation and its Subantarctic relations." In: Troll, C. and W. Lauer (eds.), *Geoecological relations between the southern temperate zone and the tropical mountains. Erdwissenschaftliche Forschung*, 11: 365–390.

Cleef, A. M. 1981. "The vegetation of the páramos of the Colombian Cordillera Oriental." *Dissertationes Botanicae*, 61: 1–320.

Colinvaux, P. A. 1996. "Quaternary environmental history and forest diversity in the Neotropics." In: Jackson, J. B. C., A. F. Budd, and A. G. Coates (eds.), *Evolution and environment in tropical America*. University of Chicago Press, Chicago, pp. 359–406.

Colinvaux, P. A. 1997. "Amazonian diversity in light of the paleoecological record." *Quaternary Research*, 34: 330–345.

Colinvaux, P. A. 1998. "A new vicariance model for Amazonian endemics." *Global Ecology and Biogeography Letters*, 7: 95–96.

Colinvaux, P. A. and P. E. de Oliveira. 1999. "A palynological history of the Amazon rainforests through glacial cycles." *Acta Palaeontologica Romanica*, 2: 99–103.

Colinvaux, P. A., P. E. de Oliveira, and M. B. Bush. 2000. "Amazonian and Neotropical plant communities on glacial time-scales: The failure of the aridity and refuge hypotheses." *Quaternary Science Review*, 19: 141–169.

Colinvaux, P. A., P. E. de Oliveira, J. E. Moreno, M. C. Miller, and M. B. Bush. 1996. "A long pollen record from lowland Amazonia: Forest and cooling in glacial times." *Science*, 274: 85–88.

Colli, G. R. 2005. "As origens e a diversificação da herpetofauna do Cerrado." In: Scariot, A., J. C. Sousa-Silva and J. M. Felfili (eds.), *Cerrado: Ecologia, biodiversidade e conservação*. Ministério do Meio Ambiente, Brasília, pp. 249–264.

CONANP. 2004. *Programa de Conservación y Manejo Reserva de la Biosfera Archipiélago de Revillagigedo*. Comisión Nacional de Áreas Naturales Protegidas, Mexico City.

Conroy, C. J., Y. Hortelano, F. A. Cervantes, and J. A. Cook. 2001. "The phylogenetic position of southern relictual species of *Microtus* (Muridae: Rodentia) in North America." *Mammalian Biology*, 56: 332–344.

Contreras-Medina, R., I. Luna-Vega, and J. J. Morrone. 2007a. "Gymnosperms and cladistic biogeography of the Mexican transition zone." *Taxon*, 56: 905–915.

Contreras-Medina, R., I. Luna-Vega, and J. J. Morrone. 2007b. "Application of parsimony analysis of endemicity to Mexican gymnosperm distributions: Grid-cells, biogeographical provinces and track analysis." *Biological Journal of the Linnean Society*, 92: 405–417.

Corona, A. and J. J. Morrone. 2005. "Track analysis of the species of *Lampetis (Spinthoptera)* Casey, 1909 (Coleoptera: Buprestidae) in North America, Central America, and the West Indies." *Caribbean Journal of Science*, 41: 37–41.

Corona, A. M., V. H. Toledo, and J. J. Morrone. 2007. "Does the Trans-Mexican Volcanic Belt represent a natural biogeographic unit?: An analysis of the distributional patterns of Coleoptera." *Journal of Biogeography*, 34: 1008–1015.

Corona, A. M., V. H. Toledo, and J. J. Morrone. 2009. "Track analysis of the Mexican species of Buprestidae (Coleoptera): Testing the complex nature of the Mexican Transition Zone." *Journal of Biogeography*, 36: 1730–1738.

Cortés, R. and P. Franco. 1997. "Análisis panbiogeográfico de la flora de Chiribiquete, Colombia." *Caldasia*, 19: 465–478.

Cortés-Ortiz, L., E. Bermingham, C. Rico, E. Rodríguez-Luna, I. Sampaio, and M. Ruiz-García. 2003. "Molecular systematics and biogeography of the Neotropical monkey genus, *Alouatta*." *Molecular Phylogenetics and Evolution*, 26: 64–81.

Cortés-Ramírez, G., A. Gordillo-Martínez, and A. G. Navarro-Sigüenza. 2012. "Patrones biogeográficos de las aves de la península de Yucatán." *Revista Mexicana de Biodiversidad*, 83: 530–542.

Cortés-Rodríguez, N., B. E. Hernández-Baños, A. G. Navarro-Sigüenza, and K. E. Omland. 2008. "Geographic variation and genetic structure in the streak-nacked oriole: Low mitochondrial DNA differentiation reveals recent divergence." *The Condor*, 110: 729–739.

Coscarón, M. del C. and J. J. Morrone. 1997. "Cladistics and biogeography of the assassin bug genus *Melanolestes* Stål (Heteroptera: Reduviidae)." *Proceedings of the Entomological Society of Washington*, 99: 55–59.

Coscarón, S. and C. L. Coscarón-Arias. 1995. "Distribution of Neotropical Simuliidae (Insecta, Diptera) and its areas of endemism." *Revista de la Academia Colombiana de Ciencias*, 19: 717–732.

Costa, L. P. 2003. "The historical bridge between the Amazon and the Atlantic Forest of Brazil: A study of molecular phylogeography with small mammals." *Journal of Biogeography*, 30: 71–86.

Costa, M., Á. L. Viloria, O. Huber, S. Attal, and A. Orellana. 2013. "Lepidoptera del Pantepui. Parte I: Endemismo y caracterización biogeográfica." *Entomotropica*, 28: 193–217.

Coulleri, J. P. and M. S. Ferrucci. 2012. "Biogeografía histórica de *Cardiospermum* y *Urvillea* (Sapindaceae) en América: Paralelismos geográficos e históricos con los bosques secos estacionales neotropicales." *Boletín de la Sociedad Argentina de Botánica*, 47: 103–117.

Couri, M. S. and C. J. B. de Carvalho. 2008. "A review of the Neotropical genus *Drepanocnemis* Stein (Diptera, Muscidae), with phylogenetic analysis and biogeographic considerations of its species." *Journal of Natural History*, 42: 2659–2678.

Cox, C. B. C. 2001. "The biogeographic regions reconsidered." *Journal of Biogeography*, 28: 511–523.

Cox, C. B. C. and P. D. Moore. 1998. *Biogeography: An ecological and evolutionary approach*. Blackwell Science, Oxford.

Cracraft, J. 1985. "Historical biogeography and patterns of differentiation within the South American avifauna: Areas of endemism." *Ornithological Monographs*, 36: 49–84.

Cracraft, J. 1988. "Deep-history biogeography: Retrieving the historical pattern of evolving continental biotas." *Systematic Zoology*, 37: 221–236.

Cracraft, J. and R. O. Prum. 1988. "Patterns and processes of diversification: Speciation and historical congruence in some Neotropical birds." *Evolution*, 42: 603–620.

Craw, R. C., J. R. Grehan, and M. J. Heads. 1999. *Panbiogeography: Tracking the history of life*. Oxford Biogeography Series 11, Oxford University Press, New York.

Crawford, A. J. and E. N. Smith. 2005. "Cenozoic biogeography and evolution in direct-developing frogs of Central America (Leptodactylidae: *Eleutherodactylus*) as inferred from a phylogenetic analysis of nuclear and mitochondrial genes." *Molecular Phylogenetics and Evolution*, 35: 536–555.

Crisci, J. V., M. S. de la Fuente, A. A. Lanteri, J. J. Morrone, E. Ortiz Jaureguizar, R. Pascual, and J. L. Prado. 1993. "Patagonia, Gondwana Occidental (GW) y Oriental (GE), un modelo de biogeografía histórica." *Ameghiniana*, 30: 104.

Croizat, L. 1958. *Panbiogeography. Vols. 1, 2a, and 2b.* Published by the author, Caracas.

Croizat, L. 1976. *Biogeografía analítica y sintética ('panbiogeografía') de las Américas.* Biblioteca de la Academia de Ciencias Físicas, Matemáticas y Naturales, Caracas.

Crother, B. I. and C. Guyer. 1996. "Caribbean historical biogeography: Was the dispersal-vicariance debate eliminated by an extraterrestrial bolide?" *Herpetologica*, 52: 440–465.

Cuatrecasas, J. 1986. "Speciation and radiation of the Espeletiinae in the Andes." In: Vuilleumier, F. and M. Monasterio (eds.), *High altitude tropical biogeography*. Oxford University Press and American Museum of Natural History, New York and Oxford, pp. 267–303.

Daniel, G. M. and F. Z. Vaz-de-Mello. 2016. "Biotic components of dung beetles (Insecta: Coleoptera: Scarabaeidae: Scarabainae) from Pantanal-Cerrado border and its implications for Chaco regionalization." *Journal of Natural History*, 50: 1159–1173.

Darlington, P. J. 1957. *Zoogeography: The geographical distribution of animals.* John Wiley and Sons, New York.

DaSilva, M. B. and R. Pinto-da-Rocha (2010). "História biogeográfica da Mata Atlântica: Opiliões (Arachnida) como modelo para su inferência." In: Carvalho, C. J. B. de and E. A. B. Almeida (eds.), *Biogeografia da América do Sul: Padroes e processos*. Editora Roca Limitada, São Paulo, pp. 221–220.

Dávalos, L. M. 2004. "Phylogeny and biogeography of Caribbean mammals." *Biological Journal of the Linnean Society*, 81: 373–394.

Daza, J. M., T. A. Castoe, and C. L. Parkinson. 2010. "Using regional comparative phylogeographic data from snake lineages to infer historical processes in Middle America." *Ecography*, 33: 343–354.

de Candolle, A. P. 1820. "Géographie botanique." In: *Dictionnaire des Sciences Naturelles*. Masson, Strasbourg and Paris, pp. 359–422.

De la Cruz, L. J. 1989. *Regionalización faunística*. Nuevo Atlas Nacional de Cuba, Instituto Cubano de Geodesia y Cartografía and Instituto Geográfico Nacional de España, Madrid.

De Marmels, J. 2007. "*Tepuibasis* gen. nov. from the Pantepui region of Venezuela, and with biogeographic, phylogenetic and taxonomic considerations on the Teinobasinae (Zigoptera: Coenagrionidae)." *Odonatologica*, 36: 117–146.

de Queiroz, A. 2014. *The monkey's voyage: How improbable journeys shaped the history of life*. Basic Books, New York

Del Río, M. G., J. J. Morrone, and A. A. Lanteri. 2015. "Evolutionary biogeography of South American weevils of the tribe Naupactini (Coleoptera: Curculionidae)." *Journal of Biogeography*, 42: 1293–1304.

Del Risco, E. del, and A. Vandama. 1989. *Regionalización florística*. Nuevo Atlas Nacional de Cuba, Instituto Cubano de Geodesia y Cartografía and Instituto Geográfico Nacional de España, Madrid.

De-Nova, J. A., R. Medina, J. C. Montero, A. Weeks, J. A. Rosell, M. E. Olson, L. E. Eguiarte, and S. Magallón. 2012. "Insights into the historical construction of species-rich Mesoamerican seasonally dry tropical forests: The diversification of *Bursera* (Burseraceae, Sapindales)." *New Phytologist*, 193: 276–287.

Désamoré, A., A. Vanderpoorten, B. Laenen, S. R. Gradstein, and P. J. R. Kok. 2010. "Biogeography of the Lost World (Pantepui region, northeastern South America): Insights from bryophytes." *Phytotaxa*, 9: 254–265.

Di Castri, F. 1968. "Esquisse écologique du Chili." In: Delamare Debouteville, C. and E. Rappoport (eds.), *Biologie de L'Amérique Australe. Vol. 4.* Editions du Centre National de la Recherche Scientifique, Paris, pp. 6–52.

Dice, L. R. 1943. *The biotic provinces of North America.* University of Michigan Press, Ann Arbor.

Diels, L. 1908. *Pflanzengeographie.* Göschensche Verlagshandlung, Leipzig.

Dillon, M. O., T. Tu, L. Xie, V. Quipuscoa Silvestre, and J. Wen. 2009. "Biogeographic diversification in *Nolana* (Solanaceae), a ubiquitous member of the Atacama and Peruvian deserts along the western coast of South America." *Journal of Systematics and Evolution*, 47: 457–476.

Dinerstein, E., D. M. Olson, D. J. Graham, A. L. Webster, S. A. Primm, M. P. Bookbinder, and G. Ledec. 1995. *A conservation assessment of the terrestrial ecoregions of Latin America and the Caribbean.* The World Bank, Washington, D.C.

Domínguez-Domínguez, O., M. Vila, R. Pérez-Rodríguez, N. Remón, and I. Doadrio. 2011. "Complex evolutionary history of the Mexican stoneroller *Campostoma ornatum* Girard, 1856 (Actinopterygii: Cyprinidae)." *BMC Evolutionary Biology*, 11: 1–20.

Donato, M. 2006. "Historical biogeography of the family Tristiridae (Orthoptera: Acridomorpha) applying dispersal-vicariance analysis." *Journal of Arid Environments*, 66: 421–434.

Donnelly, T. W. 1988. "Geologic constraints on Caribbean biogeography." In: Liebherr, J. K. (ed.), *Zoogeography of Caribbean insects.* Cornell University Press, Ithaca and London, pp. 15–37.

Donnelly, T. W. 1992. "Geological setting and tectonic history of Mesoamerica." In: Quinteros, D. and A. Aiello (eds.), *Insects of Panama and Mesoamerica: Selected studies.* Oxford Science Publication, Oxford, pp. 1–13.

Donoghue, M. J. and B. R. Moore. 2003. "Toward an integrative historical biogeography." *Integrative and Comparative Biology*, 43: 261–270.

Dos Santos, D. A., D. Emmerich, C. Molineri, C. Nieto, and E. Domínguez. 2016. "On the position of Uruguay in the South American puzzle: Insights from Ephemeroptera (Insecta)." *Journal of Biogeography*, 43: 361–371.

Drude, O. 1884. *Die Florenreiche der Erde: Darstellung der gegenwärtigen Verbreitung verhältnisse der Pflanzen: Ein Beitrag zur vergleichenden Erdkunde.* Petermanns Mitteilungen, Justus Perthes, Gotha.

Drude, O. 1890. *Handbuch der Pflanzengeographie.* Verlag J. Engelhorn, Stuttgart.

Dugès, E. 1902. "Algo sobre la distribución geográfica de algunas aves." *Memorias y Revista de la Sociedad Científica Antonio Alzate*, 18: 44–46.

Duno-de Stefano, R., L. L. Can-Itza, A. Rivera-Ruiz, and L. M. Calvo-Irabién. 2012. "Regionalización y relaciones biogeográficas de la Península de Yucatán con base en los patrones de distribución de la familia Leguminosae." *Revista Mexicana de Biodiversidad*, 83: 1053–1072.

Durante, S. P., N. C. Cabrera, and L. E. Gómez de la Vega. 2008. "Megachilidae." In: Claps, L. E., G. Debandi, and S. Roig-Juñent (eds.), *Biodiversidad de artrópodos argentinos, Vol. 2.* Sociedad Entomológica Argentina, San Miguel de Tucumán, pp. 421–433.

Ebach, M. C. and C. J. Humphries. 2003. "Ontology of biogeography." *Journal of Biogeography*, 30: 959–962.

Ebach, M. C., J. J. Morrone, L. R. Parenti, and Á. L. Viloria. 2008. "International Code of Area Nomenclature." *Journal of Biogeography*, 35: 1153–1157.

Eberhard, J. R. and E. Bermingham. 2004. "Phylogeny and biogeography of the *Amazona ochrocephala* (Aves: Psittacidae) complex." *The Auk*, 121: 318–332.

Echeverría-Londoño, S. and D. R. Miranda-Esquivel. 2011. "MartiTracks: A geometrical approach for identifying geographical patterns of distribution." *PLOS ONE*, 6: 1–7.

Echeverry, A. 2011. "Biogeografía y geología: Una reflexión sobre su interacción a partir de tres casos caribeños." *Revista de Geografía Norte Grande*, 48: 27–43.

Echeverry, A. and J. J. Morrone. 2010. "Parsimony analysis of endemicity as a panbiogeographical tool: An analysis of Caribbean plant taxa." *Biological Journal of the Linnean Society*, 101: 961–976.

Echeverry, A. and J. J. Morrone. 2013. "Generalized tracks, area cladograms and tectonics in the Caribbean." *Journal of Biogeography*, 40: 1619–1637.

Edwards, C. W. and R. D. Bradley. 2002a. "Molecular systematics and historical phylobiogeography of the *Neotoma mexicana* species group." *Journal of Mammalogy*, 83: 20–30.

Edwards, C. W. and R. D. Bradley. 2002b. "Molecular systematics of the genus *Neotoma*." *Molecular Phylogenetics and Evolution*, 25: 489–500.

Ellenrieder, N. von. 2001. "Species composition and distribution of the Argentinean Aeshnidae (Odonata: Anisoptera)." *Revista de la Sociedad Entomológica Argentina*, 60: 39–60.

Endler, J. 1982. "Pleistocene forest refuges: Fact or fancy?" In: Prance, T. P. (ed.), *Biological diversification in the tropics*. Columbia University Press, New York, pp. 179–200.

Engler, A. 1879. *Versuch einer Entwicklungsgeschichte der Pflanzenwelt, insbesondere der Florengebiete seit der Tertiärperiode. Vol 1. Die Extratropischen Gebiete der Nördlichen Hemisphäre*. Verlag von W. Engelmann, Leipzig.

Engler, A. 1882. *Versuch einer Entwicklungsgeschichte der Pflanzenwelt, insbesondere der Florengebiete seit der Tertiärperiode. Vol. 2. Die extratropischen Gebiete der Südlichen Hemisphäre und die Tropischen Gebiete*. Verlag von W. Engelmann, Leipzig.

Engler, A. 1899. *Die Entwicklung der Pflanzengeographie in den letzten hundert Jahren und weitere Aufgaben derselben: Wissenschaftliche Beiträge zum Gedächtnis der hundertjährigen Wiederkehr des Antritts von Alexander von Humboldt's Reise nach Amerika*. Gesellschaft für Erdkunde zu Berlin, Berlin.

Erwin, T. E. and M. G. Pogue. 1988. "*Agra*, arboreal beetles of Neotropical forests: Biogeography and the forest refugium hypothesis (Carabidae)." In: Heyer, W. R. and E. Vanzolini (eds.), *Proceedings of a Workshop on Neotropical distribution patterns*. Academia Brasileira de Ciencias, Rio de Janeiro, pp. 161–188.

Escalante, P., A. G. Navarro, and A. T. Peterson. 1998. "Un análisis geográfico, ecológico e histórico de la diversidad de aves terrestres de México." In: Ramamoorthy, T. P., R. Bye, A. Lot, and A. Fa (eds.), *Diversidad biológica de México: Orígenes y distribución*. Instituto de Biología, UNAM, Mexico City, pp. 279–304.

Escalante, T. 2009. "Un ensayo sobre regionalización biogeográfica." *Revista Mexicana de Biodiversidad*, 80: 551–560.

Escalante, T., D. Espinosa, and J. J. Morrone. 2003. "Using parsimony analysis of endemicity to analyze the distribution of Mexican land mammals." *Southwestern Naturalist*, 48: 563–578.

Escalante, T., E. A. Martínez-Salazar, J. Falcón-Ordaz, M. Linaje, and R, Guerrero. 2011. "Análisis panbiogeográfico de *Vexillata* (Nematoda: Ornithostrongylidae) y sus huéspedes (Mammalia: Rodentia)." *Acta Zoológica Mexicana (nueva serie)*, 27: 25–46.

Escalante, T., J. J. Morrone, and G. Rodríguez-Tapia. 2013. "Biogeographic regions of North American mammals based on endemism." *Biological Journal of the Linnean Society*, 110: 485–499.

Escalante, T., G. Rodríguez, N. Cao, M. C. Ebach, and J. J. Morrone. 2007a. "Cladistic biogeographic analysis suggests an early Caribbean diversification in Mexico." *Naturwissenschaften*, 94: 561–565.

Escalante, T., G. Rodríguez, N. Gámez, L. León-Paniagua, O. Barrera, and V. Sánchez-Cordero. 2007b. "Biogeografía y conservación de los mamíferos." In: Luna, I., J. J. Morrone, and D. Espinosa (eds.), *Biodiversidad de la Faja Volcánica Transmexicana*. Las Prensas de Ciencias, UNAM, Mexico City, pp. 485–502.

Escalante, T., G. Rodríguez, and J. J. Morrone. 2004. "The diversification of Nearctic mammals in the Mexican transition zone." *Biological Journal of the Linnean Society*, 83: 327–339.

Escalante, T., G. Rodríguez, and J. J. Morrone. 2005. "Las provincias biogeográficas del Componente Mexicano de Montaña desde la perspectiva de los mamíferos continentales." *Revista Mexicana de Biodiversidad*, 76: 199–205.

Escalante, T., C. Szumik, and J. J. Morrone. 2009. "Areas of endemism of Mexican mammals: Reanalysis applying the optimality criterion." *Biological Journal of the Linnean Society*, 98: 468–478.

Espadas Manrique, C., R. Durán, and J. Argáez. 2003. "Phytogeographic analysis of taxa endemic to the Yucatán Peninsula using geographic information systems, the domain heuristic method and parsimony analysis of endemicity." *Diversity and Distributions*, 9: 313–330.

Espinosa, D., C. Aguilar, and S. Ocegueda. 2004. "Identidad biogeográfica de la Sierra Madre Oriental y posibles subdivisiones bióticas." In: Luna, I., J. J. Morrone, and D. Espinosa (eds.), *Biodiversidad de la Sierra Madre Oriental*. Las Prensas de Ciencias, UNAM, Mexico City, pp. 487–500.

Espinosa, D., J. Llorente, and J. J. Morrone. 2006. "Historical biogeographic patterns of the species of *Bursera* (Burseraceae) and their taxonomical implications." *Journal of Biogeography*, 33: 1945–1958.

Espinosa, D., J. J. Morrone, C. Aguilar, and J. Llorente. 2000. "Regionalización biogeográfica de México: Provincias bióticas." In: Llorente, J., E. González, and N. Papavero (eds.), *Biodiversidad, taxonomía y biogeografía de artrópodos de México: Hacia una síntesis de su conocimiento. Vol. II*. UNAM, Mexico City, pp. 61–94.

Espinosa, D. and S. Ocegueda. 2007. "Introducción." In: Luna, I., J. J. Morrone, and D. Espinosa (eds.), *Biodiversidad de la Faja Volcánica Transmexicana*. Las Prensas de Ciencias, UNAM, Mexico City, pp. 5–6.

Espinosa Organista, D., S. Ocegueda Cruz, C. Aguilar Zúñiga, Ó. Flores Villela, and J. Llorente-Bousquets. 2008. "El conocimiento biogeográfico de las especies y su regionalización natural." In: Sarukhán, J. (ed.), *Capital natural de México. Vol. I. Conocimiento actual de la biodiversidad*. Conabio, Mexico City, pp. 33–65.

Feldman, C. R., O. Flores-Villela, and T. J. Papenfuss. 2011. "Phylogeny, biogeography, and display evolution in the tree and brush lizard genus *Urosaurus* (Squamata: Phrynosomatidae)." *Molecular Phylogenetics and Evolution*, 61: 714–725.

Fernandes, A. 2006. *Fitogeografia brasileira: Provincias florísticas*. Realce Editora e Indústria Gráfica, Fortaleza.

Fernandes, A. and Bezerra, P. 1990. *Estudo fitogeográfico do Brasil*. Stylus Comunicaçoes, Fortaleza.

Fernández, H. R. and M. G. Cuezzo. 1997. *La región Neotropical: Algunos aspectos históricos*. Facultad de Ciencias Naturales e Instituto Miguel Lillo, Universidad Nacional de Tucumán, Serie Monográfica y Didáctica nro. 35.

Ferrari, L., T. Orozco-Esquivel, V. Manea, and M. Manea. 2012. "The dynamic history of the Trans-Mexican Volcanic Belt and the Mexico subduction zone." *Tectonophysics*, 522–523: 122–149.

Ferretti, N., A. González, and F. Pérez-Miles. 2012. "Historical biogeography of mygalomorph spiders from the peripampasic orogenic arc based on track analysis and PAE as a panbiogeographical tool." *Systematics and Biodiversity*, 10: 179–193.

Ferretti, N., A. González, and F. Pérez-Miles. 2014. "Identification of priority areas for conservation in Argentina: Quantitative biogeography insights from mygalomorph spiders (Araneae: Mygalomorphae)." *Journal of Insect Conservation*, 18: 1087–1096.

Ferro, I. 2013. "Rodent endemism, turnover and biogeographical transitions on elevation gradients in the northwestern Argentinian Andes." *Mammalian Biology*, 78: 322–331.

Ferro, I. and J. J. Morrone. 2014. "Biogeographic transition zones: A search for conceptual synthesis." *Biological Journal of the Linnean Society*, 113: 1–12.

Ferrusquía-Villafranca, I. 1990. "Regionalización biogeográfica. Mapa IV. 8. 10." In: *Atlas Nacional de México. Vol. III*. Instituto de Geografía, UNAM, Mexico City, map.

Ferrusquía-Villafranca, I. 2007. "Ensayo sobre la caracterización y significación biológica." In: Luna, I., J. J. Morrone, and D. Espinosa (eds.), *Biodiversidad de la Faja Volcánica Transmexicana*. Las Prensas de Ciencias, UNAM, Mexico City, pp. 7–23.

Fiaschi, P. and J. R. Pirani. 2009. "Review of plant biogeographic studies in Brazil." *Journal of Systematics and Evolution*, 47: 477–496.

Fittkau, E. J. 1969. "The fauna of South America." In: Fittkau, E., J. J. Illies, H. Klinge, G. H. Schwabe, and H. Sioli (eds.), *Biogeography and ecology in South America. Vol. 2*. Junk, The Hague, pp. 624–650.

Fjeldså, J. 1992. "Biogeographic patterns and evolution of the avifauna of relict high-altitude woodlands of the Andes." *Steenstrupia*, 18: 9–62.

Fleming, C. A. 1987. "Comments on Udvardy's biogeographical realm Antarctica." *Journal of the Royal Society of New Zealand*, 17: 195–200.

Flores-Villela, O. and I. Goyenechea. 2001. "A comparison of hypotheses of historical area relationships for Mexico and Central America, or in search for the lost pattern." In: Johnson, J., R. G. Webb, and O. Flores-Villela (eds.), *Mesoamerican herpetology: Systematics, zoogeography, and conservation*. Centennial Museum, Special Publication 1, University of Texas, El Paso, pp. 171–181.

Flores-Villela, O. and E. A. Martínez-Salazar. 2009. "Historical explanation of the origin of the herpetofauna of México." *Revista Mexicana de Biodiversidad*, 80: 817–833.

Fontenla, J. L. 2003. "Biogeography of Antillean butterflies (Lepidoptera: Rhopalocera): Patterns of association among areas of endemism." *Transactions of the American Entomological Society*, 129: 399–410.

Fontenla Rizo, J. L. and A. López Almiral. 2008. *Archipiélago cubano: Biogeografía histórica y complejidad*. Cubalibri, Washington, D.C.

Ford, S. M. 2005. "The biogeographic history of Mesoamerican primates." In: Estrada, A., P. A. Garber, M. S. M. Pavelka, and L. Luecke (eds.), *New perspectives in the study of Mesoamerican primates: Distribution, ecology, behavior, and conservation*. Springer, New York, pp. 81–114.

Frailey, C. D. 2002. "Neogene paleogeography of the Amazon basin." In: Dort, W. Jr (ed.), *TERQUA Symposium Series 3*. Institute for Tertiary-Quaternary Studies, Lincoln, pp. 71–97.

Frailey, C. D., E. L. Lavina, A. Rancy, and J. P. Souza Filho. 1988. "A proposed Pleistocene/Holocene lake in the Amazon basin and its significance to Amazonian geology and biogeography." *Acta Amazonica*, 18: 119–143.

Francisco-Ortega, J., E. Santiago-Valentín, P. Acevedo-Rodríguez, C. Lewis, J. Pipoly III, A. W. Meerow, and M. Maunder. 2007. "Seed plant genera endemic to the Caribbean island biodiversity hotspot: A review and a molecular phylogenetic perspective." *The Botanical Review*, 73: 183–234.

Franco-Rosselli, P. and C. C. Berg. 1997. "Distributional patterns of *Cecropia* (Cecropiaceae): A panbiogeographic analysis." *Caldasia*, 19: 285–296.

Galán de Mera, A. and G. Navarro. 1992. "Comunidades vegetales acuáticas del Paraguay occidental." *Caldasia*, 17: 35–46.

Gámez, N., T. Escalante, G. Rodríguez, M. Linaje, and J. J. Morrone. 2012. "Caracterización biogeográfica de la Faja Volcánica Transmexicana y análisis de los patrones de distribución de su mastofauna." *Revista Mexicana de Biodiversidad*, 83: 258–272.

García, A. 2010. "Reptiles y anfibios." In: Ceballos, G., L. Martínez, A. García, E. Espinoza, J. B. Creel, and R. Dirzo (eds.), *Diversidad, amenazas y áreas prioritarias para la conservación de las selvas secas del Pacífico de México*. Fondo de Cultura Económica, Mexico City, pp. 165–178.

García-Marmolejo, G., T. Escalante, and J. J. Morrone. 2008. "Establecimiento de prioridades para la conservación de mamíferos terrestres neotropicales de México." *Mastozoología Neotropical*, 15: 41–65.

García-Moreno, J., A. G. Navarro-Sigüenza, A. T. Peterson, and L. A. Sánchez-González. 2004. "Genetic variation coincides with geographic structure in the common bush-tanager *(Chlorospingus ophthalmicus)* complex from Mexico." *Molecular Phylogenetics and Evolution*, 33: 186–196.

García-Trejo, E. A. and A. G. Navarro. 2004. "Patrones biogeográficos de la riqueza de especies y el endemismo de la avifauna en el oeste de México." *Acta Zoológica Mexicana (nueva serie)*, 20: 167–185.

Garzione, C. N., G. D. Hoke, J. C. Libarkin, S. Withers, B. MacFadden, J. Eiler, P. Hosh, and A. Mulch. 2008. "Rise of the Andes." *Science*, 320: 1304–1307.

Gentry, A. H. 1982. "Neotropical floristic diversity: Phytogeographical connections between Central and South America, Pleistocene climatic fluctuations, or an accident of the Andean orogeny?" *Annals of the Missouri Botanical Garden*, 69: 557–593.

Gill, T. 1885. "The principles of zoogeography." *Proceedings of the Biological Society of Washington*, 2: 1–23.

Goldani, A. and G. S. Carvalho. 2003. "Análise de parcimônia de endemismo de cercopídeos neotropicais (Hemiptera, Cercopidae)." *Revista Brasileira de Entomologia*, 47: 437–442.

Goldani, A., G. S. Carvalho, and J. C. Bicca-Marques. 2006. "Distribution patterns of Neotropical primates (Platyrrhini) based on parsimony analysis of endemicity." *Brazilian Journal of Biology*, 66: 61–74.

Goldani, A., A. Ferrari, G. S. Carvalho, and A. J. Creão-Duarte. 2002. "Análise de parcimônia de endemismo de membracídeos neotropicais (Hemiptera, Membracidae, Hoplophorionini)." *Revista Brasileira de Zoologia*, 19: 187–193.

Goldman, E. A. and Moore, R. T. 1945. The biotic provinces of Mexico. *Journal of Mammalogy*, 26: 347–360.

Goloboff, P., J. S. Farris, and K. C. Nixon. 2008. "TNT, a free program for phylogenetic analysis." *Cladistics*, 24: 774–786.

González Soriano, E. and R. Novelo Gutiérrez. 1996. "Odonata." In: Llorente Bousquets, J., A. N. García Aldrete, and E. Gónzalez Soriano (eds.), *Biodiversidad, taxonomía y biogeografía de artrópodos de México: Hacia una síntesis de su conocimiento*. Universidad Nacional Autónoma de México, Mexico City, pp. 147–167.

González-Zamora, A., I. Luna-Vega, J. L. Villaseñor, and C. A. Ruiz-Jiménez. 2007. "Distributional patterns and conservation of species of Asteraceae (asters etc.) endemic to eastern Mexico: A panbiogeographical approach." *Systematics and Biodiversity*, 5: 135–144.

Good, R. 1947. *The geography of the flowering plants*. Longman, London.

Gradstein, S. R. 1998. "Hepatic diversity in the Neotropical páramos." In: *Proceedings of the VI Congreso Latinoamericano de Botánica (1994)*. Monographs in Systematic Botany from the Missouri Botanical Garden 68, Missouri, pp. 69–85.

Graham, A. 2003. "Historical phytogeography of the Greater Antilles." *Brittonia*, 55: 357–383.

Graham, A. 2004. *A natural history of the New World: The ecology and evolution of plants in the Americas*. The University of Chicago Press, Chicago and London.

Graham, S. A. 2003. "Biogeographic patterns of the Antillean Lythraceae." *Systematic Botany*, 28: 410–420.

Grehan, J. R. 2001. "Biogeography and evolution of the Galapagos: Integration of the biological and geological evidence." *Biological Journal of the Linnean Society*, 74: 267–287.

Guerrero, P. C., M. Rosas, M. T. K. Arroyo, and J. J. Wiens. 2013. "Evolutionary lag times and recent origin of the biota of an ancient desert (Atacama-Sechura)." *Proceedings of the National Academy of Sciences*, 110: 11469–11474.

Gutiérrez-García, T. A. and E. Vázquez-Domínguez. 2012. Biogeographically dynamic genetic structure bridging two continents in the monotypic Central American rodent *Ototylomys phyllotis*. *Biological Journal of the Linnean Society*, 107: 593–610.

Gutiérrez-García, T. A. and E. Vázquez-Domínguez. 2013. "Consensus between genes and stones in the biogeographic and evolutionary history of Central America." *Quaternary Research*, 79: 311–324.

Gutiérrez-Velázquez, A., O. Rojas-Soto, P. Reyes-Castillo, and G. Halffter. 2013. "The classic theory of Mexican Transition Zone revisited: The distributional congruence patterns of Passalidae (Coleoptera)." *Invertebrate Systematics*, 27: 282–293.

Haffer, J. 1969. "Speciation in Amazonian forest birds." *Science*, 165: 131–137.

Haffer, J. 1974. *Avian speciation in tropical South America, with a systematic survey of the toucans (Ramphastidae) and jacamars (Galbulidae)*. Nuttall Ornithological Club, Cambridge, Massachusetts.

Haffer, J. 1981. "Aspects of Neotropical bird speciation during the Cenozoic." In: Nelson, G. and D. E. Rosen (eds.), *Vicariance biogeography: A critique*. Columbia University Press, New York, pp. 371–391.

Haffer, J. 1985. "Avian zoogeography of the Neotropical lowlands." *Ornithological Monographs*, 36: 13–145.

Haffer, J. 1993. "On the 'river effect' in some forest birds of southern Amazonia." *Boletim do Museu Paraense E. Goeldi, Zoologia*, 8: 217–245.

Haffer, J. 1997. "Alternative models of vertebrate speciation in Amazonia: An overview." *Biodiversity and Conservation*, 6: 451–476.

Halffter, G. 1962. "Explicación preliminar de la distribución geográfica de los Scarabaeidae mexicanos." *Acta Zoológica Mexicana*, 5: 1–17.

Halffter, G. 1964. "Las regiones Neártica y Neotropical, desde el punto de vista de su entomo-fauna." *Anais do II Congresso Latinoamericano de Zoologia, São Paulo, 1962*, 1: 51–61.

Halffter, G. 1965. "Algunas ideas acerca de la zoogeografía de América." *Revista de la Sociedad Mexicana de Historia Natural*, 26: 1–16.

Halffter, G. 1974. "Eléments anciens de l'entomofaune neotropicale: Ses implications bio-géographiques." *Quaestiones Entomologicae*, 10: 223–262.

Halffter, G. 1976. "Distribución de los insectos en la Zona de Transición Mexicana: Relaciones con la entomofauna de Norteamérica." *Folia Entomológica Mexicana*, 35: 1–64.

Halffter, G. 1978. "Un nuevo patrón de dispersión en la zona de transición mexicana: El mesoamericano de montaña." *Folia Entomológica Mexicana*, 39–40: 219–222.

Halffter, G. 1987. "Biogeography of the montane entomofauna of Mexico and Central America." *Annual Review of Entomology*, 32: 95–114.

Halffter, G., M. E. Favila, and L. Arellano. 1995. "Spatial distribution of three groups of Coleoptera along an altitudinal transect in the Mexican Transition Zone and its biogeographical implications." *Elytron*, 9: 151–185.

Halffter, G., J. Llorente-Bousquets, and J. J. Morrone. 2008. "La perspectiva biogeográfica histórica." In: Sarukhán, J. (ed.), *Capital natural de México. Vol. I. Conocimiento actual de la biodiversidad*. Conabio, Mexico City, pp. 67–86.

Harris, D., D. S. Rogers, and J. Sullivan, J. 2000. "Phylogeography of *Peromyscus furvus* (Rodentia; Muridae) based on cytochrome b sequence data." *Molecular Ecology*, 9: 2129–2135.

Heads, M. 2012. *Molecular panbiogeography of the tropics*. University of California Press, Berkeley and Los Angeles.

Hechem, V., L. Acheritobehere, and J. J. Morrone. 2011. "Patrones de distribución de las especies de *Cynanchum, Diplolepis* y *Tweedia* (Apocynaceae: Asclepiadoideae) de America del Sur austral." *Revista de Geografía Norte Grande*, 48: 45–60.

Hedges, S. B. 1982. "Caribbean biogeography: Implications of recent plate tectonic studies. *Systematic Zoology*, 31: 518–522.

Hedges, S. B. 1996a. "Vicariance and dispersal in Caribbean biogeography." *Herpetologica*, 52: 466–473.

Hedges, S. B. 1996b. "Historical biogeography of West Indian vertebrates." *Annual Review of Ecology and Systematics*, 27: 163–196.

Hedges, S. B. 2001. "Biogeography of the West Indies: An overview." In: Woods, C. A. and F. E. Sergile (eds.), *Biogeography of the West Indies: Patterns and perspectives, 2nd edition*. CRC Press, Boca Raton, pp. 15–34.

Hedges, S. B., C. A. Hass, and L. R. Maxson. 1992. "Caribbean biogeography: Molecular evidence for dispersal in West Indian terrestrial vertebrates." *Proceedings of the National Academy of Sciences of the USA*, 89: 1909–1913.

Hauman, L. 1920. "Ganadería y geobotánica en la Argentina." *Revista del Centro de Estudiantes de Agronomía y Veterinaria de la Universidad de Buenos Aires*, 102: 45–65.

Hauman, L. 1931. "Esquisse phytogéographique de l'Argentine subtropicale es de ses relations avec la geóbotanie sud-américaine." *Bulletin de la Société Royale de Botanique de Belgique*, 64: 20–80.

Heilprin, A. 1887. *The geographical and geological distribution of animals*. International Scientific Series, New York and London.

Helgen, K. M. and D. E. Wilson. 2005. "A systematic and zoogeographic overview of the raccoons of Mexico and Central America." In: Sánchez-Cordero, V. and R. A. Medellín (eds.), *Contribuciones mastozoológicas en homenaje a Bernardo Villa*. Instituto de Biología, UNAM, and Conabio, Mexico City, pp. 221–236.

Hernández, J., A. Hurtado, R. Ortiz, and T. Walschburger. 1992a. "Unidades biogeográficas de Colombia." In: Halffter, G. (ed.), *La diversidad biológica de Iberoamérica, Acta Zoológica Mexicana. Volumen Especial 1992.* Cyted-D, Programa Iberoamericano de Ciencia y Tecnología para el Desarrollo, Instituto de Ecología, A. C., Xalapa, pp. 105–151.

Hernández, J., A. Hurtado, R. Ortiz, and T. Walschburger. 1992b. "Centros de endemismo en Colombia." In: Halffter, G. (ed.), *La diversidad biológica de Iberoamérica, Acta Zoologica Mexicana. Vol. Esp. 1992.* Cyted-D, Programa Iberoamericano de Ciencia y Tecnología para el Desarrollo, Instituto de Ecología, A. C., Xalapa, pp. 175–190.

Hernández, J., T. Walschburger, R. Ortiz, and A. Hurtado. 1992c. "Origen y distribución de la biota suramericana y colombiana." In: Halffter, G. (ed.), *La diversidad biológica de Iberoamérica, Acta Zoológica Mexicana. Vol. Esp. 1992.* Cyted-D, Programa Iberoamericano de Ciencia y Tecnología para el Desarrollo. Instituto de Ecología, A. C., Xalapa, pp. 55–104.

Hernández, J. and H. Sánchez. 1992. "Biomas terrestres de Colombia." In: Halffter, G. (ed.), *La diversidad biológica de Iberoamérica, Acta Zoológica Mexicana, Vol. Esp. 1992.* Cyted-D, Programa Iberoamericano de Ciencia y Tecnología para el Desarrollo, Instituto de Ecología, A.C., Xalapa, pp. 153–173.

Hershkovitz, P. 1969. "The evolution of mammals on southern continents. VI: The Recent mammals of the Neotropical region: A zoogeographic and ecological review." *Quarterly Review of Biology*, 44: 1–70.

Hoffmann, C. C. 1936. "Relaciones zoogeográficas de los lepidópteros mexicanos." *Anales del Instituto de Biología de la UNAM*, 7: 47–58.

Holmberg, E. L. 1898. "La flora de la República Argentina." *Segundo Censo de la República Argentina*, 1: 385–474.

Holt, B. G., J. P. Lessard, M. K. Borregaard, S. A. Fritz, M. B. Araújo, D. Dimitrov, P. H. Fabre, C. H. Graham, G. R. Graves, K. A. Jønsson, D. Nogués-Bravo, Z. Wang, R. J. Whittaker, R. J. Fjeldså, and C. Rahbek. 2013. "An update of Wallace's zoogeographic regions of the world." *Science*, 339: 74–78.

Hoorn, C., F. P. Wesselingh, H. ter Steege, M. A. Bermúdez, A. Mora, J. Sevink, I. Sanmartín, A. Sánchez-Meseguer, C. L. Anderson, J. P. Figueiredo, C. Jaramillo, D. Riff, F. R. Negri, H. Hooghiemstra, J. Lundberg, T. Stadler, T. Särkinen, and A. Antonelli. 2010. "Amazonia through time: Andean uplift, climate change, landscape evolution, and biodiversity." *Science*, 330: 927–931.

Huber, O. 1994. "Recent advances in the phytogeography of the Guayana region, South America." *Mémoires de la Societé de Biogéographie (3me. série)*, 4: 53–63.

Huber, O. and Alarcón, C. 1988. *Mapa de vegetación de Venezuela. Ministerio del Ambiente y de los Recursos Naturales Renovables.* The Nature Conservancy, Caracas, map.

Huber, O. and R. Riina. 1997. *Glosario ecológico de las Américas. Vol. I. América del Sur: Países hispanoparlantes.* UNESCO, Paris.

Huber, O. and R. Riina. 2003. *Glosario ecológico de las Américas. Vol. 2. México, América Central e islas del Caribe: Países hispanoparlantes.* UNESCO, Caracas.

Hueck, K. 1953. "Distribuçao e habitat natural do pinheiro do Paraná (*Araucaria angustifolia*)." *Boletin da Facultade de Filosofia e Letras, Botánica*, 156: 3–24.

Hueck, K. 1957. "Las regiones forestales de Sudamérica." *Boletín del Instituto Forestal Latinoamericano de Investigación y Capacitación* (Mérida), 2: 1–40.

Hueck, K. 1966. *Die Wälder Südamerikas.* Fischer, Stuttgart.

Hughes, C. and R. Eastwood. 2006. "Island radiation on a continental scale: Exceptional rates of plant diversification after uplift of the Andes." *Proceedings of the National Academy of Science*, 103: 10334–10339.

Huidobro, L., J. J. Morrone, J. L. Villalobos, and F. Álvarez. 2006. "Distributional patterns of freshwater taxa (fishes, crustaceans and plants) from the Mexican transition zone." *Journal of Biogeography*, 33: 731–741.

Hulsey, C. D. and H. López-Hernández. 2011. "Nuclear Central America." In: Albert, J. S. and R. E. Reis (eds.), *Historical biogeography of Neotropical freshwater fishes*. University of California Press, Berkeley and Los Angeles, pp. 279–291.

Huxley, T. H. 1868. "On the classification and distribution of Alectoromorphae and Heteromorphae." *Proceedings of the Zoological Society of London*, 1868: 294–319.

Ibarra-Manríquez, G., J. L. Villaseñor, R. Durán, and J. Meave. 2002. "Biogeographical analysis of the tree flora of the Yucatán Peninsula." *Journal of Biogeography*, 29: 17–29.

Ippi, S. and V. Flores. 2001. "Las tortugas neotropicales y sus áreas de endemismo." *Acta Zoológica Mexicana (nueva serie)*, 84: 49–63.

Iturralde-Vinent, M. A. and R. D. E. MacPhee. 1999. "Paleogeography of the Caribbean region: Implications for Cenozoic biogeography." *Bulletin of the American Museum of Natural History*, 238: 1–95.

Jeannel, R. 1938. "Les Migadopides (Coleoptera, Adephaga), une lignée subantarctique." *Revue Française d'Entomologie*, 5: 1–55.

Jeannel, R. 1942. *La genese des faunes terrestres: Élements de biogéographie*. Paris: Presses Universitaires de France.

Johnson, M. P. and P. H. Raven. 1973. "Species number and endemism: The Galapagos archipelago revisited." *Science*, 179: 893–895.

Katinas, L., J. V. Crisci, W. L. Wagner, and P. C. Hoch. 2004. "Geographical diversification of tribes Epilobiae, Gongylocarpae, and Onagreae (Onagraceae) in North America, based on parsimony analysis of endemicity and track compatibility analysis." *Annals of the Missouri Botanical Garden*, 91: 159–185.

Katinas, L., J. J. Morrone, and J. V. Crisci. 1997. "Track analysis reveals the composite nature of the Andean biota." *Australian Systematic Botany*, 47: 111–130.

Kennan, L. 2000. "Large-scale geomorphology in the Central Andes of Peru and Bolivia: Relation to tectonic, magmatic and climatic processes." In: Summerfield, M. (ed.), *Geomorphology and global tectonics*. Wiley, London, pp. 167–192.

Kennan, L. and J. L. Pindell. 2009. "Dextral shear, terrane accretion and basin formation in the northern Andes: Best explained by interaction with a Pacific-derived Caribbean plate?" *Geological Society Special Publications*, 328: 487–531.

Kirby, M. X. and B. MacFadden. 2005. "Was southern Central America an archipelago or a peninsula in the middle Miocene? A test using land-mammal body size." *Palaeogeography, Palaeoclimatoligy, Palaeoecology*, 228: 193–202.

Kirby, W. F. 1872. "On the geographical distribution of the diurnal Lepidoptera as compared with that of birds." *Journal of the Linnean Society of London*, 11: 431–439.

Kirchhoff, P. 1943. "Mesoamerica." *Acta Americana*, 1: 92–107.

Klassa, B. and C. M. D. Santos. 2015. "Areas of endemism in the Neotropical region based on the geographical distribution of Tabanomorpha (Diptera: Brachycera)." *Zootaxa*, 4058: 519–534.

Kobelkowsky-Vidrio, T., C. A. Ríos-Muñoz, and A. G. Navarro-Sigüenza. 2014. "Biodiversity and biogeography of the avifauna of the Sierra Madre Occidental, Mexico." *Biodiversity and Conservation*, 23: 2087–2105.

Kohlmann, B. and G. Halffter. 1988. "Cladistic and biogeographical analysis of *Ateuchus* (Coleoptera: Scarabaeidae) of Mexico and the United States." *Folia Entomológica Mexicana*, 74: 109–130.

Kohlmann, B. and G. Halffter. 1990. "Reconstruction of a specific example of insect invasion waves: The cladistic analysis of *Canthon* (Coleoptera: Scarabaeidae) and related genera in North America." *Quaestiones Entomologicae*, 26: 1–28.

Kohlmann, B. and M. J. Wilkinson. 2007. "The Tárcoles line: Biogeographic effects of the Talamanca ridge in lower Central America." *Giornale Italiano di Entomologia*, 12: 1–30.

Kreft, H. and W. Jetz. 2010. "A framework for delineating biogeographical regions based on species distributions." *Journal of Biogeography*, 37: 2029–2053.

Kress, J. 1990. "The diversity and distribution of *Heliconia* in Brazil." *Acta Botanica Brasilica*, 4: 159–167.

Kury, A. B. 2013. "Order Opiliones Sundevall, 1833." *Zootaxa*, 3703: 27–33.

Kuschel, G. 1960. "Terrestrial zoology in southern Chile." *Proceedings of the Royal Society of London, series B*, 152: 540–550.

Kuschel, G. 1961. "Composition and relationship of the terrestrial faunas of Easter, Juan Fernandez, Desventuradas, and Galapagos Islands." In: *Tenth Pacific Science Congress*. Pacific Science Association, Honolulu, pp. 79–95.

Kuschel, G. 1964. "Problems concerning an Austral region." In: Gressitt, J. L., C. H. Lindroth, F. R. Fosberg, C. A. Fleming, and E. G. Turbott (eds.), *Pacific Basin biogeography: A symposium, 1963 [1964]*. Bishop Museum Press, Honolulu, pp. 443–449.

Kuschel, G. 1969. "Biogeography and ecology of South American Coleoptera." In: Fittkau, E., J. J. Illies, H. Klinge, G. H. Schwabe, and H. Sioli (eds.), *Biogeography and ecology in South America. Vol. 2*. Junk, The Hague, pp. 709–722.

Lamas, C. J. E., S. S. Nihei, A. M. Cunha, and M. S. Couri. 2014. "Phylogeny and biogeography of *Heterostylum* (Diptera: Bombyliidae): Evidence for an ancient Caribbean diversification model." *Florida Entomologist*, 97: 952–966.

Lamas, G. 1982. "A preliminary zoogeographical division of Peru, based on butterfly distributions (Lepidoptera, Papilionoidea)." In: Prance, T. P. (ed.), *Biological diversification in the tropics*. Columbia University Press, New York, pp. 336–357.

Lane, J. 1943. "The geographic distribution of Sabethini (Dipt., Culicidae)." *Revista de Entomología*, 4: 440–429.

Lanteri, A. A. 1990. "Revisión sistemática del género *Cyrtomon* Schönherr (Coleoptera, Curculionidae)." *Revista Brasileira de Entomologia*, 34: 387–402.

Lanteri, A. A. 1992. "Systematics, cladistics and biogeography of a new weevil genus, *Galapaganus* (Coleoptera: Curculionidae) from the Galapagos Islands, and coasts of Ecuador and Peru." *Transactions of the American Entomological Society*, 118: 227–267.

Lanteri, A. A. 2001. "Biogeografía de las islas Galapagos: Principales aportes de los estudios filogenéticos." In: Llorente, J. and J. J. Morrone (eds.), *Introducción a la biogeografía en Latinoamérica: Conceptos, teorías, métodos y aplicaciones*. Las Prensas de Ciencias, UNAM, Mexico City, pp. 141–151.

Laubenfels, D. J. de. 1970. *A geography of plants and animals*. The Brown Foundations of Geography Series, W. M. C. Brown Company Publishers, Dubuque.

Leite, Y. R. N., P. J. R. Kok, and M. Weksler. 2015. "Evolutionary affinities of the 'Lost World' mouse suggest a late Pliocene connection between the Guiana and Brazilian shields." *Journal of Biogeography*, 42: 706–715.

Leite, Y. R. N. and D. S. Rogers. 2013. "Revisiting Amazonian phylogeography: Insights into diversification hypotheses and novel perspectives." *Organisms Diversity and Evolution*, 13: 639–664.

León, H. 1946. *Flora de Cuba. Vol. 1*. Cultural, S. A., La Habana.

León-Paniagua, L. and J. J. Morrone. 2009. "Do the Oaxacan Highlands represent a natural biotic unit? A cladistic biogeographical test based on vertebrate taxa." *Journal of Biogeography*, 36: 1939–1944.

León-Paniagua, L., A. Navarro, B. Hernández, and J. C. Morales. 2007. "Diversification of arboreal mice of genus *Habromys* (Rodentia: Cricetidae: Neotominae)." *Molecular Phylogenetics and Evolution*, 62: 653–664.

Lieberman, B. S. 2004. "Range expansion, extinction, and biogeographic congruence: A deep time perspective." In: Lomolino, M. V. and L. R. Heaney (eds.), *Frontiers of biogeography: New directions in the geography of nature*. Sinauer Associates Inc., Sunderland, MA, pp. 111–124.

Liebherr, J. K. 1986. "*Barylaus*, new genus (Coleoptera: Carabidae) endemic to the West Indies with Old World affinities." *Journal of the New York Entomological Society*, 94: 83–97.

Liebherr, J. K. 1991. "A general area cladogram for montane Mexico based on distributions in the Platynine genera *Elliptoleus* and *Calathus* (Coleoptera: Carabidae)." *Proceedings of the Entomological Society of Washington*, 93: 390–406.

Liebherr, J. K. 1994a. "Biogeographic patterns of montane Mexican and Central American Carabidae (Coleoptera)." *The Canadian Entomologist*, 126: 841–860.

Liebherr, J. K. 1994b. "Identification of New World *Agonum*, review of the Mexican fauna, and description of *Incagonum*, new genus, from South America (Coleoptera: Carabidae: Platynini)." *Journal of the New York Entomological Society*, 102: 1–55.

Liria, J. 2008. "Sistemas de información geográfica y análisis espaciales: Un método combinado para realizar estudios panbiogeográficos." *Revista Mexicana de Biodiversidad*, 79: 281–284.

Llorente Bousquets, J. 1996. "Biogeografía de artrópodos de México: ¿Hacia un nuevo enfoque?" In: Llorente Bousquets, J., A. N. García Aldrete, and E. Gónzalez Soriano (eds.), *Biodiversidad, taxonomía y biogeografía de artrópodos de México: Hacia una síntesis de su conocimiento*. Universidad Nacional Autónoma de México, Mexico City, pp. 41–56.

Londoño, C., A. Cleef and S. Madriñán. 2014. "Angiosperm flora and biogeography of the páramo region of Colombia, northern Andes." *Flora*, 209: 81–87.

López, H. L., R. C. Menni, M. Donato, and A. M. Miquelarena. 2008. "Biogeographical revision of Argentina (Andean and Neotropical regions): An analysis using freshwater fishes." *Journal of Biogeography*, 35: 1564–1579.

López Almiral, A. 2005. "Nueva perspectiva para la regionalización de Cuba: Definición de los sectores." In: Llorente Bousquets, J. and J. J. Morrone (eds.), *Regionalización biogeográfica en Iberoamérica y tópicos afines—Primeras Jornadas Biogeográficas de la Red Iberoamericana de Biogeografía y Entomología Sistemática (RIBES XII. I-CYTED)*. Las Prensas de Ciencias, UNAM, Mexico City, pp. 417–428.

López Ruf, M., J. J. Morrone, and E. P. Hernández. 2006. "Patrones de distribución de las Naucoridae argentinas (Insecta: Heteroptera)." *Revista de la Sociedad Entomológica Argentina*, 65: 111–121.

Lott, E. J. and T. H. Atkinson. 2010. "Diversidad florística." In: Ceballos, G., L. Martínez, A. García, E. Espinoza, J. B. Creel, and R. Dirzo (eds.), *Diversidad, amenazas y áreas prioritarias para la conservación de las selvas secas del Pacífico de México*. Fondo de Cultura Económica, Mexico City, pp. 63–76.

Lourenço, W. R. 1986. "Diversité de la faune scorpionique de la region Amazonienne; centers d'endémisme; nouvel appui à la théorie des refuges forestiers du Pléistocène." *Amazoniana*, 9: 559–580.

Lovette, I. J., B. S. Arbogast, R. L. Curry, R. M. Zink, C. A. Botero, J. P. Sullivan, A. L. Talaba, R. B. Harris, D. R. Rubenstein, R. E. Ricklefs, and E. Bermingham. 2012. "Phylogenetic relationships of the mockingbirds and trashers (Aves: Mimidae)." *Molecular Phylogenetics and Evolution*, 63: 219–229.

Löwenberg-Neto, P. 2014. "Neotropical region: A shapefile of Morrone's (2014) biogeographical regionalization." *Zootaxa*, 3802: 300.

Löwenberg-Neto, P. and C. J. B. de Carvalho. 2009. "Areas of endemism and spatial diversification of the Muscidae (Insecta: Diptera) in the Andean and Neotropical regions." *Journal of Biogeography*, 36: 1750–1759.

Löwenberg-Neto, P., C. J. B. de Carvalho, and J. A. F. Diniz-Filho. 2008. "Spatial congruence between biotic history and species richness of Muscidae (Diptera, Insecta) in the Andean and Neotropical regions." *Journal of Zoological Systematics and Evolutionary Research*, 46: 374–380.

Löwenberg-Neto, P., C. J. B. de Carvalho, and B. A. Hawkins. 2011. "Tropical niche conservatism as a historical narrative hypothesis for the Neotropics: A case study using the fly family Muscidae." *Journal of Biogeography*, 38: 1936–1947.

Lücking, R. 2003. "Takhtajan's floristic regions and foliicolous lichen biogeography: A compatibility analysis." *Lichenologist*, 35: 33–54.

Luebert, F. 2011. "Hacia una fitogeografía histórica del desierto de Atacama." *Revista de Geografía Norte Grande*, 50: 105–133.

Luebert, F. and P. Pliscoff. 2006. *Sinopsis bioclimática y vegetacional de Chile*. Editorial Universitaria, Santiago de Chile.

Luna, I., J. J. Morrone, and D. Espinosa (eds.). 2004. *Biodiversidad de la Sierra Madre Oriental*. Las Prensas de Ciencias, UNAM, Mexico City.

Luna, I., J. J. Morrone, and D. Espinosa (eds.). 2007. *Biodiversidad de la Faja Volcánica Transmexicana*. FES Zaragoza, UNAM, Mexico City.

Luna Vega, I., O. Alcántara Ayala, D. Espinosa Organista, and J. J. Morrone. 1999. "Historical relationships of the Mexican cloud forests: A preliminary vicariance model applying parsimony analysis of endemicity to vascular plant taxa." *Journal of Biogeography*, 26: 1299–1305.

Luna-Vega, I., D. Espinosa, and R. Contreras-Medina (eds.). 2016. *Biodiversidad de la Sierra Madre del Sur: Una síntesis preliminar*. UNAM, Mexico City.

Lundberg, J. G., L. G. Marshall, J. Guerrero, B. Horton, M. C. S. L. Malabarba, and F. Wesselingh. 1998. "The stage for Neotropical fish diversification: A history of tropical South American rivers." In: Malabarba, L. R., R. E. Reis, R. P. Vari, Z. M. S. Lucena, and C. A. S. Lucena (eds.), *Phylogeny and classification of Neotropical fishes*. EDIPUCRS, Porto Alegre, pp. 13–48.

Luteyn, J. L. 1992. "Paramos: Why study them?" In: Balslev, H. and J. L. Luteyn (eds.), *Paramo: An Andean ecosystem under human influence*. Academic Press, London, pp. 1–14.

Lydekker, B. A. 1896. *A geographical history of mammals*. Cambridge University Press, Cambridge.

MacDonald, G. M. 2003. *Biogeography: Space, time, and life*. John Wiley and Sons, New York.

MacFadden, B. 1981. "Comments on Pregill's appraisal of historical biogeography of Caribbean vertebrates: Vicariance, dispersal or both?" *Systematic Zoology*, 30: 370–272.

MacPhee, R. D. E. and M. Iturralde-Vinent. 2005. "The interpretation of Caribbean paleogeography: Reply to Hedges." In: Alcover, J. A. and P. Bover (eds.), Proceedings of the International Symposium "Insular Vertebrate Evolution: The Palaeontological Approach." *Monografies de la Societat d'Història Natural de les Balears*, 12: 175–184.

Magallón, S. A. 2004. "Dating lineages: Molecular and paleontological approaches to the temporal framework of clades." *International Journal of Plant Sciences*, 165: S7–S21.

Maguire, B. 1970. "On the flora of the Guayana highland." *Biotropica*, 2: 85–100.

Maguire, B. 1979. "Guayana, region of the Roraima Sandstone Formation." In: Larsen, K. and L. Holm-Nielsen (eds.), *Tropical botany*. Academic Press, London, pp. 223–238.

Malone, C. L., T. Wheeler, J. F. Taylor, and S. K. Davis. 2000. "Phylogeography of the Caribbean rock iguana *(Cyclura)*: Implications for conservation and insights on the biogeographic history of the West Indies." *Molecular Phylogenetics and Evolution*, 17: 269–279.

Marinho-Filho, J., R. B. Machado, and R. P. B. Henriques. 2010. "Evolução do conhecimento e da conservação do Cerrado brasileiro." In: Diniz, I. R., J. M. Filho, R. B. Machado, and R. B. Cavalcanti (eds.), *Cerrado: Conhecimiento científico quantitativo como subsídio para açoes de conservação*. Universidade de Brasília, Brasília, pp. 13–31.

Marino, P. I., G. R. Spinelli, and P. Posadas. 2001. "Distributional patterns of species of Ceratopogonidae (Diptera) in southern South America." *Biogeographica*, 77: 113–122.

Mariño-Pérez, R., H. Brailovsky, and J. J. Morrone. 2007. "Análisis panbiogeográfico de las especies mexicanas de *Pselliopus* Bergroth (Hemiptera: Heteroptera: Reduviidae: Harpactorinae)." *Acta Zoológica Mexicana (nueva serie)*, 23: 77–88.

Marks, B. D., S. J. Hackett, and A. P. Capparella. 2002. "Historical relationships among Neotropical forest areas of endemism as determined by mitochondrial DNA sequence variation within the wedge-billed woodcreeper (Aves: Dendrocolaptidae: *Glyphorhynchus spirurus).*" *Molecular Phylogenetics and Evolution*, 24: 153–167.

Márquez, J. and J. J. Morrone. 2003. "Análisis panbiogeográfico de las especies de *Heterolinus* y *Homalolinus* (Coleoptera: Staphylinidae: Xantholinini)." *Acta Zoológica Mexicana (nueva serie)*, 90: 15–25.

Marshall, C. J. and J. K. Liebherr. 2000. "Cladistic biogeography of the Mexican transition zone." *Journal of Biogeography*, 27: 203–216.

Martínez-Aquino, A., R. Aguilar-Aguilar, H. Santa Anna del Conde-Juárez, and R. Contreras-Medina. 2007. "Empleo de herramientas panbiogeográficas para detectar áreas para conservar: Un ejemplo con taxones dulceacuícolas." In: Luna, I., J. J. Morrone, and D. Espinosa (eds.), *Biodiversidad de la Faja Volcánica Transmexicana*. Las Prensas de Ciencias, UNAM, Mexico City, pp. 449–460.

Martínez Carretero, E. 1995. "La Puna Argentina: Delimitación general y división en distritos florísticos." *Boletín de la Sociedad Argentina de Botánica*, 31: 27–40.

Martínez-Gordillo, M. and J. J. Morrone. 2005. "Patrones de endemismo y disyunción de los géneros de Euphorbiaceae *sensu lato*: Un análisis panbiogeográfico." *Boletín de la Sociedad Mexicana de Botánica*, 77: 21–34.

Mateos, M., O. I. Sanjur, and R. C. Vrijenhoek. 2002. "Historical biogeography of the live-bearing fish genus *Poeciliopsis* (Poeciliidae: Cyprinodontiformes)." *Evolution*, 56: 972–984.

Matthew, W. D. 1915. "Climate and evolution." *Annals of the New York Academy of Sciences*, 24: 171–318.

Mattick, F. 1964. "Übersicht über die Florenreiche und Florengebiete der Erde. A." In: Melchior, H. (ed.), *Engler's Syllabus der Pflanzenfamilien. 12 ed. Vol. 2*. Angiospermen, Gerbrüder Borntraeger, Berlin, pp. 626–629.

Maury, E. A., R. Pinto da Rocha, and J. J. Morrone. 1996. "Distribution of *Acropsopilio chilensis* Silvestri, 1904 in southern South America (Opiliones, Palpatores, Caddidae)." *Biogeographica*, 72: 127–132.

Maxson, L. 1992. "Tempo and pattern in Anuran speciation and phylogeny: An albumin perspective." In: Adler, K. (ed.), *Herpetology: Current research on the biology of Amphibians and Reptiles*. Society for the Study of Amphibians and Reptiles, Oxford (Ohio), pp. 41–57.

Maya-Martínez, A., J. J. Schmitter-Soto, and C. Pozo. 2011. "Panbiogeography of the Yucatan Peninsula based on Charaxinae (Lepidoptera: Nymphalidae)." *Florida Entomologist*, 94: 527–533.

Mayr, E. and W. H. Phelps Jr. 1967. "The origin of the bird fauna of the South Venezuelan highlands." *Bulletin of the American Museum of Natural History*, 136: 269–328.

Mello-Leitão, C. de. 1937. *Zoo-geografia do Brasil*. Biblioteca Pedagógica Brasileira, Brasiliana, São Paulo.

Mello-Leitão, C. de. 1943. "Los alacranes y la zoogeografía de Sudamérica." *Revista Argentina de Zoogeografía*, 2: 125–131.

Menezes, R. S. T., S. G. Brady, A. F. Carvalho, M. A. del Lama, and M. A. Costa. 2015. "Molecular phylogeny and historical biogeography of the Neotropical swarm-founding social wasp genus *Synoeca* (Hymenoptera: Vespidae)." *PLOS ONE*, 10: 1–15.

Mercado-Salas, N. F., C. Pozo, J. J. Morrone, and E. Suárez-Morales. 2012. "Distribution patterns of the American species of the freshwater genus *Eucyclops* (Copepoda: Cyclopoida)." *Journal of Crustacean Biology*, 32: 457–464.

Mermudes, J. R. and D. S. Napp. 2006. "Revision and cladistic analysis of the genus *Ptychoderes* Schoenherr, 1823 (Coleoptera, Anthribidae, Anthribinae, Ptychoderini)." *Zootaxa*, 1182: 1–130.

Merriam, C. H. 1892. "The geographic distribution of life in North America with special reference to the Mammalia." *Proceedings of the Biological Society of Washington*, 7: 1–64.

Miguez-Gutiérrez, A., J. Castillo, J. Márquez, and I. Goyenechea. 2013. "Biogeografía cladística de la zona de transición Mexicana con base en un análisis de árboles reconciliados." *Revista Mexicana de Biodiversidad*, 84: 215–224.

Miller, R. R. 1966. "Geographical distribution of Central American freshwater fishes." *Copeia*, 4: 773–802.

Monasterio, M. 1986. "Adaptive strategies of *Espeletia* in the Andean desert páramo." In: Vuilleumier, F. and M. Monasterio (eds.), *High altitude tropical biogeography*. Oxford University Press and American Museum of Natural History, New York and Oxford, pp. 49–80.

Monrós, F. 1958. "Consideraciones sobre la fauna del sur de Chile y revisión de la tribus Stenomelini (Coleoptera, Chrysomelidae)." *Acta Zoológica Lilloana*, 15: 143–153.

Moore, R. T. 1945. "The Transverse volcanic biotic province of central Mexico and its relationship to adjacent provinces." *Transactions of the San Diego Society of Natural History*, 10: 217–236.

Morain, S. A. 1984. *Systematic and regional biogeography*. Van Nostrand Reinhold Company, New York.

Morales, J. M., M. Sirombra, and A. D. Brown. 1995. "Riqueza de árboles en las Yungas argentinas." In: Brown, A. D. and H. R. Grau (eds.), *Investigación, conservación y desarrollo en selvas subtropicales de montaña*. LIEY, San Miguel de Tucumán, pp. 163–174.

Moreira, G. R. P., A. Ferrari, C. A. Mondin, and A. C. Cervi. 2011. "Panbiogeographical analysis of passion vines at their southern limit of distribution in the Neotropics." *Revista Brasileira de Biociencias*, 9: 28–40.

Moreira-Muñoz, A. 2007. "The Austral floristic realm revisited." *Journal of Biogeography*, 34: 1649–1660.

Moreira-Muñoz, A. 2011. *Plant geography of Chile*. Springer, New York.

Morello, J. 1955. "Estudios botánicos en las regiones áridas de la Argentina. II." *Revista de Agricultura del Noroeste Argentino*, 1: 385–524.

Morello, J. 1958. "La provincia biogeográfica del Monte." *Opera Lilloana*, 2: 1–155.

Morello, J. 1984. *Perfil ecológico de Sudamérica: Características estructurales de Sudamérica y su relación con espacios semejantes del planeta*. Ediciones Cultura Hispánica, Instituto de Cooperación Iberamericana, Barcelona.

Morón, M. A. 1996. "Scarabaeidae (Coleoptera)." In: Llorente Bousquets, J., A. N. García Aldrete, and E. González Soriano (eds.), *Biodiversidad, taxonomía y biogeografía de artrópodos de México: Hacia una síntesis de su conocimiento*. Universidad Nacional Autónoma de México, Mexico City, pp. 309–328.

Morrone, J. J. 1993. "Cladistic and biogeographic analyses of the weevil genus *Listroderes* Schoenherr (Coleoptera: Curculionidae)." *Cladistics*, 9: 397–411.

Morrone, J. J. 1994a. "Systematics, cladistics, and biogeography of the Andean weevil genera *Macrostyphlus, Adioristidius, Puranius*, and *Amathynetoides*, new genus (Coleoptera: Curculionidae)." *American Museum Novitates*, 3104: 1–63.

Morrone, J. J. 1994b. "Distributional patterns of species of Rhytirrhinini (Coleoptera: Curculionidae) and the historical relationships of the Andean provinces." *Global Ecology and Biogeography Letters*, 4: 188–194.

Morrone, J. J. 1996. "The biogeographical Andean subregion: A proposal exemplified by Arthropod taxa (Arachnida, Crustacea, and Hexapoda)." *Neotropica*, 42: 103–114.

Morrone, J. J. 1999. "Presentación preliminar de un nuevo esquema biogeográfico de América del Sur." *Biogeographica*, 75: 1–16.

Morrone, J. J. 2000a. "What is the Chacoan subregion?" *Neotropica*, 46: 51–68.

Morrone, J. J. 2000b. "A new regional biogeography of the Amazonian subregion, based mainly on animal taxa." *Anales del Instituto de Biología de la UNAM, serie Zoología*, 71: 99–123.

Morrone, J. J. 2001a. *Biogeografía de América Latina y el Caribe. Vol. 3*. MandT-Manuales and Tesis SEA, Sociedad Entomológica Aragonesa, Zaragoza.

Morrone, J. J. 2001b. "The Paraná subregion and its provinces." *Physis* (Buenos Aires), 58: 1–7.

Morrone, J. J. 2001c. "Toward a formal definition of the Paramo-Punan subregion and its provinces." *Revista del Museo Argentino de Ciencias Naturales, nueva serie*, 3: 1–12.

Morrone, J. J. 2001d. "A proposal concerning formal definitions of the Neotropical and Andean regions." *Biogeographica*, 77: 65–82.

Morrone, J. J. 2001e. "Toward a cladistic model for the Caribbean subregion: Delimitation of areas of endemism." *Caldasia*, 23: 43–76.

Morrone, J. J. 2002. "Biogeographic regions under track and cladistic scrutiny." *Journal of Biogeography*, 29: 149–152.

Morrone, J. J. 2004a. "Panbiogeografía, componentes bióticos y zonas de transición." *Revista Brasileira de Entomologia*, 48: 149–162.

Morrone, J. J. 2004b. "La zona de transición Sudamericana: Caracterización y relevancia evolutiva." *Acta Entomológica Chilena*, 28: 41–50.

Morrone, J. J. 2005. "Hacia una síntesis biogeográfica de México." *Revista Mexicana de Biodiversidad*, 76: 207–252.

Morrone, J. J. 2006. "Biogeographic areas and transition zones of Latin America and the Caribbean Islands based on panbiogeographic and cladistic analyses of the entomofauna." *Annual Review of Entomology*, 51: 467–494.

Morrone, J. J. 2008. "Endemism." In: Jorgensen, S. E. and B. D. Fath (eds.), *Encyclopedia of ecology*. Elsevier, Oxford, pp. 1254–1259.

Morrone, J.J. 2009. *Evolutionary biogeography: An integrative approach with case studies*. Columbia University Press, New York.

Morrone, J. J. 2010a. "América do Sul e geografia da vida: Comparação de algumas propostas de regionalização." In: Carvalho, C. J. B. de and E. A. B. Almeida (eds.), *Biogeografia da América do Sul: Padroes e processos*. Editora Roca Limitada, São Paulo, pp. 14–40.

Morrone, J. J. 2010b. "Fundamental biogeographic patterns across the Mexican transition zone: An evolutionary approach." *Ecography*, 33: 355–361.

Morrone, J. J. 2014a. "Parsimony analysis of endemicity (PAE) revisited." *Journal of Biogeography*, 41: 842–854.

Morrone, J.J. 2014b. "Biogeographical regionalisation of the Neotropical region." *Zootaxa*, 3782: 1–110.

Morrone, J. J. 2014c. "Cladistic biogeography of the Neotropical region: Identifying the main events in the diversification of the terrestrial biota." *Cladistics*, 30: 202–214.

Morrone, J. J. 2015a. "Track analysis beyond panbiogeography." *Journal of Biogeography*, 42: 413–425.

Morrone, J. J. 2015b. "Biogeographic regionalisation of the world: A reappraisal." *Australian Systematic Botany*, 28: 81–90.

Morrone, J. J. 2015c. "Halffter's Mexican transition zone (1962–2014), cenocrons and evolutionary biogeography." *Journal of Zoological Systematics and Evolutionary Research*, 53: 249–257.

Morrone, J.J. 2015d. "Biogeographical regionalisation of the Andean region." *Zootaxa*, 3936: 207–236.

Morrone, J. J. and J. V. Crisci. 1995. "Historical biogeography: Introduction to methods." *Annual Review of Ecology and Systematics*, 26: 373–401.

Morrone, J. J. and M. del C. Coscarón. 1996. "Distributional patterns of the American Peiratinae (Heteroptera: Reduviidae)." *Zoologische Medelingen Leiden*, 70: 1–15.

Morrone, J. J. and M. del C. Coscarón, 1998. "Cladistics and biogeography of the assassin bug genus *Rasahus* Amyot and Serville (Heteroptera: Reduviidae: Peiratinae)." *Zoologische Medelingen Leiden*, 72: 73–87.

Morrone, J. J., D. Espinosa, C. Aguilar Zúñiga, and J. Llorente Bousquets. 1999. "Preliminary classification of the Mexican biogeographic provinces: A parsimony analysis of endemicity based on plant, insect, and bird taxa." *Southwestern Naturalist*, 44: 507–514.

Morrone, J. J., D. Espinosa, and J. Llorente. 2002. "Mexican biogeographic provinces: Preliminary scheme, general characterizations, and synonymies." *Acta Zoológica Mexicana (nueva serie)*, 85: 83–108.

Morrone, J. J. and C. Ezcurra. 2016. "On the Prepuna biogeographic province: A nomenclatural clarification." *Zootaxa*, 4132: 287–289.

Morrone, J. J. and A. Gutiérrez. 2005. "Do fleas (Insecta: Siphonaptera) parallel their mammal host diversification in the Mexican transition zone?" *Journal of Biogeography*, 32: 1315–1325.

Morrone, J. J. and E. C. Lopretto. 1994. "Distributional patterns of freshwater Decapoda (Crustacea: Malacostraca) in southern South America: A panbiogeographic approach." *Journal of Biogeography*, 21: 97–109.

Morrone, J. J. and J. Márquez. 2001. Halffter's Mexican Transition Zone, beetle generalised tracks, and geographical homology." *Journal of Biogeography*, 28: 635–650.

Morrone, J. J. and J. Márquez. 2003. "Aproximación a un Atlas Biogeográfico de México: Componentes bióticos principales y provincias biogeográficas." In: Morrone, J. J. and J. Llorente (eds.), *Una perspectiva latinoamericana de la biogeografía*. Las Prensas de Ciencias, UNAM, Mexico City, pp. 217–220.

Morrone, J. J. and J. Márquez. 2008. "Biodiversity of Mexican arthropods (Arachnida and Hexapoda): A biogeographical puzzle." *Acta Zoológica Mexicana (nueva serie)*, 24: 15–41.

Morrone, J. J. and E. Urtubey. 1997. "Historical biogeography of the northern Andes: A cladistic analysis based on five genera of Rhytirrhinini (Coleoptera: Curculionidae) and Barnadesia (Asteraceae)." *Biogeographica*, 73: 115–121.

Müller, P. 1973. *The dispersal centers of terrestrial vertebrates in the Neotropical realm: A study in the evolution of the Neotropical biota and its native landscapes*. Junk, The Hague.

Muñiz, O. 1996. "The phytogeographical subdivision of Cuba." In: Borhidi, A. (ed.), *Phytogeography and vegetation ecology of Cuba. 2a ed.* Akademiai Kiado, Budapest, pp. 283–385.

Murphy, P. G. and A. E. Lugo. 1995. "Dry forests of Central America and the Caribbean." In: Bullock, S. H., H. A. Mooney, and E. Medina (eds.), *Seasonally dry tropical forests*. Cambridge University Press, Cambrige, UK, pp. 9–34.

Murray, A. 1866. *The geographical distribution of mammals*. Day and Son Limited, London.

Musser, G. G. and M. M. Williams. 1985. "Systematic studies of Oryzomyine rodents (Muridae): Definitions of *Oryzomys villosus* and *Oryzomys talamancae*." *American Museum Novitates*, 2810: 1–22.

Naka, L. N. 2011. Avian distribution patterns in the Guiana Shield: Implications for the delimitation of Amazonian areas of endemism." *Journal of Biogeography*, 38: 681–696.

Navarro, A. G., A. T. Peterson, E. López-Medrano, and H. Benítez. 2001. "Species limits in Mesoamerican *Aulacorhynchus* toucanets." *Wilson Bulletin*, 113: 363–372.

Navarro, F. R., F. Cuezzo, P. A. Goloboff, C. Szumik, M. Lizarralde de Grosso, and M. G. Quintana. 2009. "Can insect data be used to infer areas of endemism? An example from the Yungas of Argentina." *Revista Chilena de Historia Natural*, 82: 507–522.

Navarro-Sigüenza, A. G., A. Lira-Noriega, A. T. Peterson, A. Oliveras de Ita, and A. Gordillo-Martínez. 2007. "Diversidad, endemismo y conservación de las aves." In: Luna, I., J. J. Morrone, and D. Espinosa (eds.), *Biodiversidad de la Faja Volcánica Transmexicana*. Las Prensas de Ciencias, UNAM, Mexico City, pp. 462–483.

Neall, V. E. and S. A. Trewick. 2008. "The age and origin of the Pacific Islands: A geological overview." *Philosophical Transactions of the Royal Society B*, 363: 3293–3308.

Newbigin, M. I. 1913. *Animal geography: The faunas of the natural regions of the world*. Clarendon Press, Oxford.

Nihei, S. S. and C. J. B. de Carvalho. 2004. Taxonomy, cladistics and biogeography of *Coenosopsia* Malloch (Diptera, Anthomyiidae) and its significance to the evolution of anthomyiids in the Neotropics." *Systematic Entomology*, 29: 260–275.

Nihei, S. S. and C. J. B. de Carvalho. 2007. "Systematics and biogeography of *Polietina* Schnabl and Dziedzicki (Diptera, Muscidae): Neotropical area relationships and Amazonia as a composite area." *Systematic Entomology*, 32: 477–501.

Nogueira, C., S. Ribeiro, G. C. Costa, and G. R. Colli. 2011. "Vicariance and endemism in a Neotropical savanna hotspot: Distribution patterns of Cerrado squamate reptiles." *Journal of Biogeography*, 38: 1097–1922.

Nores, M., 1992. "Bird speciation in subtropical South America in relation to forest expansion and retraction." *The Auk*, 109: 346–357.

Nori, J., J. M. Díaz Gómez, and G. C. Leynaud. 2011. "Biogeographic regions of central Argentina based on snake distribution: Evaluating two different methodological approaches." *Journal of Natural History*, 45: 1005–1020.

O'Brien, C. W. 1971. "The biogeography of Chile through entomofaunal regions." *Entomological News*, 82: 197–207.

O'Brien, C. W. and W. Tang. 2015. "Revision of the New World cycad weevils of the subtribe Allocorynina, with description of two new genera and three new subgenera (Coleoptera: Belidae: Oxycoryninae)." *Zootaxa*, 3970: 1–87.

Ocampo, F. C. and E. Ruiz Manzano. 2008. "Scarabaeidae." In: Claps, L. E., G. Debandi and S. Roig-Juñent (eds.), *Biodiversidad de artrópodos argentinos. Vol. 2.* Sociedad Entomológica Argentina, San Miguel de Tucumán, pp. 535–557.

Ojeda, A. A., A. Novillo, R. A. Ojeda, and S. Roig-Juñent. 2013. "Geographical distribution and ecological diversification of South American octodontid rodents." *Journal of Zoology*, 289: 285–293.

Ojeda, R. A., C. E. Borghi, and V. G. Roig. 2002. "Mamíferos de Argentina." In: G. Ceballos and J. A. Simonetti (eds.), *Diversidad y conservación de mamíferos neotropicales.* Conabio and UNAM, Mexico City, pp. 23–63.

Olson, D. M., E. Dinerstein, E. D. Wikramanayake, N. D. Burgess, G. V. N. Powell, E. C. Underwood, J. A. D'Amico, I. E. Itoua, H. E. Strand, J. C. Morrison, C. J. Loucks, T. F. Allnutt, T. H. Ricketts, Y. Kura, J. F. Lamoreux, W. W. Wettengel, P. Hedao, and K. R. Kassem. "Terrestrial ecoregions of the world: A new map of life on Earth." *BioScience*, 51: 933–938.

Orfila, R. N. 1941. "Apuntaciones ornitológicas para la zoogeografía neotropical. I. Distrito Sabánico." *Revista Argentina de Zoogeografía*, 1: 85–92.

Ornelas, J. F., V. Sosa, D. E. Soltis, J. M. Daza, C. González, P. S. Soltis, C. Gutiérrez-Rodríguez, A. Espinosa de los Monteros, T. A. Castoe, C. Bell, and E. Ruiz-Sánchez. 2013. "Comparative phylogeographic analyses illustrate complex evolutionary history of threatened cloud forests of northern Mesoamerica." *PLOS ONE*, 8: 1–11.

Ortega, J. and H. T. Arita. 1998. "Neotropical-Nearctic limits in Middle America as determined by distributions of bats." *Journal of Mammalogy*, 79: 772–781.

Ortega Baes, P., S. Sühring, and G. Ceballos. 2002. "Mamíferos de Uruguay." In: Ceballos, G. and J. A. Simonetti (eds.), *Diversidad y conservación de mamíferos neotropicales.* Conabio and UNAM, Mexico City, pp. 551–565.

Page, R. D. M. and C. Lydeard. 1994. "Towards a cladistic biogeography of the Caribbean." *Cladistics*, 10: 21–41.

Paggi, S. J. de. 1990. "Ecological and biogeographical remarks on the rotifer fauna of Argentina." *Revue d'Hydrobiologie Tropicale*, 23: 297–311.

Palacios, W., C. Cerón, R. Valencia, and R. Sierra. 1999. "Las formaciones naturales de la Amazonía del Ecuador." In: Sierra, R. (ed.), *Propuesta preliminar de un sistema de clasificación de vegetación para el Ecuador continental.* Proyecto INEFAN/GEF-BIRF and EcoCiencia, Quito, pp. 109–119.

Palma, R. E., P. A. Marquet, and D. Boric-Bargetto. 2005. "Inter- and intraspecific phylogeography of small mammals in the Atacama desert and adjacent areas of northern Chile." *Journal of Biogeography*, 32: 1931–1941.

Panfilov, D. V. 1970. *Regionalización zoogeográfica.* Atlas Nacional de Cuba, Instituto de Geografía, Academia de Ciencias de Cuba, and Institut geografii, Akademiia Nauk SSSR, La Habana.

Parent, C. E., A. Caccone, and K. Petren. 2008. "Colonization and diversification of Galapagos terrestrial fauna: A phylogenetic and biogeographical synthesis." *Philosophical Transactions of the Royal Society B*, 363: 3347–3361.

Parodi, L. R. 1934. "Las plantas indígenas no alimenticias cultivadas en la Argentina." *Revista Argentina de Agronomía*, 1: 165–212.

Parodi, L. R. 1945. "Las regiones fitogeográficas argentinas y sus relaciones con la industria forestal." In: Verdoorn, F. (ed.), *Plants and plant science in Latin America*. Waltham, Massachusetts, pp. 127–132.

Patton, J. L., M. N. F. da Silva, and J. R. Malcolm. 2000. "Mammals of the rio Juruá and the evolutionary and ecological diversification of Amazonia." *Bulletin of the American Museum of Natural History*, 244: 1–306.

Paula-Souza, J. and J. R. Pirani. 2014. "A biogeographical overwiew of the 'lianescent clade' of Violaceae in the Neotropical region." In: Greer, F. E. (ed.), *Dry forests: Ecology, species diversity and sustainable management*. Nova Publishers, New York, pp. 1–28.

Peck, S. B. 1991. "Beetle (Coleoptera) faunas of tropical oceanic islands: With emphasis on the Galapagos archipelago, Ecuador." In: Zunino, M., X. Bellés and M. Blas (eds.), *Advances in coleopterology*. European Association of Coleopterology, Torino, pp. 177–192.

Peck, S. B. and J. Kukalová-Peck. 1990. "Origin and biogeography of the beetles (Coleoptera) of the Galapagos archipelago, Ecuador." *Canadian Journal of Zoology*, 68: 1617–1638.

Pellegrino, K. C. M., M. T. Rodrigues, A. N. Waite, M. Morando, Y. Y. Yassuda, and J. W. Sites Jr. 2005. "Phylogeography and species limits in the *Gymnodactylus darwinii* complex (Gekkonidae, Squamata): Genetic structure coincides with river systems in the Brazilian Atlantic forest." *Biological Journal of the Linnean Society*, 85: 13–26.

Pennington, R. T., M. Lavin, D. E. Prado, C. A. Pendry, S. K. Pell, and C. A. Butterworth. 2004. "Historical climate change and speciation: Neotropical seasonally dry forest plants show patterns of both Tertiary and Quaternary diversification." *Philosophical Transactions of the Royal Society of London B*, 359: 515–538.

Peña, L. E. 1966a. "A preliminary attempt to divide Chile into entomofaunal regions, based on the Tenebrionidae (Coleoptera)." *Postilla*, 97: 1–17.

Peña, L. E. 1966b. "Ensayo preliminar para dividir Chile en regiones entomofaunísticas, basadas especialmente en la familia Tenebrionidae (Col.)." *Revista Universitaria, Universidad Católica de Chile*, 28–29: 209–220.

Pérez-Hernández, R. and D. Lew. 2001. "Las clasificaciones e hipótesis biogeográficas para la Guayana venezolana." *Interciencia*, 26: 373–382.

Perret, M., A. Chautems, and R. Spichiger. 2006. "Dispersal-vicariance analyses in the tribe Sinningieae (Gesneriaceae): A clue to understanding biogeographical history of the Brazilian Atlantic Forest." *Annals of the Missouri Botanical Garden*, 93: 340–358.

Pielou, E. C. 1992. *Biogeography*. Krieger Publishing Company, Malabar.

Pindell, J. L. and S. Barrett. 1990. "Geological evolution of the Caribbean region: A plate-tectonic perspective." In: Dengo, G. and J. E. Case (eds.), *The geology of North America. Vol. H. The Caribbean region*. Geological Society of America, Boulder, Colorado.

Pinto-da-Rocha, R. 1997. "Systematic revision of the Neotropical family Stygnidae (Opiliones, Laniatores, Gonyleptoidea)." *Arquivos de Zoologia, São Paulo*, 33: 163–342.

Pinto-da-Rocha, R. and M. B. da Silva. 2005. "Faunistic similarity and historic biogeography of the harvestmen of southern and southeastern Atlantic rain forest of Brazil." *The Journal of Arachnology*, 33: 290–299.

Pires, A. C. and L. Marinoni. 2010. "Historical relationships among Neotropical endemic areas based on *Sepedonea* (Diptera: Sciomyzidae) phylogenetic and distribution data." *Zoologia*, 27: 681–690.

Pires, A. C. and L. Marinoni. 2011. "Distributional patterns of the Neotropical genus *Thecomyia* Perty (Diptera, Sciomyzidae) and phylogenetic support." *Revista Brasileira de Entomologia*, 55: 6–14.

Pitman, W. C., S. Cande, J. LaBrecque, and J. Pindell. 1993. "Fragmentation of Gondwana: The separation of Africa and South America." In: Goldblatt, P. (ed.), *Biological relationships between Africa and South America*. Yale University Press, New Haven, pp. 15–34.

Porzecanski, A. L. and J. Cracraft. 2005. "Cladistic analysis of distributions and endemism (CADE): Using raw distributions of birds to unravel the biogeography of the South American aridlands." *Journal of Biogeography*, 32: 261–275.

Posadas, P. E., J. M. Estévez, and J. J. Morrone. 1997. "Distributional patterns and endemism areas of vascular plants in the Andean subregion." *Fontqueria*, 48: 1–10.

Poux, C., P. Chevret, D. Huchon, W. E. de Jong, and E. J. P. Douzery. 2006. "Arrival and diversification of Caviomorph rodents and Platyrrhine primates in South America." *Systematic Biology*, 55: 228–244.

Prado, D. E. 1993a. "What is the Gran Chaco vegetation in South America? I. A review. Contribution to the study of the flora and vegetation of the Chaco. V." *Candollea*, 48: 145–172.

Prado, D. E. 1993b. "What is the Gran Chaco vegetation in South America? II. A redefinition. Contribution to the study of the flora and vegetation of the Chaco. VII." *Candollea*, 48: 615–629.

Prado, D. E. and P. E. Gibbs. 1993. "Patterns of species distributions in the dry seasonal forests of South America." *Annals of the Missouri Botanical Garden*, 80: 902–927.

Prance, G. T. (ed.). 1982. *Biological diversification in the tropics*. Columbia University Press, New York.

Pregill, G. K. 1981. "An appraisal of the vicariance hypothesis of Caribbean biogeography and its application to West Indian terrestrial vertebrates." *Systematic Zoology*, 30: 147–155.

Procheş, Ş. 2005. "The world's biogeographical regions: Cluster analyses based on bat distributions." *Journal of Biogeography*, 32: 607–614.

Procheş, Ş. and S. Ramdhani. 2012. "The world's zoogeographical regions confirmed by cross-taxon analyses." *BioScience*, 62: 260–270.

Proença, C. E. B., L. H. Soares-Silva, V. L. Rivera, M. F. Simon, R. C. de Oliveira, I. A. dos Santos, J. N. Batista, C. L. Ramalho, Z. J. G. Miranda, C. F. R. Cardoso, M. A. Barboza, L. B. Bianchetti, E. G. Gonçalves, R. F. Singer, S. M. Gomes, S. R. Silva, R. C. Martins, C. B. R. Munhoz, and S. F. de Carvalho. 2010. "Regionalização, centros de endemismos e conservaçao com base em espécies de angiospermas indicadoras da biodiversidade do Cerrado brasileiro." In: Diniz, I. R., J. M. Filho, R. B. Machado, and R. B. Cavalcanti (eds.), *Cerrado: Conhecimiento científico quantitativo como subsídio para açoes de conservação*. Universidade de Brasília, Brasília, pp. 89–148.

Puga-Jiménez, A. L., A. R. Andrés-Hernández, H. Carrillo-Ruiz, D. Espinosa-Organista, and S. P. Rivas-Arancibia. 2013. "Patrones de distribución del género *Zanthoxylum* L. (Rutaceae) en México." *Revista Mexicana de Biodiversidad*, 84: 1179–1188.

Quijano-Abril, M. A., R. Callejas-Posada, and D. R. Miranda-Esquivel. 2006. "Areas of endemism and distribution patterns for Neotropical *Piper* species (Piperaceae)." *Journal of Biogeography*, 33: 1266–1278.

Racheli, L. and T. Racheli. 2003. "Historical relationships of Amazonian areas of endemism based on raw distributions of parrots (Psittacidae)." *Tropical Zoology*, 16: 33–46.

Racheli, L. and T. Racheli. 2004. "Patterns of Amazonian area relationships based on raw distributions of Papilionid butterflies (Lepidoptera: Papilionidae)." *Biological Journal of the Linnean Society*, 82: 345–357.

Ragonese, A. E. 1967. *Vegetación y ganadería en la República Argentina.* Colección Científica del INTA, Buenos Aires.

Ramella, L. and R. Spichiger. 1989. "Interpretación preliminar del medio físico y de la vegetación del Chaco Boreal. Contribución al estudio de la flora y de la vegetación del Chaco. I." *Candollea*, 44: 639–680.

Ramírez-Barahona, S., A. Torres-Miranda, M. Palacios-Ríos, and I. Luna-Vega. 2009. "Historical biogeography of the Yucatan Peninsula, Mexico: A perspective from ferns (Monilophyta) and lycopods (Lycophyta)." *Biological Journal of the Linnean Society*, 98: 775–786.

Ramírez-Pulido, J. and Castro-Campillo, A. 1990. "Regionalización mastofaunística (mamíferos). Mapa IV. 8. 8. A." In: *Atlas Nacional de México. Vol. III.* Instituto de Geografía, UNAM, Mexico City, map.

Ramírez-Pulido, J., A. Castro-Campillo, and A. Salame-Méndez. 2005. "Relación de algunas especies del género *Reithrodontomys* (Rodentia: Muridae) en la colección de mamíferos de la UAMI." In: Sánchez-Cordero, V. and R. A. Medellín (eds.), *Contribuciones mastozoológicas en homenaje a Bernardo Villa.* Instituto de Biología, UNAM and Conabio, Mexico City, pp. 399–422.

Ramos, K. S. and G. A. R. Melo. 2010. "Taxonomic revision and phylogenetic relationships of the bee genus *Parapsaenythia* Friese (Hymenoptera, Apidae, Protandrenini), with biogeographic inferences for the South American Chacoan subregion." *Systematic Entomology*, 35: 449–474.

Rangel, J. O. (ed.). 2000a. *Colombia Diversidad Biótica III: La región de vida paramuna de Colombia.* Universidad Nacional de Colombia, Santafé de Bogotá.

Rangel, J. O. 2000b. "La diversidad beta: Tipos de vegetación." In: Rangel, J. O. (ed.), *Colombia Diversidad Biótica III: La región de vida paramuna de Colombia.* Universidad Nacional de Colombia, Santafé de Bogotá, pp. 658–718.

Rangel, J. O. (ed.). 2004a. *Colombia Diversidad Biótica IV: El Chocó biogeográfico/Costa Pacífica.* Universidad Nacional de Colombia, Santafé de Bogotá.

Rangel, J. O. 2004b. "La diversidad beta: Tipos de vegetación." In: Rangel, J. O. (ed.), *Colombia Diversidad Biótica IV: El Chocó biogeográfico/Costa Pacífica.* Universidad Nacional de Colombia, Santafé de Bogotá, pp. 769–815.

Rangel, J. O., M. Aguilar, H. Sánchez, and P. Lowy. 1995a. "Región Costa Pacífica." In: Rangel, J. O. (ed.), *Colombia: Diversidad biótica I.* Instituto de Ciencias Naturales, Convenio Inderena-Universidad Nacional de Colombia, Santafé de Bogotá, pp. 121–139.

Rangel, J. O., M. Aguilar, H. Sánchez, P. Lowy, A. Garzón, and L. A. Sánchez. 1995b. "Región de la Amazonia." In: Rangel, J. O. (ed.), *Colombia: Diversidad biótica I.* Instituto de Ciencias Naturales, Convenio Inderena-Universidad Nacional de Colombia, Santafé de Bogotá, pp. 82–103.

Rangel, J. O., A. Garzón, and P. Lowy. 1995c. "Sierra Nevada de Santa Marta-Colombia (con énfasis en la parte norte Transecto del río Buritaca-La Cumbre)." In: Rangel, J. O. (ed.), *Colombia: Diversidad biótica I.* Instituto de Ciencias Naturales, Convenio Inderena-Universidad Nacional de Colombia, Santafé de Bogotá, pp. 155–170.

Rangel, J. O., P. Lowy, and M. Aguilar. 1995d. "Marco general y alcances del estudio." In: Rangel, J. O. (ed.), *Colombia: Diversidad biótica I.* Instituto de Ciencias Naturales, Convenio Inderena-Universidad Nacional de Colombia, Santafé de Bogotá, pp. 17–24.

Rangel, J. O., P. D. Lowy, M. Aguilar, and A. Garzón. 1997. "Tipos de vegetación en Colombia: Una aproximación al conocimiento de la terminología fitosociológica, fitoecológica y de uso común." In: Rangel, J. O., P. D. Lowy, and M. Aguilar (eds.), *Colombia Diversidad Biótica II: Tipos de vegetación en Colombia*. Instituto de Ciencias Naturales, Universidad Nacional de Colombia, Santafé de Bogotá, pp. 89–381.

Rapoport, E. H. 1968. "Algunos problemas biogeográficos del Nuevo Mundo con especial referencia a la región Neotropical." In: Delamare Debouteville, C. and E. H. Rapoport (eds.), *Biologie de l'Amerique Australe. Vol. 4.* CNRS, Paris, pp. 55–110.

Rapoport, E. H. 1971. "The Nearctic-Neotropical frontier." Proceedings of the XIII International Congress of Entomology, Nauka, Leningrad, pp. 190–191.

Ratcliffe, B. C., R. D. Cave, and E. B. Cano. 2013. "The Dynastine scarab beetles of Mexico, Guatemala, and Belize (Coleoptera: Scarabaeidae: Dynastinae)." *Bulletin of the University of Nebraska State Museum*, 27: 1–661.

Rauchenberger, M. 1988. "Historical biogeography of Poeciliid fishes in the Caribbean." *Systematic Zoology*, 37: 356–365.

Restall, R., C. Rodner, and M. Lentino. 2006. *Birds of northern South America*. Helm, London.

Reyes, R. and N. H. Campos. 1992. "Moluscos, anélidos y crustáceos asociados a las raíces de *Rhizophora mangle* Linnaeus, en la región de Santa Marta, Caribe colombiano." *Caldasia*, 17: 133–148.

Ribas, C. C., R. G. Moyle, C. Y. Miyaki, and J. Cracraft. 2007. "The assembly of montane biotas: Linking Andean tectonics and climatic oscillations to independent regimes of diversification in *Pionus* parrots." *Proceedings of the Royal Society B*, 274: 2399–2408.

Ribeiro, M. C., J. P. Metzger, A. C. Martensen, F. J. Ponzoni, and M. M. Hirota. 2009. "The Brazilian Atlantic forest: How much is left, and how is the remaining forest distributed? Implications for conservation." *Biological Conservation*, 142: 1141–1153.

Ribichich, A. M. 2002. "El modelo clásico de la fitogeografía de Argentina: Un análisis crítico." *Interciencia*, 27: 669–675.

Ricklefs, R. and E. Bermingham. 2008. "The West Indies as a laboratory of biogeography and evolution." *Philosophical Transactions of the Royal Society B*, 363: 2393–2413.

Riddle, B. R., M. N. Dawson, E. A. Hadly, D. J. Hafner, M. J. Hickerson, S. J. Mantooth, and A. D. Yoder. 2008. "The role of molecular genetics in sculpting the future of integrative biogeography." *Progress in Physical Geography*, 32: 173–202.

Ringuelet, R. A. 1955. "Panorama zoogeográfico de la provincia de Buenos Aires." *Notas del Museo de La Plata*, 18: 1–15.

Ringuelet, R. A. 1961. "Rasgos fundamentales de la zoogeografía de la Argentina." *Physis* (Buenos Aires), 22: 151–170.

Ringuelet, R. A. 1975. "Zoogeografía y ecología de los peces de aguas continentales de la Argentina y consideraciones sobre las áreas ictiológicas de América del Sur." *Ecosur*, 2: 1–122.

Ringuelet, R. A. 1978. "Dinamismo histórico de la fauna brasílica en la Argentina." *Ameghiniana*, 15: 255–262.

Ringuelet, R. A. 1981. *El ecotono faunístico subtropical-pampásico y sus cambios históricos*. IV Jornadas Argentinas de Zoología, Symposia, La Plata, pp. 75–80.

Ríos-Muñoz, C. A. 2013. "¿Es posible reconocer una unidad biótica entre América del Norte y del Sur?" *Revista Mexicana de Biodiversidad*, 84: 1022–1030.

Ríos-Muñoz, C. A. and A. G. Navarro-Sigüenza. 2012. "Patterns of species richness and biogeographic regionalization of the avifaunas of the seasonally dry tropical forest in Mesoamerica." *Studies on Neotropical Fauna and Environment*, 47: 171–182.

Rivas-Martínez, S. and G. Navarro. 1994. *Mapa biogeográfico de Suramérica.* Published by the authors, Madrid.

Rivas-Martínez, S., G. Navarro, Á. Penas, and M. Costa. 2011. "Biogeographic map of South America. A preliminary survey." *International Journal of Geobotanical Research,* 1: 21–40.

Rivas-Martínez, S. and O. Tovar. 1983. "Síntesis biogeográfica de los Andes." *Collectanea Botanica* (Barcelona), 14: 515–521.

Rizzini, C. T. 1963. "Nota prévia sôbre a divisão fitogeográfica (florístico-sociológica) do Brasil." *Revista Brasileira de Geografia,* 25: 3–64.

Rizzini, C. T. 1997. *Tratado de fitogeografia do Brasil: Aspectos ecológicos, sociológicos e florísticos.* Âmbito Cultural Edições Ltda., Rio de Janeiro.

Rodríguez-Olarte, D., J. I. Mojica Corzo, and D. C. Taphorn Baechle. 2011. "Northern South America: Magdalena and Maracaibo basins." In: Albert, J. S. and R. E. Reis (eds.), *Historical biogeography of Neotropical freshwater fishes.* University of California Press, Berkeley and Los Angeles, pp. 243–257.

Roig, F. A. 1998. "La vegetación de la Patagonia." In: Correa, M. N. (ed.), *Flora Patagónica, tomo VIII(1).* INTA, Colección Científica, Buenos Aires, pp. 48–166.

Roig, F. A. and E. Martínez Carretero. 1998. "La vegetación puneña en la provincia de Mendoza, Argentina." *Phytocoenologia,* 28: 565–608.

Roig, F. A., S. Roig-Juñent, and V. Corbalán. 2009. "Biogeography of the Monte desert." *Journal of Arid Environments,* 73: 164–172.

Roig Juñent, S. 1994. "Historia biogeográfica de América del Sur austral." *Multequina* (Mendoza), 3: 167–203.

Roig-Juñent, S., G. Flores, S. Claver, G. Debandi, and A. Marvaldi. 2001. "Monte desert (Argentina): Insect biodiversity and natural areas." *Journal of Arid Environments,* 47: 77–94.

Roig-Juñent, S., G. E. Flores, and C. Mattoni. 2003. "Consideraciones biogeográficas de la Precordillera (Argentina), con base en artrópodos epígeos." In: Morrone, J. J. and J. Llorente (eds.), *Una perspectiva latinoamericana de la biogeografía.* Las Prensas de Ciencias, UNAM, Mexico City, pp. 275–288.

Roig-Juñent, S., M. F. Tognelli, and J. J. Morrone. 2008. "Aspectos biogeográficos de los insectos de la Argentina." In: Claps, L. E., G. Debandi, and S. Roig-Juñent (eds.), *Biodiversidad de artrópodos argentinos, Vol. 2.* Sociedad Entomológica Argentina, San Miguel de Tucumán, pp. 11–29.

Rojas-Parra, C. A. 2007. "Una herramienta automatizada para realizar análisis panbiogeográficos." *Biogeografía,* 1: 31–33.

Ron, S. R. 2000. "Biogeographic area relationships of lowland Neotropical rainforest based on raw distributions of vertebrate groups." *Biological Journal of the Linnean Society,* 71: 379–402.

Rosas, M. V., M. G. del Río, A. A. Lanteri, and J. J. Morrone. 2011. "Track analysis of the North and Central American species of the *Pantomorus-Naupactus* complex (Coleoptera: Curculionidae)." *Journal of Zoological Systematics and Evolutionary Research,* 49: 309–314.

Rosen, D. E. 1976. "A vicariance model of Caribbean biogeography." *Systematic Zoology,* 24: 431–464.

Rosen, D. E. 1985. "Geological hierarchies and biogeographic congruence in the Caribbean." *Annals of the Missouri Botanical Garden,* 72: 636–659.

Rueda, M., M. A. Rodríguez, and B. A. Hawkins. 2013. "Identifying global zoogeographical regions: Lessons from Wallace." *Journal of Biogeography,* 40: 2215–2225.

Ruggiero, A. and C. Ezcurra. 2003. "Regiones y transiciones biogeográficas: Complementariedad de los análisis en biogeografía histórica y ecológica." In: Morrone, J. J. and J. Llorente (eds.), *Una perspectiva latinoamericana de la biogeografía*. Las Prensas de Ciencias, UNAM, Mexico City, pp. 141–154.

Ruiz-Sánchez, E. and C. D. Specht. 2013. "Influence of the geological history of the Trans-Mexican Volcanic Belt on the diversification of *Nolina parviflora* (Asparagaceae: Nolinoideae)." *Journal of Biogeography*, 40: 1336–1347.

Rull, V. 2004a. "An evaluation of the Lost World and Vertical Displacement hypotheses in the Chimantá Massif, Venezuelan Guyana." *Global Ecology and Biogeography Letters*, 13: 141–148.

Rull, V. 2004b. "Biogeography of the 'Lost World': A paleoecological perspective." *Earth-Science Reviews*, 67: 125–137.

Rull, V. 2005. "Biotic diversification in the Guayana highlands: A proposal." *Journal of Biogeography*, 32: 921–927.

Ryan, R. M. 1963. "The biotic provinces of Central America." *Acta Zoológica Mexicana*, 6: 1–55.

Rzedowski, J. 1963. "Contribuciones a la fitogeografía florística e histórica de México. I. Algunas consideraciones acerca del elemento endémico en la flora mexicana." *Boletín de la Sociedad Botánica de México*, 27: 52–65.

Rzedowski, J. 1978. *La vegetación de México*. Editorial Limusa, Mexico City.

Rzedowski, J. 1991. "Diversidad y orígenes de la flora fanerogámica de México." *Acta Botanica Mexicana*, 14: 3–21.

Rzedowski, J. and T. Reyna-Trujillo. 1990. "Tópicos biogeográficos. Mapa IV. 8. 3." In: *Atlas Nacional de México, Vol. III*. Instituto de Geografía, UNAM, Mexico City.

Salazar Bravo, J., T. L. Yates, and L. M. Zalles. 2002. "Mamíferos de Bolivia." In: Ceballos, G. and J. A. Simonetti (eds.), *Diversidad y conservación de mamíferos neotropicales*. Conabio and UNAM, Mexico City, pp. 65–113.

Samek, V. 1973. *Regiones fitogeográficas de Cuba*. Academia de Ciencias de Cuba, Serie Forestal, La Habana.

Samek, V., E. del Risco, and R. Vandana. 1988. "Fitorregionalización del Caribe." *Revista del Jardín Botánico Nacional* (La Habana), 9: 25–38.

Sampaio, E. V. S. B. 1995. "Overview of the Brazilian caatinga." In: Bullock, S. H., H. A. Mooney, and E. Medina (eds.), *Seasonally dry tropical forests*. Cambridge University Press, Cambridge, UK, pp. 35–63.

Sánchez-González, L. A., J. J. Morrone, and A. Navarro-Sigüenza. 2008. "Distributional patterns of the Neotropical humid montane forest avifaunas." *Biological Journal of the Linnean Society*, 94: 175–194.

Sánchez-González and A. Navarro-Sigüenza. 2009. "History meets ecology: A geographical analysis of ecological restriction in the Neotropical humid montane forests avifaunas." *Diversity and Distributions*, 15: 1–11.

Sánchez-González, L. A., A. G. Navarro-Sigüenza, J. F. Ornelas, and J. J. Morrone. 2013. "What's in a name?: Mesoamerica." *Revista Mexicana de Biodiversidad*, 84: 1305–1308.

Sánchez Osés, C. and R. Pérez-Hernández. 1998. "Revisión histórica de las subdivisiones biogeográficas de la región Neotropical, con especial énfasis en Suramérica." *Montalbán*, 31: 169–210.

Sánchez Osés, C. and R. Pérez-Hernández. 2005. "Historia y tabla de equivalencias de las propuestas de subdivisiones biogeográficas de la región Neotropical." In: Llorente Bousquets, J. and J. J. Morrone (eds.), *Regionalización biogeográfica en Iberoamérica y tópicos afines—Primeras Jornadas Biogeográficas de la Red Iberoamericana de Biogeografía y Entomología Sistemática (RIBES XII. I-CYTED).* Las Prensas de Ciencias, UNAM, Mexico City, pp. 495–508.

Sandoval, M. L. and R. M. Barquez. 2013. "The Chacoan bat fauna identity: Patterns of distributional congruence and conservation implications." *Revista Chilena de Historia Natural*, 86: 75–94.

Sanginés-Franco, C., I. Luna-Vega, O. Alcántara Ayala, and R. Contreras-Medina. 2011. "Distributional patterns and biogeographic analysis of ferns in the Sierra Madre Oriental, Mexico." *American Fern Journal*, 101: 81–104.

Santa Anna del Conde Juárez, H., R. Contreras-Medina, and I. Luna-Vega. 2009. "Biogeographic analysis of endemic cacti of the Sierra Madre Oriental, Mexico." *Biological Journal of the Linnean Society*, 97: 373–389.

Santiago-Alvarado, M., G. Montaño-Arias, and D. Espinosa. 2016. "Áreas de endemismo de la Sierra Madre del Sur." In: Luna-Vega, I., D. Espinosa, and R. Contreras-Medina (eds.), *Biodiversidad de la Sierra Madre del Sur: Una síntesis preliminar.* UNAM, Mexico City, pp. 431–448.

Santiago-Valentin, E. and R. G. Olmstead. 2004. "Historical biogeography of Caribbean plants: Introduction to current knowledge and possibilities from a phylogenetic perspective." *Taxon*, 53: 299–319.

Savage, J. M. 1966. "The origins and history of Central America herpetofauna." *Copeia*, 4: 719–766.

Savage, J. M. 1982. "The enigma of the Central American herpetofauna: Dispersals or vicariance?" *Annals of the Missouri Botanical Garden*, 69: 464–547.

Schmidt, K. P. 1954. "Faunal realms, regions, and provinces." *Quarterly Review of Biology*, 29: 322–331.

Schmidt, K. P. and R. F. Inger. 1951. "Amphibians and reptiles of Hopkins-Branner expedition to Brazil." *Fieldiana, Zoology*, 31: 439–465.

Schuh, R. T. and M. D. Schwartz. 1985. "Revision of the plant bug genus *Rhinacloa* Reuter with a phylogenetic analysis (Hemiptera: Miridae)." *Bulletin of the American Museum of Natural History*, 179: 382–470.

Sclater, P. L. 1858. "On the general geographic distribution of the members of the class Aves." *Proceedings of the Linnean Society of London, Zoology*, 2: 130–145.

Sclater, W. L. 1894. "The geography of mammals. I. Introductory." *The Geographical Journal*, 3: 95–105.

Sclater, W. L. and P. L. Sclater. 1899. *The geography of mammals.* Kegan Paul, Trench, Trübner and Co., London.

Sequeira, A. S., A. A. Lanteri, M. A. Scataglini, V. A. Confalonieri, and B. D. Farrell. 2000. "Are flightless *Galapaganus* weevils older than the Galapagos Islands they inhabit?" *Heredity*, 85: 20–29.

Shannon, R. C. 1927. "Contribución a los estudios de las zonas biológicas de la República Argentina." *Revista de la Sociedad Entomológica Argentina*, 4: 1–14.

Sick, W. D. 1969. "Geographical substance." *Monographiae Biologicae*, 19: 449–474.

Sigrist, M. S. and C. J. B. de Carvalho. 2009. "Historical relationships among areas of endemism in the tropical South America using Brooks Parsimony Analysis (BPA)." *Biota Neotropica*, 9: 79–90.

Silva, J. M. C. and D. C. Oren. 1996. "Application of parsimony analysis of endemicity in Amazonian biogeography: An example with primates." *Biological Journal of the Linnean Society*, 59: 427–437.

Silva, J. M. C., A. B. Rylands, and G. A. B. da Fonseca. 2005. "The fate of the Amazonian areas of endemism." *Conservation Biology*, 19(3): 689–694.

Simpson, G. G. 1940. "Mammals and land bridges." *Journal of the Washington Academy of Sciences*, 30: 137–163.

Simpson, G. G. 1950. "History of the fauna of Latin America." *American Scientist*, 38: 361–389.

Simpson, G. G. 1953. *Evolution and geography: An essay on historical biogeography with special reference to mammals.* Condon Lecture Series, Oregon State System of Higher Education, Eugene.

Sklenář, P., E. Dušková, and H. Balslev. 2011. "Tropical and temperate: Evolutionary history of páramo flora." *Botanical Review*, 77: 71–108.

Smith, C. H. 1983. "A system of world mammal faunal regions. I. Logical and statistical derivation of the regions." *Journal of Biogeography*, 10: 455–466.

Smith, H. 1941. "Las provincias bióticas de México, según la distribución geográfica de las lagartijas del género *Sceloporus.*" *Anales de la Escuela Nacional de Ciencias Biológicas*, 2: 103–110.

Smith, M. L., S. B. Hedges, W. Buck, A. Hemphill, S. Inchaustegui, M. A. Ivie, D. Martina, M. Maunder, and J. Francisco-Ortega. 2004. "Caribbean Islands." In: Mittermeier, R. A., R. R. Gil, M., Hoffman, J. Pilgrim, T. Brooks, C. G. Mittermeier, J. Lamoreux, and G. A. B. da Fonseca (eds.), *Hotspots revisited: Earth's biologically richest and most threatened terrestrial ecoregions.* CEMEX, Mexico City, pp. 112–118.

Smith, S. A. and E. Bermingham. 2005. "The biogeography of lower Mesoamerican freshwater fishes." *Journal of Biogeography*, 32: 1835–1854.

Sobral-Souza, T., M. S. Lima-Ribeiro, and V. N. Solferini. 2015. "Biogeography of Neotropical rainforests: Past connections between Amazon and Atlantic Forest detected by ecological niche modeling." *Evolutionary Ecology*, 29: 543–655.

Soderstrom, T. R., E. J. Judziewicz, and L. G. Clark. 1988. "Distribution patterns of Neotropical bamboos." In: Heyer, W. R. and E. Vanzolini (eds.), *Proceedings of a Workshop on Neotropical distribution patterns.* Academia Brasileira de Ciencias, Rio de Janeiro, pp. 121–157.

Solbrig, O. T., W. F. Blair, F. A. Enders, A. C. Hulse, J. H. Hunt, M. A. Mares, J. Neff, D. Otte, B. B. Simpson, and C. S. Tomoff. 1977. "The biota: The dependent variable." In: Orians, G. H. and O. T. Solbrig (eds.), *Convegent evolution in warm deserts.* US/IBP Synthesis Series 3, Dowden, Hutchinson and Ross, Stroudsburg, Pennsylvania, pp. 50–66.

Soriano, A. 1949. "El límite entre las provincias botánicas Patagónica y Central en el territorio del Chubut." *Lilloa*, 20: 193–202.

Soriano, A. 1950. "La vegetación del Chubut." *Revista Argentina de Agricultura*, 17: 30–66.

Souza-Dias, P. G. B., L. D. De Campos, and S. S. Nihei. 2015. "Two new species of *Eidmanacris* (Orthoptera: Grylloidea: Phalangopsidae) from the Atlanti forest of São Paulo State, Brazil." *Florida Entomologist*, 98: 547–555.

Spichiger, R., L. Ramella, R. Palese, and F. Mereles. 1991. "Proposición de leyenda para cartografía de las formaciones vegetales del Chaco paraguayo. Contribución al estudio de la flora y de la vegetación del Chaco. III." *Candollea*, 46: 541–564.

Stange, L. A., A. L. Terán, and A. Willink. 1976. "Entomofauna de la provincia biogeográfica del Monte." *Acta Zoológica Lilloana*, 32: 73–119.

Stehli, F. G. and S. C. Webb (eds.). 1985. *The great American biotic interchange.* Topics in Geobiology Vol. 4, Plenum Press, New York.

Steyermark, J. A. 1986. "Speciation and endemism in the flora of the Venezuelan tepuis." In: Vuilleumier, F. and M. Monasterio (eds.), *High altitude tropical biogeography.* Oxford University Press and American Museum of Natural History, New York and Oxford, pp. 317–373.

Stuart, L. C. 1964. "Fauna of Middle America." In: West, R. C. (ed.), *Handbook of Middle American Indians. Vol. 1.* University of Texas Press, Austin, pp. 316–363.

Sturm, H. 1990. "Contribución al conocimiento de las relaciones entre los frailejones (Espeletiinae, Asteraceae) y los animales en la región del páramo andino." *Revista de la Academia Colombiana de Ciencias,* 17: 667–685.

Suárez-Mota, M., O. Téllez-Valdés, R. Lira-Saade, and J. L. Villaseñor. 2013. "Una regionalización de la Faja Volcánica Transmexicana con base en su riqueza florística." *Botanical Sciences,* 91: 93–105.

Swofford, D. L. 2003. *PAUP*: Phylogenetic Analysis Using Parsimony (*and other methods). Version 4.* Sinauer Associates, Sunderland, Massachusetts. http://paup.csit.fsu .edu/.

Szumik, C. A. and P. A. Goloboff. 2015. "Higher taxa and the identification of areas of endemism." *Cladistics,* 31: 568–572.

Takhtajan, A. 1986. *Floristic regions of the world.* University of California Press, Berkeley.

Thomé, M. T. C., F. Sequeira, F. Brusquetti, B. Carstens, C. F. B. Haddad, M. T. Rodrigues, and J. Alexandrino. 2016. "Recurrent connections between Amazon and Atlantic forests shaped diversity in Caatinga four-eyed frogs." *Journal of Biogeography,* 43: 1045–1056.

Tiemann-Boege, I., C. W. Kilpatrick, D. J. Schmidly, and R. D. Bradley. 2000. "Molecular phylogenetics of the *Peromyscus boylii* species group (Rodentia: Muridae) based on mitochondrial cytochrome b sequences." *Molecular Phylogenetics and Evolution,* 16: 366–378.

Toledo, V. H., A. M. Corona, and J. J. Morrone. 2007. "Track analysis of the Mexican species of Cerambycidae (Insecta, Coleoptera)." *Revista Brasileira de Entomologia,* 51: 131–137.

Torres Miranda, A. and I. Luna. 2007. "Hacia una síntesis panbiogeográfica." In: Luna, I., J. J. Morrone and D. Espinosa (eds.), *Biodiversidad de la Faja Volcánica Transmexicana.* Las Prensas de Ciencias, UNAM, Mexico City, pp. 503–514.

Torres Miranda, A. and I. Luna Vega. 2006. "Análisis de trazos para establecer áreas de conservación en la Faja Volcánica Transmexicana." *Interciencia,* 31: 849–855.

Townsend, J. H. 2014. "Characterizing the Chortís Block biogeographic province: Geological, physiographic, and ecological associations and herpetofaunal diversity." *Mesoamerican Herpetology,* 1: 204–251.

Trejo-Torres, J. C. and J. D. Ackerman. 2001. "Biogeography of the Antilles based on a parsimony analysis of orchid distributions." *Journal of Biogeography,* 28: 775–794.

Udvardy, M. D. F. 1975. *A classification of the biogeographical provinces of the world.* International Union for Conservation of Nature and Natural Resources Occasional Paper 18, Morges.

Upham, N. S. and B. D. Patterson. 2012. "Diversification and biogeography of the Neotropical caviomorph lineage Octodontoidea (Rodentia: Hystricognathi)." *Molecular Phylogenetics and Evolution,* 63: 417–429.

Urtubey, E., T. F. Stuessy, K. Tremetsberger, and J. J. Morrone. 2010. "The South American biogeographic transition zone: An analysis from Asteraceae." *Taxon,* 59: 505–509.

Valencia, R., C. Cerón, W. Palacios, and R. Sierra. 1999. "Las formaciones naturales de la sierra del Ecuador." In: Sierra, R. (ed.), *Propuesta preliminar de un sistema de clasificación de vegetación para el Ecuador continental.* Proyecto INEFAN/GEF-BIRF and EcoCiencia, Quito, pp. 79–108.

van der Hammen, T. 1974. "The Pleistocene changes of vegetation and climate in tropical South America." *Journal of Biogeography*, 1: 3–26.

van der Hammen, T. and A. M. Cleef. 1986. "Development of the High Andean páramo flora and vegetation." In: Vuilleumier, F. and M. Monasterio (eds.), *High altitude tropical biogeography.* Oxford University Press and American Museum of Natural History, New York and Oxford, pp. 153–201.

Vanin, S. A. 1976. "Taxonomic revision of the South American Belidae (Coleoptera)." *Arquivos de Zoologia, São Paulo*, 28: 1–75.

Vanzolini, P. E. 1963. "Problemas faunisticos do Cerrado." In: Ferri, M. G. (ed.), *Simpósio sôbre o Cerrado.* Universidade de São Paulo, São Paulo, pp. 307–320.

Vanzolini, P. E. and E. E. Williams. 1970. "South American anoles: Geographic differentiation and evolution of the *Anolis chrysolepis* species group (Sauria, Iguanidae)." *Arquivos de Zoologia, São Paulo*, 19: 1–298.

Vari, R. P. 1992. "Systematics of the Neotropical Characiform genus *Cyphocharax* Fowler (Pisces: Ostariophysi)." *Smithsonian Contributions to Zoology*, 529: 1–137.

Vázquez-Miranda, H., A. G. Navarro-Sigüenza, and J. J. Morrone. 2007. "Biogeographical patterns of the avifaunas of the Caribbean Basin islands: A parsimony perspective." *Cladistics*, 23: 180–200.

Vidal, M. A., E. R. Soto, and A. Veloso. 2009. "Biogeography of Chilean herpetofauna: Distributional patterns of species richness and endemism." *Amphibia-Reptilia*, 30: 151–171.

Viloria, A. L. 2005. "Las mariposas (Lepidoptera: Papilionoidea) y la regionalización biogeográfica de Venezuela." In: Llorente Bousquets, J. and J. J. Morrone (eds.), *Regionalización biogeográfica en Iberoamérica y tópicos afines—Primeras Jornadas Biogeográficas de la Red Iberoamericana de Biogeografía y Entomología Sistemática (RIBES XII. I-CYTED).* Las Prensas de Ciencias, UNAM, Mexico City, pp. 441–459.

Vivó, J. A. 1943. "Los límites biogeográficos en América y la zona cultural mesoamericana." *Revista Geográfica*, 3: 109–131.

Voronov, A. G. 1970. *Regionalización geobotánica.* Atlas Nacional de Cuba, La Habana.

Vuilleumier, F. 1986. "Origins of the tropical avifaunas of the high Andes." In: Vuilleumier, F. and M. Monasterio (eds.), *High altitude tropical biogeography.* Oxford University Press and American Museum of Natural History, New York and Oxford, pp. 586–622.

Vuilleumier, F. 1993. "Biogeografía de aves en el Neotrópico: Jerarquías conceptuales y perspectivas para futuras investigaciones." *Revista Chilena de Historia Natural*, 66: 11–51.

Wallace, A. R. 1852. "On the monkeys of the Amazon." *Proceedings of the Zoological Society of London*, 20: 107–110.

Wallace, A. R. 1876. *The geographical distribution of animals. Vol. I and II.* Harper and Brothers, New York.

Wen, J. and S. M. Ickert-Bond. 2009. "Evolution of the Madrean-Tethyan disjunctions and the North and South American amphitropical disjunctions in plants." *Journal of Systematics and Evolution*, 47: 331–348.

Werneck, F. P. 2011. "The diversification of eastern South American open vegetation biomes: Historical biogeography and perspectives." *Quaternary Science Reviews*, 30: 1630–1648.

Werneck, F. P., C. Nogueira, G. R. Colli, J. W. Sites Jr., and G. C. Costa. 2012. "Climatic stability in the Brazilian Cerrado: Implications for biogeographical connections of South American savannas, species richness and conservation in a biodiversity hotspot." *Journal of Biogeography*, 39: 1695–1706.

West, R. C. 1964. "The natural regions of Middle America." In: West, R. C. (ed.), *Handbook of Middle American Indians. Vol. 1.* University of Texas Press, Austin, pp. 363–383.

Whittaker, R. J., B. R. Riddle, B. A. Hawkins, and R. J. Ladle. 2013. "The geographical distribution of life and the problem of regionalization: 100 years after Alfred Russel Wallace." *Journal of Biogeography*, 40: 2209–2214.

Wiles, J. S. and V. M. Sarich. 1983. "Are the Galapagos iguanas older than the Galapagos?" In: Bowman, R. I., M. Berson, and A. E. Levington (eds.), *Patterns of evolution in Galapagos organisms*. California Academy of Sciences, San Francisco, pp. 177–186.

Williams, E. E. 1989. "Old problems and new opportunities in West Indian biogeography." In: Woods, C. A. (ed.), *Biogeography of the West Indies: Past, present, and future*. Sandhill Crane, Gainesville, pp. 1–46.

Willink, A. 1991. "Contribución a la zoogeografía de insectos argentinos." *Boletín de la Academia Nacional de Ciencias de Córdoba*, 59: 125–147.

Winker, K. 2011. "Middle America, not Mesoamerica, is the accurate term for biogeography." *Condor*, 113: 5–6.

Woodburne, M. O. 2010. "The great American biotic interchange: Dispersals, tectonics, climate, sea level and holding pens." *Journal of Mammalian Evolution*, 17: 245–264.

Woodman, N. 2005. "Evolution and biogeography of Mexican small-eared shrews of the *Cryptotis mexicana*-group (Insecta: Soricidae)." In: Sánchez-Cordero, V. and R. A. Medellín (eds.), *Contribuciones mastozoológicas en homenaje a Bernardo Villa*. Instituto de Biología, UNAM and Conabio, Mexico City, pp. 523–534.

Wüster, W., J. E. Ferguson, J. A. Quijada-Mascareñas, C. E. Pook, M. G. Salomão, and R. S. Thorpe. 2005. "Tracing an invasion: Landbridges, refugia, and the phylogeography of the Neotropical rattlesnake (Serpentes: Viperidae: *Crotalus durissus*)." *Molecular Ecology*, 14: 1095–1108.

Yamaguti, H. Y. and R. Pinto-da-Rocha. 2009. "Taxonomic review of Bourguyiinae, cladistic analysis, and a new hypothesis of biogeographic relationships of the Brazilian Atlantic rainforest (Arachnida: Opiliones, Gonyleptidae)." *Zoological Journal of the Linnean Society*, 156: 319–362.

Zanella, F. C. V. 2010. "Evolução da biota da diagonal de formações abertas secas da América do Sul." In: Carvalho, C. J. B. de and E. A. B. Almeida (eds.), *Biogeografia da América do Sul: Padroes e processos*. Editora Roca Limitada, São Paulo, pp. 198–220.

Zuckerland, E. and L. Pauling. 1965. "Molecular disease, evolution and genetic heterogeneity." In: Kasha, M. and B. Pullman (eds.), *Horizons in biochemistry*. Academic Press, London and New York, pp. 189–225.

Zuloaga, F. O., O. Morrone, and D. Rodríguez. 1999. "Análisis de la biodiversidad en plantas vasculares de la Argentina." *Kurtziana*, 27: 17–167.

Zunino, M. 2003. "Nuevos conceptos en la biogeografía histórica: Implicancias teóricas y metodológicas." In: Morrone, J. J. and J. Llorente (eds.), *Una perspectiva latinoamericana de la biogeografía*. Las Prensas de Ciencias, UNAM, Mexico City, pp. 159–162.

Zunino, M. and G. Halffter. 1988. "Análisis taxonómico, ecológico y biogeográfico de un grupo americano de *Onthophagus* (Coleoptera: Scarabaeidae)." *Monografie del Museo Regionale di Science Naturale*, 9: 1–211.

Index

Printed and bound by CPI Group (UK) Ltd, Croydon, CR0 4YY

24/10/2024

01778306-0006